T0320467

Mobile Applications and Solutions for Social Inclusion

Sara Paiva
Instituto Politécnico de Viana do Castelo, Portugal

A volume in the Advances in Wireless Technologies and Telecommunication (AWTT) Book Series

Published in the United States of America by
IGI Global
Information Science Reference (an imprint of IGI Global)
701 E. Chocolate Avenue
Hershey PA, USA 17033
Tel: 717-533-8845
Fax: 717-533-8661
E-mail: cust@igi-global.com
Web site: http://www.igi-global.com

Library of Congress Cataloging-in-Publication Data

Names: Paiva, Sara, 1979- editor.
Title: Mobile applications and solutions for social inclusion / Sara Paiva, editor.
Description: Hershey, PA : Information Science Reference, [2018] | Includes bibliographical references.
Identifiers: LCCN 2017038365| ISBN 9781522552703 (hardcover) | ISBN 9781522552710 (ebook)
Subjects: LCSH: Mobile computing. | Telematics. | Online social networks.
Classification: LCC QA76.59 .M6245 2018 | DDC 006.7/54--dc23 LC record available at https://lccn.loc.gov/2017038365

This book is published in the IGI Global book series Advances in Wireless Technologies and Telecommunication (AWTT) (ISSN: 2327-3305; eISSN: 2327-3313)

British Cataloguing in Publication Data
A Cataloguing in Publication record for this book is available from the British Library.

All work contributed to this book is new, previously-unpublished material.
The views expressed in this book are those of the authors, but not necessarily of the publisher.

For electronic access to this publication, please contact: eresources@igi-global.com.

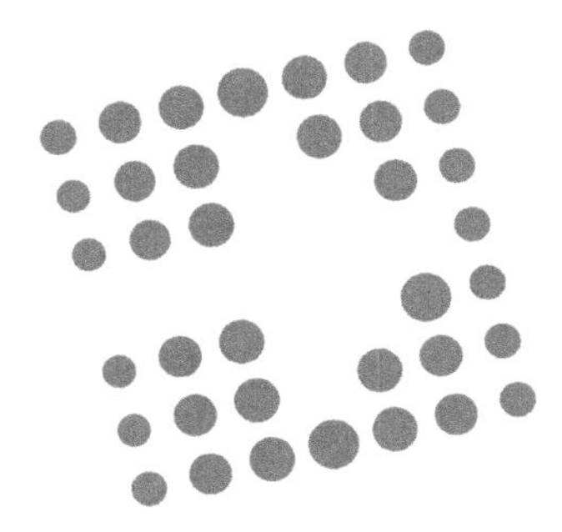

Advances in Wireless Technologies and Telecommunication (AWTT) Book Series

ISSN:2327-3305
EISSN:2327-3313

Editor-in-Chief: Xiaoge Xu, Xiamen University Malaysia, Malaysia

MISSION

The wireless computing industry is constantly evolving, redesigning the ways in which individuals share information. Wireless technology and telecommunication remain one of the most important technologies in business organizations. The utilization of these technologies has enhanced business efficiency by enabling dynamic resources in all aspects of society.

The **Advances in Wireless Technologies and Telecommunication Book Series** aims to provide researchers and academic communities with quality research on the concepts and developments in the wireless technology fields. Developers, engineers, students, research strategists, and IT managers will find this series useful to gain insight into next generation wireless technologies and telecommunication.

COVERAGE

- Digital Communication
- Wireless Technologies
- Telecommunications
- Cellular Networks
- Wireless Sensor Networks
- Radio Communication
- Virtual Network Operations
- Mobile Technology
- Mobile Web Services
- Grid Communications

IGI Global is currently accepting manuscripts for publication within this series. To submit a proposal for a volume in this series, please contact our Acquisition Editors at Acquisitions@igi-global.com or visit: http://www.igi-global.com/publish/.

The Advances in Wireless Technologies and Telecommunication (AWTT) Book Series (ISSN 2327-3305) is published by IGI Global, 701 E. Chocolate Avenue, Hershey, PA 17033-1240, USA, www.igi-global.com. This series is composed of titles available for purchase individually; each title is edited to be contextually exclusive from any other title within the series. For pricing and ordering information please visit http://www.igi-global.com/book-series/advances-wireless-technologies-telecommunication/73684. Postmaster: Send all address changes to above address.

Titles in this Series

For a list of additional titles in this series, please visit:
https://www.igi-global.com/book-series/advances-wireless-technologies-telecommunication/73684

Handbook of Research on Environmental Policies for Emergency Management and ...
Augustine Nduka Eneanya (University of Lagos, Nigeria)
Engineering Science Reference • ©2018 • 393pp • H/C (ISBN: 9781522531944) • US $295.00

Positioning and Navigation in Complex Environments
Kegen Yu (Wuhan University, China)
Information Science Reference • ©2018 • 577pp • H/C (ISBN: 9781522535287) • US $195.00

Centrality Metrics for Complex Network Analysis Emerging Research and Opportunities
Natarajan Meghanathan (Jackson State University, USA)
Information Science Reference • ©2018 • 213pp • H/C (ISBN: 9781522538028) • US $175.00

Examining Cloud Computing Technologies Through the Internet of Things
Pradeep Tomar (Gautam Buddha University, India) and Gurjit Kaur (Gautam Buddha University, India)
Information Science Reference • ©2018 • 311pp • H/C (ISBN: 9781522534457) • US $215.00

Advanced Mobile Technologies for Secure Transaction Processing Emerging Research ...
Raghvendra Kumar (LNCT Group of Colleges, India) Preeta Sharan (The Oxford College of Engineering, India) and Aruna Devi (Surabhi Software, India)
Information Science Reference • ©2018 • 177pp • H/C (ISBN: 9781522527596) • US $130.00

Examining Developments and Applications of Wearable Devices in Modern Society
Saul Emanuel Delabrida Silva (Federal University of Ouro Preto, Brazil) Ricardo Augusto Rabelo Oliveira (Federal University of Ouro Preto, Brazil) and Antonio Alfredo Ferreira Loureiro (Federal University of Minas Gerais (UFMG), Brazil)
Information Science Reference • ©2018 • 330pp • H/C (ISBN: 9781522532903) • US $195.00

Graph Theoretic Approaches for Analyzing Large-Scale Social Networks
Natarajan Meghanathan (Jackson State University, USA)
Information Science Reference • ©2018 • 355pp • H/C (ISBN: 9781522528142) • US $225.00

For an entire list of titles in this series, please visit:
https://www.igi-global.com/book-series/advances-wireless-technologies-telecommunication/73684

701 East Chocolate Avenue, Hershey, PA 17033, USA
Tel: 717-533-8845 x100 • Fax: 717-533-8661
E-Mail: cust@igi-global.com • www.igi-global.com

Table of Contents

Detailed Table of Contents

Social exclusion can occur in a variety of ways, one of which is lack of social interaction. The recognition of the social relations that occur in a group is fundamental to identify possible exclusions. This chapter proposes SocialCount, a mobile application that identifies social interactions performed by the user. In order not to interfere in the naturalness of relationships, the application was designed to infer social interactions without user intervention. The data of the interactions generated sociograms that represented the structure of the relations in a group in a simple way. Through the sociogram it was possible to visualize the users who may be socially at risk and alert the professionals responsible to solve the situation.

Currently, tourists spend a lot of time planning their trips because they need to make the most of every moment. In this sense, technology has been a great ally, especially when performing and adapting this planning in the event of some unforeseen event during the journey. And the emergence of distinct types of mobile devices was presented as an opportunity to improve the experience of tourism significantly. In this context, this chapter aims to identify the main computing needs to a mobile application to support the promotion of tourist sites for the traveler. The authors adopted a literature review as the research methodology. The main result of this chapter is the proposal of the Urbis prototype, an application that aims to help tourists to know better the cities they are visiting, even in the absence of local information or a specialized tour guide.

This chapter focuses on the development of mobile applications in a research strategy that combines computer sciences and architecture and urbanism. The main goal of the research is to develop mobile applications that help specific target people in daily life situations and that clearly contribute for advances in the fields of computer sciences and social responses. The authors discuss a group of mobile apps that were developed for smartphones and tablets and that respond to the following broad goals: 1) mapping of the physical space in order to adapt it to respond better to the users' needs, 2) adaptation of the physical space to the users' needs, and finally, 3) give the users a better knowledge about the physical space they are in. For each app developed, the authors describe the research problem involved, the goals, the development process, and the developed solution as well as the tests conducted to measure their performance. Usability and satisfaction tests revealed that the developed apps have a good acceptance by the target users.

The economic and social challenges felt in recent years because of the financial crisis impact wave were somehow attenuated by the silent work provided by the third sector institutions. Therefore, the incessant instability of the markets, as well as the population life-expectancy increasing and the implications thereof require new approaches towards pointing strategies to mitigate these problematic situations. For that reason, the development of technological solutions and applications for the private social solidarity institutions is an utmost challenge towards guaranteeing their sustainability and efficiency over time. The adoption of such solutions should be properly conceived to enhance their efficiency of the daily routines and to fulfill and inclusion of all users, while trying to reduce the technological literacy. The development of a technological framework to support the adoption of the practices, selection of technical requirements, and functionalities is seen as a great contribution for setting the roadmap that should be followed. This chapter explores the development of a framework for technological embedding in private social solidarity institutions.

This chapter aims to present a case study about the use mobile devices as a tool to practice in science classes in subjects related to the parts of the human body and digestive system. This case study was carried out with 24 students of the fifth grade of a public school located in a region of high social vulnerability of the city of Araranguá, Santa Catarina, Brazil, in partnership with the Laboratory of Remote Experimentation (RexLab). For the practice, the following applications were used: Human Body Systems 3D, Human Anatomy and Puzzle Anatomy. At the end, students were asked to respond to a questionnaire about the level of satisfaction related to the use of tablets in the classroom. Through the answers, a positive reaction related to the integration of digital technologies in the classroom was perceived, characterizing, thus, an assertive opportunity to continue the use of mobile devices in the school environment, giving an improvement in the quality of teaching of the sciences provided by the applications, considering the benefits of digital inclusion.

Education, including the subset of language learning, has been greatly influenced by information and communication technologies. This influence manifests itself in the form of various paradigms, starting from distance or digital learning (d-learning) to electronic learning (e-learning) then mobile learning (m-learning) and eventually ubiquitous learning (u-learning). The integration of these paradigms with supportive techniques to enhance inclusion, engagement, and to overcome the classic problem of lack of motivation led to a series of innovations, culminated in the notion of educational mobile game applications. This chapter focuses on the roots of this emergent trend, including the elements of mobile technology and the aspect of gaming, and how instrumental are they in empowering and motivating learners. The relationship of mobile games with the concept of gamification is examined, and a few major challenges to building effective mobile game applications for language learners are highlighted for future attention.

This chapter describes mobile payment, a mobile financial activity born of digital revolution, which is the combination of electronic money and mobile technology. The underlying technologies of mobile payment, its big players, and its status quo and future trend are discussed. In addition, this chapter discusses how mobile payment is related to social equality and social inclusion. Through presenting the historical, technical, economic, and social aspects of mobile payment, this chapter intends to provide readers with a holistic view of one of the fast-evolving financial activities that are transforming business, individuals, and the society.

In spite of the great potential for the development of mobile banking in Brazil, since the banking index reaches more than half of the population and the number of internet users is even higher, this potential is not evenly distributed among the age groups in the country. Taking into account the tendency to resist new technologies as one ages, this chapter aimed to identify the factors that lead the Brazilian population aged 45 years or over to use the internet and, within this spectrum, identify the barriers to the adoption of mobile banking technology. A questionnaire was applied and 113 responses were analyzed and categorized between functional and psychological aspects in these barriers. This chapter presents the results of this research.

Non-communicable diseases (NCDs) are the global leading cause of morbidity and mortality and disproportionately affect more in the less developed countries. Mobile technologies are being used for a variety of purposes in healthcare. Most importantly, they are enabling new ways for NCDs management by providing powerful tools to both doctors and patients for effective prevention and treatment. As the common risk factors of NCDs are related to human behavior; therefore, mobile phone-based health solutions can be used to combat with rising burden of NCDs by focusing on behavioral change programs to promote a healthy lifestyle. This chapter discusses the common NCDs, their burden, and future estimated projections, and shows how mobile phone technologies can provide effective NCDs management in developing countries—which have a lot of issues in their healthcare systems.

A health mobile application (app) has enabled users to access personal health records at any time and place. As an app provides health service to users, it is crucial for an app to be accessible to every user. However, often an app does not provide proper visual aid for users who are visually impaired. The authors restrict the range of visually impaired to retinitis pigmentosa (RP) patients in this chapter. RP is a rare type of progressive retinal disease that is hard to cure. Unfortunately, there are no established guidelines to assist RP patients in using their degrading sense of vision.

In this chapter, the authors review WCAG (web content accessibility guidelines) specified by W3C, analyze the UX designs of 140 popular health apps chosen based on the number of download counts in app stores, and propose a set of standard-compliant UX design guidelines to assist the visually impaired (RP) in accessing visual data and evaluate its compliance compared to WCAG.

Preface

The last few years have brought a considerable breakthrough of mobile devices at multiple levels. We have assisted to an increase of processing power, image resolution as well as the incorporated physical sensors of smartphones. This evolution was accompanied by a development of the society that adapted and shaped itself to this new reality (Moreira, Ferreira, Santos, & Durão, 2017), through a big number of sold equipment, number of people that use smartphones in their daily routines and number of *apps* that are published daily. It is also clear the change that has taken place in many daily routines performed by people, that started using apps to control and monitor certain aspects of their lives. Examples are the calculation of number of steps taken per day, calories wasted, instantaneous possibility of sharing photos in social networks, amongst others.

Social inclusion is defined, by the United Nations (2016), "as the process of improving the terms of participation in society, particularly for people who are disadvantaged, through enhancing opportunities, access to resources, voice and respect for rights." Mobile applications play an important role in the present and future society by helping to fight social exclusion. Some examples are in the field of education (Monjelat, 2017), health (Duretti et al., 2015), or elder and disabled people (Lunardini et al., 2017).

This is the scenario as we enter the fourth industrial revolution where lifestyles and industrial and economic development will count on mobile development innovation. Worldwide, two-thirds of the population uses mobile services and 60% of them use mobile internet. 39% of mobile users use their mobile phone to search for work or access services to improve their health or look for educational content. It is predicted that by 2020 mobile phones will reach 90% in developed countries and 70% in developing countries (Team, 2016).

THE CHALLENGES

The challenges mobile development will face for the next years are mainly related to the innovation they will be able to bring to promote social inclusion and the heterogeneity of the domains they will target. Digital technology will increase its presence in most of the object we use daily and people in general will have to be prepared to use these devices more and more to the most trivial tasks (Pachler, Bachmair, & Cook, 2010).

Since the last decade, schools, workplaces and higher education institutions are embracing mobile learning. Education will benefit from the ubiquity and portability of mobile devices (Goundar, 2011) and as such we will assist to the increase of these equipment in learning methodologies (Sung, Chang, & Liu, 2016; Fisher, 2010). Contents will need to be reformulated and teachers will have to adjust to this new emergent reality.

Another discipline where mobile development will contribute to is monetary transactions, making them safer and easier. For this purpose, plenty of services are being created to enable finance activities to be performed via mobile devices (Barnes & Corbitt, 2003) and protocols are being studied (Zhang & Zeng, 2017) to assure their safe execution.

Elder and disabled people are a perfect example where mobile apps can contribute to increase social inclusion as many obstacles exist to them in their daily tasks. Improving their mobility (Brito, Viana, Lourenço, Sousa, & Paiva, 2017; Mendes, Martins, & Paiva, 2016), monitoring health conditions (Brunschwiler et al., 2017), assisting them outdoors (Masulli et al., 2017), or helping them with cognitive impairments (Jecan, Rusu, Arba, & Mican, 2017) are just some of the challenges in this field.

Health monitoring and prevention bring together several researchers as these are problems that affects everyone and improvements are necessary. This is a field where plenty of challenges still remain. Some current research include studies for COPD patients (Brunschwiler et al., 2017), intervention to enhance home carer's disposal of medical waste (Ndayizigamiye & Hangulu, 2017), prevention and treatment of cholera (Imaja & Ndayizigamiye, 2017), and studies to support health application in low-end cellphones (Emamian & Li, 2017).

ORGANIZATION OF THE BOOK

The book is organized into 12 chapters. A brief description of each of the chapters follows:

Chapter 1 focuses on social exclusion and in the variety of ways it can occur, such as lack of social interaction. The chapter presents SocialCount, a mobile application that identifies social interactions performed by the user.

Chapter 2 presents an application for tourists, allowing them to easily integrate in the foreign city and promote social inclusion for them. The main result of this paper is the proposal of the Urbis prototype, an application that aims to help tourists to know better the cities they're visiting, even in the absence of local information or a specialized tour guide.

Chapter 3 focuses on the development of mobile applications in a research strategy that combines computer sciences, architecture and urbanism. The main goal of this research is to develop mobile applications that help specific target people in daily life situations and that clearly contribute for advances in the fields of computer sciences and social responses.

Chapter 4 targets the development of technological solutions and applications for the Private Social Solidarity Institutions is an utmost challenge towards guaranteeing their sustainability and efficiency over time.

Chapter 5 presents a case study about the use of mobile devices as a tool to practice in science classes in subjects related to the parts of the human body and digestive system. This paper reinforces the use of mobile devices in school environment, claiming it is an improvement in the quality of teaching of the sciences provided by the applications, considering the benefits of digital inclusion.

Chapter 6 refers to the integration of learning paradigms with supportive techniques to enhance inclusion, engagement and to overcome the classic problem of lack of motivation, explaining how it led to a series of innovations that culminated in the notion of educational mobile game applications.

Chapter 7 describes mobile payment, a mobile financial activity born of digital revolution, which is the combination of electronic money and mobile technology. The underlying technologies of mobile payment, its big players, and its status quo and future trend are discussed. In addition, this chapter discusses how mobile payment is related to social equality and social inclusion.

Chapter 8 presents a study on mobile banking in Brazil that aimed to identify the factors, which lead the Brazilian population aged 45 years, or over to use the Internet and, within this spectrum, identify the barriers to the adoption of mobile banking technology.

Chapter 9 introduces the peripheral arterial disease (PAD) which is characterized by a leg pain during walking to which a recommended treatment is to perform supervised physical activity. In this chapter, it is presented a system that monitories the physical activity containing one application for smartwatch, one application for smartphone and a back-end webservice.

Chapter 10 refers to chronic stress which is a spreading disease that affects millions of individuals with an enormous economic and social impact. The present study aims to understand how stress can be continuously monitored with the goal of predicting and alerting the occurrence of chronic or pathological stress and burnout situations.

Chapter 11 reviews non-communicable diseases (NCDs) which are the global leading cause of morbidity and mortality and disproportionately affecting more to the less developed countries. This chapter discusses the common NCDs, their burden and future estimated projections and show how mobile phone technologies can provide effective NCDs management in developing countries.

Chapter 12 review Web Content Accessibility Guidelines, analyse the UX designs of 140 popular health apps and propose a set of standard-compliant UX design guidelines to assist the visually impaired with retinitis pigmentosa in accessing visual data and evaluate its compliance compared to WCAG.

REFERENCES

Barnes, S., & Corbitt, B. (2003). Mobile banking: Concept and potential. *International Journal of Mobile Communications*, *1*(3), 273–288. doi:10.1504/IJMC.2003.003494

Brito, D., Viana, T., Lourenço, A., Sousa, D., & Paiva, S. (2017). A Mobile Solution to Help Visually Impaired People in Public Transports and in Pedestrian Walks. *International Journal of Sustainable Development and Planning*, *13*(2).

Brunschwiler, T., Straessle, R., Weiss, J., Michel, B., Van Kessel, T., & Ko, B. J. … Muehlner, U. (2017). CAir : Mobile-Health Intervention for COPD Patients. In *IEEE 19th International Conference on e-Health Networking, Applications and Services (Healthcom)* (pp. 17–19). IEEE.

Duretti, S., Marchioro, C. E., Marasso, L., Vicari, C., Fiorano, L., Papas, E. G., … Falda, S. (2015). ALL4ALL: IoT and telecare project for social inclusion. *2015 IEEE 1st International Forum on Research and Technologies for Society and Industry, RTSI 2015 - Proceedings, 80*, 17–22. 10.1109/RTSI.2015.7325065

Emamian, P., & Li, J. (2017). SimpleHealth – a Mobile Cloud Platform to Support Lightweight Mobile Health Applications for Low-end Cellphones. In *IEEE 19th International Conference on e-Health Networking, Applications and Services (Healthcom)* (pp. 1–6). IEEE.

Fisher, K. D. (2010). Technology-enabled active learning environments: An appraisal. CELE Exchange. Centre for Effective Learning Environments, 2010(6–10), 1–8.

Goundar, S. (2011, December). What is the Potential Impact of Using Mobile Devices in Education? *SIG GlobDev*, *4*, 1–30.

Imaja, I. M., & Ndayizigamiye, P. (2017). A design of a mobile health intervention for the prevention and treatment of Cholera in South Kivu in the Democratic Republic of Congo. In *IEEE Global Humanitarian Technology Conference (GHTC)* (pp. 1–5). IEEE. 10.1109/GHTC.2017.8239251

Jecan, S., Rusu, L., Arba, R., & Mican, D. (2017). Mobile application for elders with cognitive impairments. In *Internet Technologies and Applications* (pp. 155–160). ITA. doi:10.1109/ITECHA.2017.8101928

Lunardini, F., Basilico, N., Ambrosini, E., Essenziale, J., Mainetti, R., & Pedrocchi, A., … Borghese, N. A. (2017). Exergaming for balance training, transparent monitoring, and social inclusion of community-dwelling elderly. *RTSI 2017 - IEEE 3rd International Forum on Research and Technologies for Society and Industry, Conference Proceedings*. 10.1109/RTSI.2017.8065964

Masulli, F., Rovetta, S., Cabri, A., Traverso, C., Capris, E., & Torretta, S. (2017). An Assistive Mobile System Supporting Blind and Visual Impaired People when Are Outdoor. In *IEEE 3rd International Forum on Research and Technologies for Society and Industry (RTSI)* (pp. 1–6). IEEE.

Mendes, D., Martins, P., & Paiva, S. (2016). Mobile platform for helping visually impaired citizens using public transportation: A case study in a Portuguese historic center. *International Journal of Emerging Research in Management and Technology*, *5*(6).

Monjelat, N. (2017). Programming Technologies for Social Inclusion: An experience in professional development with elementary teachers. In *2017 Twelfth Latin American Conference on Learning Technologies (LACLO)* (pp. 1–8). Academic Press. 10.1109/LACLO.2017.8120901

Moreira, F., Ferreira, M. J., Santos, C. P., & Durão, N. (2017). Evolution and use of mobile devices in higher education: A case study in Portuguese Higher Education Institutions between 2009/2010 and 2014/2015. *Telematics and Informatics*, *34*(6), 838–852. doi:10.1016/j.tele.2016.08.010

Nations, U. (2016). *Identifying social inclusion and exclusion*. Academic Press.

Ndayizigamiye, P., & Hangulu, L. (2017). A Design of a Mobile Health Intervention to Enhance Home- Carers ' Disposal of Medical Waste in South Africa. In *IEEE Global Humanitarian Technology Conference (GHTC)* (pp. 1–6). IEEE. 10.1109/GHTC.2017.8239241

Pachler, N., Bachmair, B., & Cook, J. (2010). Mobile Learning Structures, Agency, Practices (G. Kress, Ed.). Academic Press.

Sung, Y.-T., Chang, K.-E., & Liu, T.-C. (2016). The effects of integrating mobile devices with teaching and learning on students' learning performance: A meta-analysis and research synthesis. *Computers & Education*, *94*(Supplement C), 252–275. doi:10.1016/j.compedu.2015.11.008

Team, G. (2016). *GSMA Intelligence. Global Mobile Engagement Index*. Retrieved from https://gsmaintelligence.com/

Zhang, X., & Zeng, H. (2017). Mobile Payment Protocol Based on Dynamic Mobile Phone Token. In *2017 Twelfth Latin American Conference on Learning Technologies (LACLO)* (pp. 680–985). Academic Press. 10.1109/ICCSN.2017.8230198

Chapter 1
An Approach for Detecting Social Interactions on Mobile Devices

Isadora Vasconcellos e Souza
Universidade Federal de Santa Maria, Brazil

William Bortoluzzi Pereira
Universidade Federal de Santa Maria, Brazil

João Carlos D Lima
Universidade Federal de Santa Maria, Brazil

ABSTRACT

Social exclusion can occur in a variety of ways, one of which is lack of social interaction. The recognition of the social relations that occur in a group is fundamental to identify possible exclusions. This chapter proposes SocialCount, a mobile application that identifies social interactions performed by the user. In order not to interfere in the naturalness of relationships, the application was designed to infer social interactions without user intervention. The data of the interactions generated sociograms that represented the structure of the relations in a group in a simple way. Through the sociogram it was possible to visualize the users who may be socially at risk and alert the professionals responsible to solve the situation.

DOI: 10.4018/978-1-5225-5270-3.ch001

INTRODUCTION

Humans are constantly developing and for this they have the need to communicate, whether through spoken language, writing or through gestures. According to De Mello and Teixeira (2011), since birth man is a social being in development and all its manifestations happen because there is another social being. Even when he does not use oral language, the individual is already interacting and becoming familiar with the environment in which he lives.

Social interaction is one of the main means of human development, especially in the early years of life. The individual begins to interact, knowing and adapting to the environment in which he lives and to the culture that is submitted. For Rabello and Passos (2010), humans are born immersed in culture, and this will be one of the main influences on his development.

In situations where the individual is deprived of social interactions, he is subject to social exclusion. Social exclusion describes a state in which individuals cannot fully participate in economic, social, political and cultural life, as well as the process that leads and sustains such a state. Participation can be hampered when people do not have access to material resources, including income, employment, land and housing, or services such as education and health care (DESA, 2016).

Burchardt et al. (2002) identified four dimensions of exclusion: consumption (the capacity to purchase goods and services), production (participation in economically or socially valuable activities), political engagement (involvement in local or national decision-making) and social interaction (integration with family, friends and community).

The researchers' proposal in this work is to implement a tool capable of solving the following question: "How to capture the dynamics of social interactions between individuals to identify possible social exclusion in different environments?". The SocialCount application aims to capture the social interactions carried out by users and through these data, create social graphs that represent the structure of relationships between individuals in the same social environment. And thus, to highlight cases in which users are socially excluded in the dimension of social interaction.

Psychology and Sociology have a field of study called sociometry. Sociometry consists of using tests to know the structure of a group of people and identifying subdivisions and positions, such as leaders, isolates, rival factions, etc. (Jennings, 1959). It is a way of measuring the degree of relationships between people (Rostampoor-Vajari, 2012). However, the tests are performed through questionnaires. Therefore, they are subject to limitations that include several sources of errors. Completing surveys and questionnaires induces partiality, unconcern, etc. (Groves, 2004) and scaling restrictions of the experiments (Palaghias, 2016).

To automate the process of measuring the degree of relationships in an unobtrusive way. The authors of this paper opted for the development of a mobile application, since the mobile devices are accessible and are part of the daily life of the majority of the population. In this way, it is likely that when an interaction occurs, the device is close to the user. To detect interactions naturally, SocialCount was developed based on ubiquitous and context-aware computing.

UBIQUITOUS COMPUTING

The term ubiquitous computing was first used by Professor Mark Weiser, who was chief scientist at the Xerox PARC Research Center (Palo Alto Research Center). In an article published in 1991 entitled "The Computer for the 21st Century," Mark Weiser states that in the future the user will focus only on the task and not the tools used to accomplish this task, the technology will be implicit in the context. For the author, the most profound technologies are those that disappear, are present in daily life, integrated into daily activities, becoming ubiquitous.

The main objective of Ubiquitous Computing is to spread the human-machine interaction, becoming a transparent interaction. It is to give the possibility of integrating natural behaviors of the human being with computing, through interconnected devices, in an omnipresent way and being inserted into the life daily life of the users (Prasad, 2012). The paradigm of Ubiquitous Computing is embedded in small, distributed and integrated computing devices, whether in a fixed or mobile environment. In this way, is incorporated into our lives effortlessly into the giving autonomy, portability, and mobility over daily needs / tasks of each user (Saha & Mukherjee, 2003).

Like Temdee and Prasad (2017), many authors consider ubiquitous computing and pervasive computing synonyms. However, according to (De Araújo, 2003), Ubiquitous Computing arises with the need to obtain mobility to a set of functionalities found in pervasive computing, giving autonomy to the computational device, dynamically assigning skills to configure services / tasks according to your need, or environment.

Already Roussos (2006) said, what differentiates ubiquitous computing from other paradigms is that ubiquitous computing has the ability to communicate via Wireless with other computing devices because communication is embedded in places, objects, and even in people, in this way it is possible to interact freely with digital resources.

The Figure 1 shows the degree of immersion and mobility of the areas of ubiquitous computing, pervasive computing, mobile computing and traditional computing methods.

Figure 1. Dimensions of ubiquitous computing
Source: Adapted from Lyytinen & Youngjin, 2002.

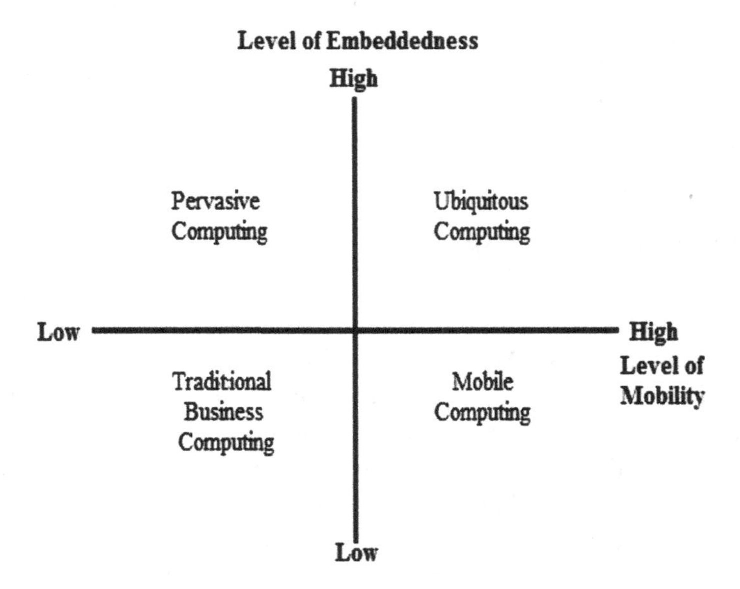

- **Ubiquitous Computing:** Located in the first quadrant has a high degree of mobility, and a high degree of integration with the environment, Ubiquitous Computing consists of small distributed/integrated computing devices organized in different environments in order to provide time independent information and services or location.
- **Pervasive Computing:** Located in the second quadrant, it has a low degree of mobility and a high degree of integration with the environment, i.e. the computer is present in the environment invisibly to the user, the computer can obtain information about the environment it is operating, adjusting to the environment, tuning applications to better meet both device and user needs.
- **Traditional Computing:** Located in the third quadrant, it has a low degree of mobility, and low degree of integration with the environment. Traditional Computing is recognized by the set of workstations, desktops, mainframes among other equipment that users perform tasks such as message exchange, create / manage documents, and these devices use peripherals (mouse, keyboard and monitor).
- **Mobile Computing:** Located in the fourth quadrant, it has a high degree of mobility, and low degree of interaction with the environment. Mobile computing has the autonomy to move through computing services, making it an ever-present device, helping the user to meet their needs related to computing services regardless of their location.

Therefore, ubiquitous computing can be defined as the union of pervasive computing and mobile computing. Thus, applications developed in the form of ubiquitous computing have a high degree of mobility and a high degree of immersion. An area that uses the Ubiquitous Computing standards is called Context-aware Computing.

CONTEXT-AWARE COMPUTING

The Context-aware is represented by the ability of the system to use the context in which it is currently providing appropriate information / responses to users. Several systems can be considered as Context-awareness, but are often called by other names such as intelligent systems, evolving system, etc.

In Ubiquitous Computing it is important that modules and sensors interact with each other, and among users, becoming sensitive, allowing interaction with users through response actions and / or reactions, and these characteristics make it a Context-Awareness.

The term Context Sensitive in Ubiquitous Computing was first used by (Schilit, 1994). According to the author, the main factors for shaping the context are: "where are you", "who you are" and "what features are nearby". For Dey (2001), Context-aware computing uses relevant information / services about entities in the world where relevance depends on the user's task.

Context-aware is a very broad term, so a context definition and awareness can be very varied, work on context-awareness usually shows its purpose through implementations with different levels of consciousness.

According Dey (2001) exists three categories of resources that a context sensitive application can adopt:

- **Presentation of Information and Services to a User:** A context aware application can perceive that the user is in a museum and offer relevant information about a work. Or realize that the user is in a different city from the cities he usually visits and recommend nearby restaurants at times that the user usually has lunch or dinner.
- **Automatic Execution of a Service:** A practical example of automatic execution of a service would be the ability to recognize that the user is in a library by studying and blocking the notifications of their mobile device so that there is no deviation of attention. In this case, you would need information about the user's current location and the activity he is performing. When the application identifies that the user's activity or location has changed, you can unlock the notifications automatically.

- **Marking a Context of Information:** Context aware application can mark contexts according to certain information. For example, with accelerometer data, heart rate, location and time may generate information that the user is exercising. Therefore, the application may assume that the user is in a gym performing physical activities.

According to Abowd (1999), there are different ways of using context sensitivity, these forms may interact together or separate. The context can be divided into five subcategories: Environmental Context, Personal Context, Social Context, Task Context and Spatio-Temporal Context (Kofod-Petersen and Cassens, 2004). According to Kofod-Petersen and Cassens, these subcategories are defined as:

- **Environmental Context:** This part captures the user's surroundings, such as things, services, people, and information accessed by the user.
- **Personal Context:** This part describes the mental and physical information about the user, such as mood, expertise and disabilities.
- **Social Context:** This describes the social aspects of the user, such as information about the different roles a user can assume.
- **Task Context:** the task context describes what the user is doing; it can describe the user's goals, tasks and activities.
- **Spatio-Temporal Context:** This type of context is concerned with attributes like: time, location and the community present.

The Figure 2 exemplifies some of the characteristics that can be identified in the context of the user. Aspects of health (blood pressure), environment (light), identity (name), current time, location and social (sociograms). This work has the focus on identifying the user's social context.

SOCIAL CONTEXT

According to Kofod-Petersen and Cassens (2006), social context describes social aspects of the user, such as information about friends, relatives and colleagues. For Kolvenbach et al. (2004) and Carter (2013) the social context is the medium by which people can relate easily, including the culture in which the individual lives and was educated and the people and institutions he interacts with. Adams et al. (2008) adds places and activities to the concept. Therefore, according to the definition of Schilit et al. (1994) cited above, the social context addresses two of the main defining aspects of context: "where you are" and "who you are".

Figure 2. Contexts representation
Source: Authors.

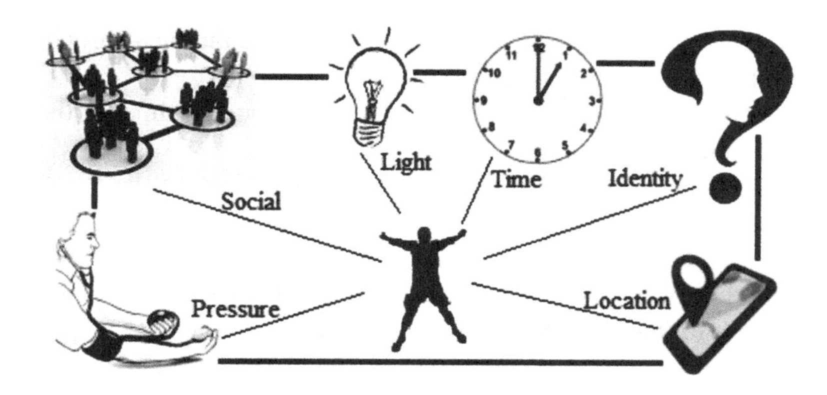

Figure 3. Definition of social context
Source: Authors.

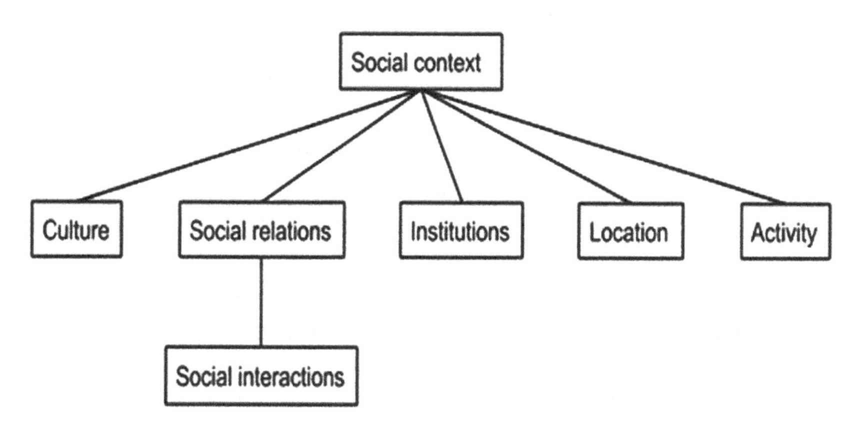

To facilitate understanding, the Figure 3 presents the relationship and hierarchy of concepts needed to define the social context. The social context is characterized by institutions, culture, social relations and people.

- **Institutions:** Are complex social forms that reproduce themselves, such as: government, family, languages, universities, hospitals, corporate corporations, and legal systems (Miller, 2010).
- **Culture:** Represents a complex of knowledge, beliefs, art, morals, laws, customs and all other habits and capacities acquired by man as a member of society (Tylor, 1871).

- **Social Relation:** It is formed by social interactions that have an intensity and intimacy (Primo, 2006). It involves communication in all its forms (Hari and Kujala, 2009). It is the way people relate to one another in a particular social context (Park et al., 1921).
- **Location:** Represents the location of the user at the time of the context identification.
- **Activity:** Represents the current physical movement of the user, such as: running, walking, sitting.

Based on the concepts presented, it can be seen that social interactions are the elementary unit for the constitution of the social context. According to Thompson (1998), social interactions can be classified as face-to-face interaction and mediated interaction.

The face-to-face interaction happens in a context of co-presence, since the participants have the same frame of time and space. The interaction has a two-way information and communication flow. In addition to dialogue, participants perform non-verbal cues to convey messages and interpret them, such as gestures and facial expressions. The nonverbal signals to help reduce ambiguity in messages.

However, mediated interaction requires a means of communication mediating this communication. It is necessary to use a technical means (paper, electric wires, electromagnetic waves, etc.) that allows the transmission of information and symbolic content to individuals located remotely in space, time or both. Participants may be in different spatial or temporal contexts. This type of interaction is subject to a greater chance of ambiguity in communication, since the nonverbal signals are compromised.

Biamino (2011) proposed a more specific interpretation that would meet the point of view of pervasive and ubiquitous computing. They suggest that the social context can be represented through networks. "In our social vision contexts are more similar to social aggregations or social groups, identified as a number of nodes in a given location, linked by some kind of ties (relations) that determine their nature" (Biamino, 2011). The authors define a social context as a 3-tuple that describes the network:

$$cxt = \left\langle Size, Density, Type\ of\ Ties \right\rangle$$

In this work the definitions of Size, Density and Type of Ties were adapted from the work of Biamino. The Size represents the number of nodes in a defined location:

- **Small (n <= 5):** A network with a small number of nodes.
- **Private (5 < n <= 20):** A network with a few nodes;

- **Open ($20 < n <= 50$):** A relatively large network;
- **Wide ($n > 50$):** A network with a very large number of nodes.

Density represents the number of connections between the nodes:

- **Clique:** A fully connected graph;
- **Easy:** A graph with easy to close triangles;
- **Hard:** A graph with many isolated nodes and hard to close triangles.

Type of Ties is defined by the main type of relations between the nodes of the network:

- **Unknown:** No relation exists between two nodes;
- **Acquaintance:** Two nodes are not close friends, but they interact with each other;
- **Friends:** Two nodes with a friendship-kind of relation.

Each interaction inferred by SocialCount contains information about the day, place, and time it was made, and the people who participated in it. This set of information is represented by sociograms and classified according to 3-tuple criteria. Sociograms are graphical representations of the social links that a person possesses. It is a graphic design that traces the structure of interpersonal relationships in a group situation. Moreno (1953) developed the sociogram to represent the data of Sociometry.

SOCIOMETRY

The word sociometry comes from the Latin "socius" which means social, and "metrum" which means measure. Sociometry is used to measure the degree of relationship between people. One of the important things about measuring relationships is to help with possible positive changes and see how much change needs to be made to make positive change. With sociometry it is also possible to reduce conflicts between people and improve communication between them, because sociometry allows a group of people to analyze their own practices (Rock & Page, 2009).

Jacob Levy Moreno created the term sociometry and defined sociometry as "the mathematical study of psychological properties of populations, the experimental technique and the results obtained by the application of quantitative methods". Moreno carried out a scope study of 1932-1938 at the New York State Training School for Girls in Hudson, New York, where she used sociometry techniques to assign people to different residences, and several other sociometry studies were applied in different settings environments such as schools, therapy groups, companies, etc. (Moreno, 1953).

The sociometry test consists, basically, of a simple questionnaire, where each member is asked to choose, in the group to which he belongs or could belong, the individuals he would like to have as partners, that is, to associate with them in situations. Questions can be formulated in a way that serves diverse purposes of interpersonal relationships. Thus, different criteria of choices can produce different structures of the same group (Xavier, 1990).

Sociometric Criteria

Every choice is made on the basis of some criterion, whether it is an intuitive feeling, like liking or disliking a person for the first impression, or an objective criterion such as whether a person has the ability to perform certain tasks. From these choices comes a description of the networks within the group. These networks are called sociograms (Moreno, 1953).

Selection Criteria

According to Bronfenbrenner (1944), the selection of the criterion manages to create or to break a sociometric intervention. All data collection in the social sciences is obtained through questions. The question should be asked in a way that is unambiguous, avoiding confusing the person by giving an answer. A good criterion should present a meaningful choice for the person in a simple format. For example: "What kind of person would you like to be part of your work team?"

The answer must be given from the individual interpretation. Each person sets his criteria. In the example of the question "What kind of person would you like to be part of your work team?" the answer could be "a person who works well in a group". In this way, the individual can define his criterion without having interpretative doubts about the question asked to him. Thus, it is possible to increase the reliability of the data produced.

In order to define the criteria, the following items must be taken into account:

Item 1: The criterion should be simple and as direct as possible;
Item 2: The person who will answer the question should have experience with the question that will be asked;
Item 3: The criterion should be specific in the question so as not to have doubts in the answer;
Item 4: The criterion should preferably be real and not hypothetical;

SocialCount aims to identify social interactions and evidence the structure of relationships between people in a group, acquiring daily activity data and without user intervention. In this way, it is possible to obtain a vision of the interactions that occur in the practice, discarding the tendentious aspects that the data raised through questionnaires can obtain. To develop the application were considered concepts of Mobile Phone Sensing.

MOBILE PHONE SENSING

The Mobile Phone Sensing is an area of Mobile Computing that aims to make the sensing of various user characteristics. According to Khan et. al (2013), the current devices has specialized sensors, such as: ambient light sensor, accelerometer, digital compass, gyroscope, GPS, proximity, camera, and microphone. In mobile sensing devices are used as a sensor server to collect, process and distribute the data around people.

According to Lane et al. (2010), sensors provide the development of applications for a variety of domains, such as health, social networks, security, environmental monitoring and transportation. Activities such as walking, running, sitting, which previously needed specialized devices to be detected, can now be identified by downloading a specific application on a mobile device.

Ali and Khusro (2016) point out that data sensing can be done in two ways: participatory and opportunistic sensing. The participatory depends on user actions, such as data entry and decision making. On the other hand, opportunistic sensing automatically collects data and makes decisions for application requests only with the usual use of the mobile device. It is worth mentioning that, considering the models of Context-Aware Computing, applications should preferably perform opportunistic sensing, that is, without user intervention.

According to Ali and Khusro (2016), the general architecture for sensing in mobile devices has the following characteristics: information domain, sensing module, mobile device module, remote server module (optional) and visualization module. The information domain represents the area where the application will collect the data (user environment, context, activities, social relations, physiological conditions). The sensing module represents the sensors that will be used. The mobile device module is the device that the application will use. The device may be responsible for the processing and inference of information, but it may be overwhelmed. What is indicated is to have a remote server module to enjoy the power of desktop hardware. Finally, the visualization module is where data from the sensing and completion of information is displayed to the user.

The area in question covers several characteristics of the user, such as the environment (light, temperature), social relations (individuals with which the user interacts), physiological conditions (feeding, rest, exercise) and activities (walking, seated). The focus of this paper is on social interactions. Within the area of mobile devices, there are two branches that study the social characteristics: Pervasive Social Context and Social Signal Processing as can be seen in Figure 4.

Pervasive Social Context

For Schuster et al. (2013), the Pervasive Social Context combines concepts from the area of Pervasive Computing and Social Computing. The focus of the user's physical environment is moved to their social environment and the focus of online interactions is geared toward interactions with people face to face. According to Dey (2001), the context is linked to entities (people, places, objects) relevant to the user and the application. For social computing, the most important entities are the other people. In this way, the focus of the social context is people.

Still according to Schuster, the term pervasive context is all the information of entities around the pervasive device. Through sensors it is possible to measure the location and proximity of other devices and identify the people who are nearby. Thus, the distinction between social context and pervasive context disappears, leading to the term pervasive social context.

Schuster has developed the STiPI taxonomy, which is an acronym for Space, Time, People and Information and covers the 5WH issues known in the area of Pervasive Computing. The 5WH questions represent: what, why, where, when, who, how. The what and why questions are not addressed by the taxonomy because they

Figure 4. Mobile sensing detection areas
Source: Authors.

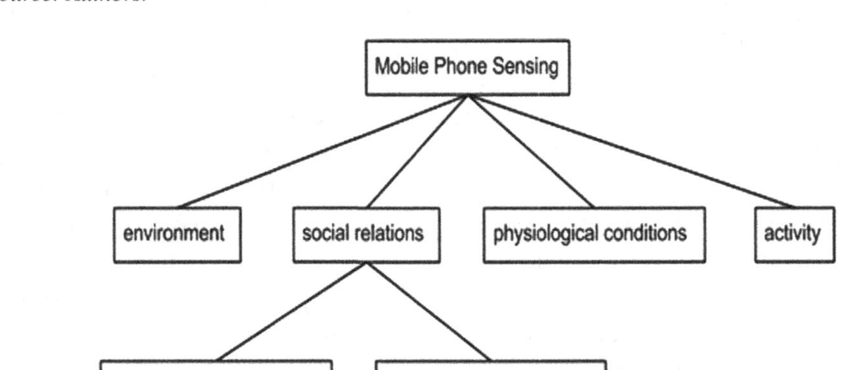

cannot be defined and vary depending on the application ("what" represents what the application is about and "why" for which the application was made).

The taxonomy represents the 5WH issues, according to the authors, as follows:

- **Space - S:** Is the spatial extent that the pervasive social context can produce and access. Answers the "where".
- **Time - Ti:** Is the time span in which the pervasive social context can produce and access. Answers the "when".
- **People - P:** Represents the people who are described in the pervasive social context. Answers the "who".
- **Information - I:** Is the source of information from which data on the pervasive social context comes, and where such data is gathered. Answers the "how".

Each part of the taxonomy has three levels. The space has the small scope (S1), the medium scope (S2), and any place (S3). Time has the short time activity (T1), medium time activity (T2) and longtime activity (T3). People have the individual levels (P1), groups (P2) and anonymous community (P3) and Information has the information level of pervasive sensors (I1), social network information (I2) and integration of the physical and virtual world (I3). Therefore, it is possible to develop several types of applications that fit certain levels of taxonomy.

The small scope (S1) addresses people who are physically close, for example in a club. The medium scope (S2) addresses a larger number of people interacting than the S1, such as a city. The S3 represents any place in society.

The short time activity (T1) includes people the user knows for a short period of time or has the potential to know, for example: a bus stop. The medium time activity (T2) includes individuals who interact for a common goal, for example, people the user knows about at work. Long time activity (T3) includes people who have a direct or indirect long-term relationship, such as family members, club members, employees, and friends.

The individual level (P1) represents the individuals that interact with the user, the social context consists of individual people (friends, friends of friends, unknown people). At the group level (P2) the social context consists of groups that interact with the user, which can be a cluster of friends on the social network. At the anonymous community level (P3) the social context consists of an anonymous community that uses the same social application.

Social Signal Processing

Psychology has long time studied human behavior. The methods most used by researchers are questionnaires, observation, audio and image recordings, among others. These methods are subject to some errors caused by the impartiality and the indifference of the people being evaluated. In addition to scalability constraints, because when the evaluation is done by humans, there is a limit of people that can be evaluated.

To automate the process of evaluating human behavior, Social Signal Processing (SSP) has emerged. The goal is to use the maximum number of sensors to collect data and automatically detect information about users' social behavior, identifying the information on the context in which the user is. SSP addresses issues of impartiality, indifference, and scalability.

Vinciarelli et al. (2010) defined some procedures for the detection of social behavior, which were later adapted to the mobile environment by Palaghias et al. (2016), who concluded that four steps are necessary to extract knowledge about social behavior in mobile devices:

Step 1: Sensing
Step 2: Social interaction detection
Step 3: Extraction of behavioral cues
Step 4: Understanding social behavior by inferring social signals

Sensing (step 1) is done by the sensors in the device, such as a gyroscope, accelerometer, camera, microphone, etc. Each sensor is responsible for generating the data of certain characteristics, the microphone will generate the data referring to the user's speech, for example.

Social interaction detection (step 2) can be performed using a single or multiple modes. In the single mode, Bluetooth or Wi-Fi connections are used to identify nearby people. In the multiple mode are used Bluetooth and Wi-Fi connections, microphone, camera, among other sensors. Data collection of social interactions allows the extraction of behavioral cues (step 3). Behavioral cues are characteristics, habits, or patterns of user interactions with people.

During a social interaction, in addition to the dialogue, there are gestures, characteristics, speech direction, etc. All these attributes are called social signs. Eagle and Pentland (2005) describe social cues as signs of nonverbal communication emitted when people are interacting socially. The union of these social signals over a period of time leads to an understanding of social behavior by inferring social signals (step 4).

RELATED WORKS

The related works were selected with the purpose of presenting the state of the art and the most used methodologies of detection of social interactions in mobile devices. The main characteristics raised for detecting interactions are: interpersonal distance, user location, relative position and conversation activity.

The most common approach used by researchers to recognize social interactions between two individuals is the Bluetooth ID search (BTID) or Wi-Fi service ID (SSID) of nearby devices. All devices/people found are classified as social interactions. This method was used in CenceMe (Miluzzo et al., 2007), SoundSense (Lu et al., 2009), E-Shadow (Teng et. al, 2014), PMSN (Zhang et. al, 2012), among others. This approach is simple and does not require specialized hardware and sensors, but the accuracy is limited by the range of Bluetooth (about 10 meters) and Wi-Fi (approximately 35 meters for indoor environments).

DARSIS (Palaghias et al., 2015) was developed to quantify social interactions in real time. The relative orientation of the users was used to obtain face direction and interpersonal distance. The proximity between the participants of the interaction is calculated through the RSSI samples of the user device's Bluetooth that are trained by learning machine (MultiBoostAB with J48). Samples are taken from three device position combinations: screen to screen, screen to back and back to back. The proximity is classified as public area, social zone, personal zone and intimate zone. The relative orientation of the user is known through uDirect (Hoseinitabatabaei et al., 2014) which identifies the relative orientation between the Earth's coordinates and the user's locomotion and predicts the direction of the face without requiring a fixed position of the device.

In Multi-modal Mobile Sensing of Social Interactions (Matic et al., 2012), they used a set of methods for the sensing of the interactions: the interpersonal distance, the relative position of the user, the direction of the face and the verification of speech activity. The interpersonal distance between the devices is captured in RSSI, where one device works as a Wi-Fi access point (Hot Spot) and another as a Wi-Fi client. The relative position is calculated by the position of the torso in relation to the coordinates of the Earth, always considering the same position of the mobile device. The speech activity is detected by an accelerometer installed on the user's chest. The device configured as a Wi-Fi access point is characterized as an intrusive process because the user generally does not use his mobile device for this purpose. Likewise, the accelerometer on the chest because is not a commonly used device.

SCAN (Social Contex-Aware smartphone Notification system) (Kim et al., 2017) detects the user's social context and blocks smartphone notifications so as not to distract the user while he or she is interacting. The system sets breakpoints to release notifications according to the following criteria: silence (when there is no conversation for 5 seconds or more), movement (when a person in the group leaves the table), user alone (when the person is alone waiting for friends) and use (when the other person participating in the interaction is using the smartphone). Social interactions are known through identification of close people and conversation. SCAN periodically searches for BLE beacons detect the presence of other people and announce their own presence, the BLE beacons have been chosen for deployment because they do not require pairing and connection actions, and have low power consumption. For the detection of conversation was used the algorithm YIN (De Cheveigné & Kawahara, 2002) that estimates the fundamental speech frequency and identifies the human voice.

SocioGlass (Xu et al., 2016) to promote additional information about the people who the user is interacting. There are 28 biographical information items that are classified into 6 groups: work, personal, education, social, leisure and family. The system uses Google Glass and an Android application that communicates via Bluetooth. Interactions are detected through facial recognition, Google Glass is responsible for providing the image of the individual who is participating in the interaction, the application receives the image, performs the processing and searches for a combination in the local database. When you do the recognition, the information related to the person in question is displayed on the Google Glass screen. The authors also implemented a smartphone-only version, where the face image is captured by the device's camera and the information is displayed on the mobile screen.

The related works are inserted in a bus stop scenario to be analyzed equally. In this case, there are many people waiting for buses, some people are talking, others are quietly looking toward the cars. The Table 1 presents a comparative board between related works and SocialCount. All the works were classified according to the STiPI and SSP taxonomy. The Multi-model and SCAN have "0" in the Time classification because they do not identify the type of the relation of the interactions.

The works that only approach the interpersonal distance to consider an interaction are submitted to have a low accuracy, mainly in situations of the real life. According to the scenario, many people are physically close and do not interact with each other. In this case several interactions would be considered wrongly. The good thing about this methodology is that Bluetooth and Wi-Fi technology are compatible with most smartphones available in the market.

DARSIS obtained a good accuracy in verifying interpersonal distance. In addition, they used the direction of the face to identify the interactions, which is a good approach considering that a person tends to direct the face to the other person

Table 1. Comparative table of related work

	Intrusive	Detection of Speech Sound	Detection From Who Is Speaking	Approach to Interaction Detection	STiPI	SSP
CenceMe, SoundSense, E-Shadow, PMSN	No	No	No	Interpersonal distance (ID Bluetooth e Wi-Fi)	S1, T1, P1, I1	1, 2
DARSIS	No	No	No	Interpersonal distance (RSSI Bluetooth), user relative orientation and face direction	S1, T3, P1, I1	1, 2, 3
Multi-model	Yes	Yes	No	Interpersonal distance (RSSI Bluetooth), user relative orientation and detection of speech sound	S1, 0, P1, I1	1, 2, 3
SCAN	No	Yes	No	Interpersonal distance (BLE beacons) and detection of speech sound	S1, 0, P1, I1	1, 2, 3
SocioGlass	Yes	No	No	Image detection with Google Glass	S1, T3, P1, I1	1, 2, 3
SocialCount	No	Yes	Yes	Interpersonal distance (ID Bluetooth), detection for speech sound and detection from who is speaking	S1, T3, P1, I1	1, 2, 3

Source: Authors.

when communicating. However, they do not consider the conversation. Considering the scenario, this can easily lead to mistakes, where two people may be standing next to each other and looking in opposite directions.

Multi-modal and SCAN consider distance and conversation. The Multi-modal uses an intrusive approach to the verification of the conversation, which damages the naturalness of the user's daily actions. SCAN uses the YIN algorithm for the same purpose, which identifies the human voice without the need for external resources to the smartphone. There may still be errors in the scene, as people around the user may be talking.

SocioGlass used images captured with Google Glass. This can be considered an intrusive procedure, since few users own the device and use it regularly. The authors have made a version that works only on the smartphone, but users need to focus the camera on the person's face, which detracts from the usability of the application.

To solve the problems mentioned in the scenario, SocialCount uses a set of approaches. Interpersonal distance is implemented through Bluetooth, the detection of the conversation by the YIN algorithm and the detection from who is speaking. This last question is fundamental for the correct consideration of interactions. In the next section the SocialCount methodology is described in detail.

SocialCount

SocialCount is a mobile application developed for the Android platform. The main objective is to detect user social interactions and provide data that describes the social context. Through these data it is possible to perform sociometry to identify people excluded from the social context. The interactions considered by the application are only face to face, that is, interactions mediated by a means of communication are not considered.

Application development is based on Social Signal Processing criteria: sensing (step 1) and social interaction detection (step 2). According to the STiPI taxonomy of the Pervasive Social Context, the SocialCount is classified as Space S1, Time T2, People P1 and Information I1. S1 because it addresses people in an enclosed room. T2 because sociometry is done with people who perform some kind of interaction with a common goal. P1 because the interactions are between users. And I1 considers only the interactions in the physical world, ignoring the virtual world.

For inference of social interactions, SocialCount detects: human voice, who is speaking, location, and nearby devices. Based on related work, the differential of SocialCount is the use of the recognition of who is speaking for the inference of social interactions. During the conversation, the application checks whether the speaker is the user or someone near him. The recognition was developed with widely known methodologies, and the focus of this work is not to elaborate a new method.

The Figure 5 shows how SocialCount works. The application is responsible for recognizing human voice in the environment, recording audio, locating nearby people and identifying the current location. The server recognizes who is speaking and stores the data in the database.

The application remains listening for the presence of the human voice in the environment. When SocialCount detects voice, it records about 8 seconds and sends the audio to the server. The YIN (De Cheveigné & Kawahara, 2002) algorithm is used to identify human voice in the environment. The algorithm is developed by the TarsosDSP (Six et al., 2014) library, which performs real-time audio processing.

Figure 5. SocialCount flowchart
Source: Authors.

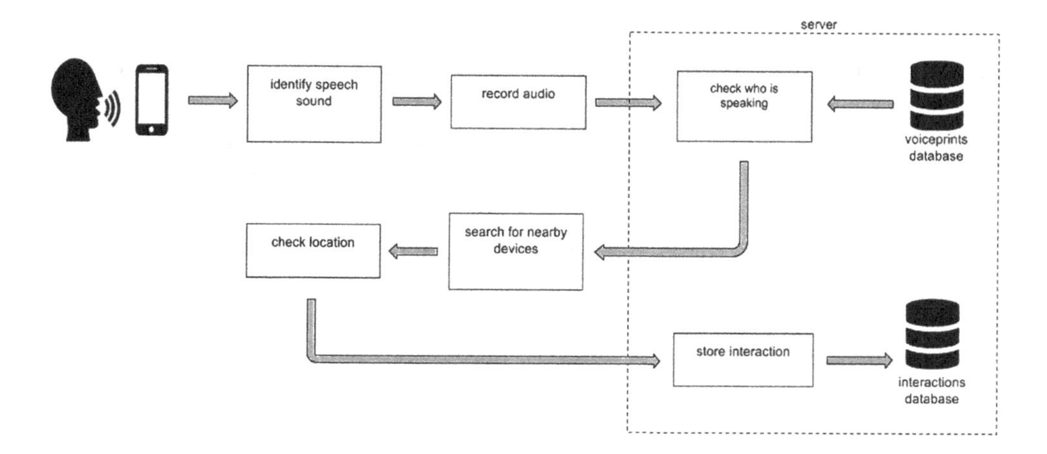

The server verifies that the user is participating in the conversation through voice prints previously stored in the database. Voice prints are a set of audios that contain speech frequency. To record the audios was elaborated phrases that contained the consonant phonemes in the native language in several vowel contexts.

The verification is developed with the Recognito (Crickx, 2014), which is a library that performs text-independent recognition of speakers on Java. The library generates a universal template with all stored voice prints. Each audio input for checking who is speaking Recognito computes the relative distance using the variables: identified voice print (VPI), unknown voice print (VPU), and universal (UM) model. VPI represents the voice prints that have already been identified, VPU is the audio input and UM is the universal model that represents the average of all stored voice prints. "If you put them on a line, you can calculate the distance between IVP/UVP and the distance between UVP/UM. Based on those numbers, you can tell how relatively close the unknown voice print is to the identified one. The UM acts as a max distance value" (Crickx, 2014).

SocialCount can classify an interaction in two ways: participation and monitoring. In participation, the server identifies who is speaking is the user, so he is participating in the interaction. In monitoring, the server identifies who is talking is not the user, it is someone who was close to the user's device. Therefore, the user may not be participating in the interaction. Data from interactions stored as monitoring are only used to increase the accuracy of the inference of the interactions stored as participation.

After finding who is talking, the server sends a response to the application. The application then searches for nearby devices. The Bluetooth ID of nearby devices is used as additional information to increase the inference accuracy of users who are participating in the interaction. If the user chooses not to turn Bluetooth on to save battery, SocialCount continues to function normally.

Finally, SocialCount detects the location of the interaction and sends all the data to the server to store. The stored data are used to generate social graphs. A sociogram can represent the interactions between users of SocialCount at a particular location over a period of time or the interactions performed by a particular user.

To classify the Type of Ties (ToT) of 3-tuple, researchers use the equation proposed by Palaghias (2016) to calculate the confidence of each social relation between a pair of users and the average of interactions performed:

$$P\left(r\right)=\frac{Q\left(r\right)}{N}$$

where Q(r) is the number of inferences of interactions that are related to social relation r and N is the total number of social interactions inferences. In order to adapt the classification according to the social characteristics of each user, the authors calculated the average number of interactions performed per node according to the equation:

$$M\left(n\right)=\frac{N}{T*N}$$

where T is the total number of nodes that have interacted with the current node n. Then, the ToT is classified as follows:

$$ToT=\begin{cases}P\left(r\right)=0, & ToT=Unknown\\ P\left(r\right)<M\left(n\right), & ToT=Acquaintance\\ P\left(r\right)\geq M\left(n\right), & ToT=Friends\end{cases}$$

To demonstrate the use of SocialCount and the classification of the social context, the researchers considered a classroom scenario.

SocialCount in a Classroom

Inasmuch as the sociometric measurement techniques have been refined and become somewhat easy to administer, school districts and classroom teaching have once again become interested in sociometry. Thus, the authors applied SocialCount in a high school classroom to identify the presence of social exclusion and inclusion.

Every teacher knows that the group of children with which he works is more that an aggregation of individuals. He knows that the group has form and structure; that there are patterns of sub-groups, cliques, and friendships. Some individuals are more accepted by the group then others. Some are more rejected. These factors play an important role in determining how the group will react to learning situations and to various types of group management employed by the teacher (Nsamenang & Tchombé, 2012).

Early identification of children likely to be experiencing social rejection and peer neglect is desirable. As in the case of early identification of developmentally handicapped children - sometimes described as "children at risk" - children who are not accepted by their peers may be thought of as being "socially at risk," too (Harris, 2011).

The Figure 6 shows students in a classroom. According to 3-tuple, the classification of the social context of the classroom is:

$$ctx = \left\langle Private, Easy, Friends \right\rangle$$

Size is Private because the number of nodes is 18, the Density is Easy because it is easy to close triangles and the Type of Ties is Friends because most of the relations are P(r)> = M(n). The contexts that have the Density classified as Hard is greater the chance to contain social exclusion. However, even in contexts where Density is Easy or Click, there may be exclusion. In this case, users 10, 15, 18 and 14 needs to receive more attention because they can be classified as "socially at risk".

Through the sociogram the teacher can easily identify the students who deserve attention in the social aspect. Thus, the responsible professionals can intervene to offer a more pleasant social experience to the student. Each case must be analyzed specifically for each student, since the low social interaction index can be derived from several sources, such as: influence of family problems, family absence, differentiated behavior (impulsiveness and aggressiveness), bullying, etc. (Silveira & Wagner, 2012).

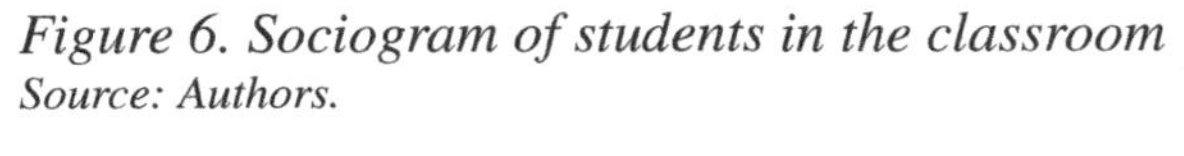

Figure 6. Sociogram of students in the classroom
Source: Authors.

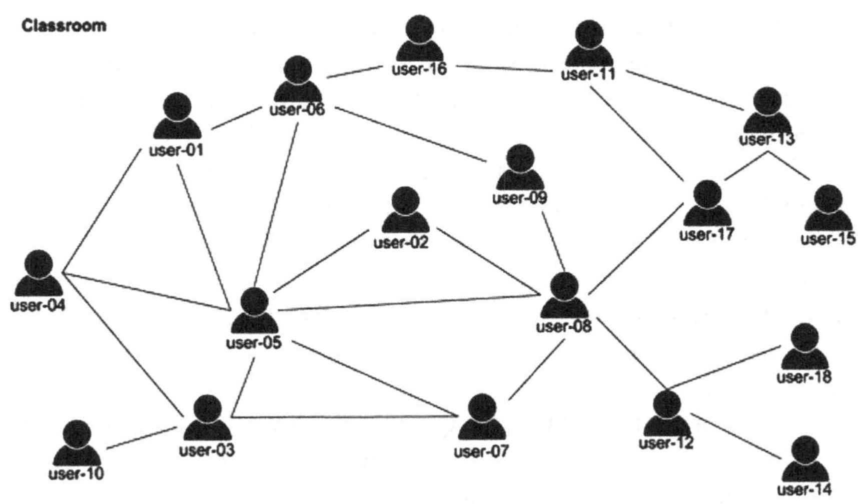

FUTURE RESEARCH DIRECTIONS

This work has as its future objectives some changes that can improve the experience with SocialCount, such as testing other approaches to identify who is speaking to increase the accuracy of classification and generate more accurate sociograms. Apply and adapt SocialCount in other areas, such as: health (infectious disease control), marketing (word-of-mouth), business (enterprise social networks). Develop of new strategies and techniques towards the register of observation and data synthesis of SocialCount in order to identify behavioral changes occurred as a consequence of the interaction with other people.

CONCLUSION

The ideal of an inclusive society varies by country and by region (Silver, 2015). However, the social exclusion can occur in at least four dimensions: consumption, production, political engagement and social interaction. Therefore, this paper proposed the solution of the following question: "How to capture the dynamics of social interactions between individuals to identify possible social exclusion in different environments?".

Psychology and sociology use a method called sociometry to map the relationships between individuals in a group. This method is usually done by questionnaires. Thus, it is subject to human error, impartiality, limitations of scalability. To address these issues, the researchers developed a mobile application called SocialCount. The application identifies the social interactions performed by users and models sociograms that demonstrate relationships in a simple way. SocialCount is developed based on context-aware computing, identifying interactions without user intervention.

SocialCount was included in a classroom and generated a sociogram that adequately represented student relationships in a classroom. Through the sociogram it was possible to visualize the students who may be "socially at risk" and alert the professionals responsible to solve the situation. SocialCount contributes to a new way of identifying social interactions within a group. Thus, it provides the realization of sociometry in different environments, including classrooms with students who would not be able to answer questionnaires.

REFERENCES

Abowd, G. D., Dey, A. K., Brown, P. J., Davies, N., Smith, M., & Steggles, P. (1999, January). Towards a better understanding of context and context-awareness. In *Handheld and ubiquitous computing* (pp. 304–307). Springer Berlin Heidelberg. doi:10.1007/3-540-48157-5_29

Ali, S., & Khusro, S. (2016). Mobile phone sensing: A new application paradigm. *Indian Journal of Science and Technology*, *9*(19). doi:10.17485/ijst/2016/v9i19/53088

Biamino, G. (2011, March). Modeling social contexts for pervasive computing environments. In *Pervasive Computing and Communications Workshops (PERCOM Workshops), 2011 IEEE International Conference on* (pp. 415-420). IEEE. 10.1109/PERCOMW.2011.5766925

Bronfenbrenner, U. (1944). A constant frame of reference for sociometric research: Part II. Experiment and inference. *Sociometry*, *7*(1), 40–75. doi:10.2307/2785536

Burchardt, T., Le Grand, J., & Piachaud, D. (2002). Degrees of exclusion: developing a dynamic, multidimensional measure. In J. Hills, J. Le Grand, & D. Piachaud (Eds.), *Understanding Social Exclusion* (pp. 30–43). Oxford, UK: Oxford University Press.

Carter, I. (2013). *Human behavior in the social environment*. AldineTransaction.

Crickx, A. (2014). *Recognito: Text Independent Speaker Recognition in Java*. Available in: <https://github.com/amaurycrickx/recognito>

De Araújo, R. B. (2003, May). Computação ubíqua: Princípios, tecnologias e desafios. In XXI Simpósio Brasileiro de Redes de Computadores (Vol. 8, pp. 11-13). Academic Press.

De Cheveigné, A., & Kawahara, H. (2002). YIN, a fundamental frequency estimator for speech and music. *The Journal of the Acoustical Society of America, 111*(4), 1917–1930. doi:10.1121/1.1458024 PMID:12002874

De Mello, E. D. F. F., & Teixeira, A. C. (2011, November). A interação social descrita por Vigotski e a sua possível ligação com a aprendizagem colaborativa através das tecnologias em rede. In Anais do Workshop de Informática na Escola (Vol. 1, No. 1, pp. 1362-1365). Academic Press.

DESA. (2016). *Leaving no one behind: the imperative of inclusive development. Report on the world social situation*. DESA.

Dey, A. K. (2001). Understanding and using context. *Personal and Ubiquitous Computing*, *5*(1), 4–7. doi:10.1007007790170019

Eagle, N., & Pentland, A. (2005). Social serendipity: Mobilizing social software. *IEEE Pervasive Computing*, *4*(2), 28–34. doi:10.1109/MPRV.2005.37

Groves, R. M. (2004). *Survey errors and survey costs* (Vol. 536). John Wiley & Sons.

Hari, R., & Kujala, M. V. (2009). Brain basis of human social interaction: From concepts to brain imaging. *Physiological Reviews*, *89*(2), 453–479. doi:10.1152/physrev.00041.2007 PMID:19342612

Harris, J. R. (2011). *The nurture assumption: Why children turn out the way they do*. Simon and Schuster.

Hoseinitabatabaei, S. A., Gluhak, A., Tafazolli, R., & Headley, W. (2014). Design, realization, and evaluation of uDirect-An approach for pervasive observation of user facing direction on mobile phones. *IEEE Transactions on Mobile Computing*, *13*(9), 1981–1994. doi:10.1109/TMC.2013.53

Jennings, H. H. (1959). *Sociometry in group relations: A manual for teachers*. American Council on Education.

Khan, W. Z., Xiang, Y., Aalsalem, M. Y., & Arshad, Q. (2013). Mobile phone sensing systems: A survey. *IEEE Communications Surveys and Tutorials*, *15*(1), 402–427. doi:10.1109/SURV.2012.031412.00077

Kim, C. P. J. L. J., & Lee, S. J. L. D. (2017). *"Don't Bother Me. I'm Socializing!": A Breakpoint-Based Smartphone Notification System*. Academic Press.

Kofod-Petersen, A., & Cassens, J. (2006). Using activity theory to model context awareness. *Lecture Notes in Computer Science*, *3946*, 1–17. doi:10.1007/11740674_1

Kolvenbach, S., Grather, W., & Klockner, K. (2004, February). Making community work aware. In *Parallel, Distributed and Network-Based Processing, 2004. Proceedings. 12th Euromicro Conference on* (pp. 358-363). IEEE. 10.1109/EMPDP.2004.1271466

Lane, N. D., Miluzzo, E., Lu, H., Peebles, D., Choudhury, T., & Campbell, A. T. (2010). A survey of mobile phone sensing. *IEEE Communications Magazine*, *48*(9), 140–150. doi:10.1109/MCOM.2010.5560598

Lu, H., Pan, W., Lane, N. D., Choudhury, T., & Campbell, A. T. (2009, June). SoundSense: scalable sound sensing for people-centric applications on mobile phones. In *Proceedings of the 7th international conference on Mobile systems, applications, and services* (pp. 165-178). ACM. 10.1145/1555816.1555834

Lyytinen, K., & Yoo, Y. (2002). Ubiquitous computing. *Communications of the ACM*, *45*(12), 63–96.

Matic, A., Osmani, V., Maxhuni, A., & Mayora, O. (2012, May). Multi-modal mobile sensing of social interactions. In *Pervasive computing technologies for healthcare (PervasiveHealth), 2012 6th international conference on* (pp. 105-114). IEEE. 10.4108/icst.pervasivehealth.2012.248689

Miller, S. (2010). *The moral foundations of social institutions: A philosophical study*. Cambridge University Press.

Miluzzo, E., Lane, N. D., Eisenman, S. B., & Campbell, A. T. (2007, October). CenceMe–injecting sensing presence into social networking applications. In *European Conference on Smart Sensing and Context* (pp. 1-28). Springer. 10.1007/978-3-540-75696-5_1

Moreno, J. L. (1953). *Who shall survive?* Academic Press.

Nsamenang, A. B., & Tchombé, T. M. (Eds.). (2012). *Handbook of African educational theories and practices: A generative teacher education curriculum*. HDRC.

Palaghias, N., Hoseinitabatabaei, S. A., Nati, M., Gluhak, A., & Moessner, K. (2015, June). Accurate detection of real-world social interactions with smartphones. In *Communications (ICC), 2015 IEEE International Conference on* (pp. 579-585). IEEE. 10.1109/ICC.2015.7248384

Palaghias, N., Hoseinitabatabaei, S. A., Nati, M., Gluhak, A., & Moessner, K. (2016). A survey on mobile social signal processing. *ACM Computing Surveys*, *48*(4), 57. doi:10.1145/2893487

Park, R. E., & Burgess, E. W. (1921). *Introduction to the Science of Sociology*. Chicago: University of Chicago Press.

Prasad, L. (2012). *Pervasive Computing goals and its Challenges for Modern Era*. Academic Press.

Primo, A. F. T. (2006). O aspecto relacional das interações na Web 2.0. In Congresso Brasileiro de Ciências da Comunicação (29.: 2006 set.: Brasília). Anais: estado e comunicação [recurso eletrônico]. Brasília, DF: Intercom: Universidade de Brasília.

Rabello, E. T., & Passos, J. S. (2010). *Vygotsky e o desenvolvimento humano. Formato do arquivo: Microsoft Powerpoint-Visualização rápida*. Retrieved from www.ceesp.com.br/arquivos/Aula

Rock, D., & Page, L. J. (2009). *Coaching with the brain in mind: Foundations for practice*. John Wiley & Sons.

Rostampoor-Vajari, M. (2012). What Is Sociometry and How We Can Apply It in Our Life. *Advances in Asian Social Science*, *2*(4), 570–573.

Roussos, G. (2006). Ubiquitous computing for electronic business. *Ubiquitous and Pervasive Commerce*, 1-12.

Saha, D., & Mukherjee, A. (2003). Pervasive computing: A paradigm for the 21st century. *Computer*, *36*(3), 25–31. doi:10.1109/MC.2003.1185214

Schilit, B., Adams, N., & Want, R. (1994, December). Context-aware computing applications. In *Mobile Computing Systems and Applications, 1994. WMCSA 1994. First Workshop on* (pp. 85-90). IEEE. 10.1109/WMCSA.1994.16

Schuster, D., Rosi, A., Mamei, M., Springer, T., Endler, M., & Zambonelli, F. (2013). Pervasive social context: Taxonomy and survey. *ACM Transactions on Intelligent Systems and Technology*, *4*(3), 46. doi:10.1145/2483669.2483679

Silveira, L. M. D. O. B., & Wagner, A. (2012). A interação família-escola diante dos problemas de comportamento da criança: Estudos de caso. *Psicologia da Educação*, (35): 95–119.

Silver, H. (2015). *The Contexts of Social Inclusion*. Academic Press.

Six, J., Cornelis, O., & Leman, M. (2014, January). TarsosDSP, a real-time audio processing framework in Java. In *Audio Engineering Society Conference: 53rd International Conference: Semantic Audio*. Audio Engineering Society.

Temdee, P., & Prasad, R. (2017). *Context-Aware Communication and Computing: Applications for Smart Environment*. Academic Press.

Teng, J., Zhang, B., Li, X., Bai, X., & Xuan, D. (2014). E-shadow: Lubricating social interaction using mobile phones. *IEEE Transactions on Computers*, *63*(6), 1422–1433. doi:10.1109/TC.2012.290

Thompson, J. B. (1998). *O advento da interação mediada. A mídia e a modernidade: uma teoria social da mídia*. Petrópolis, RJ: Vozes.

Tylor, E. B. (1871). *Primitive culture: researches into the development of mythology, philosophy, religion, art, and custom* (Vol. 2). J. Murray.

United Nations Department of Economic and Social Affairs. (2016). New York: UN DESA.

Vinciarelli, A., Murray-Smith, R., & Bourlard, H. (2010, September). Mobile social signal processing: vision and research issues. In *Proceedings of the 12th international conference on Human computer interaction with mobile devices and services* (pp. 513-516). ACM. 10.1145/1851600.1851731

Weiser, M. (1991). The Computer for the 21 st Century. *Scientific American*, *265*(3), 94–105. doi:10.1038cientificamerican0991-94 PMID:1675486

Xavier, O. S. (1990). A sociometria na administração de recursos humanos. *Revista de Administração de Empresas*, *30*(1), 45–54. doi:10.1590/S0034-75901990000100005

Xu, Q., Chia, S. C., Mandal, B., Li, L., Lim, J. H., Mukawa, M. A., & Tan, C. (2016). SocioGlass: Social interaction assistance with face recognition on google glass. *Scientific Phone Apps and Mobile Devices*, *2*(1), 1–4. doi:10.118641070-016-0011-8

Zhang, R., Zhang, Y., Sun, J., & Yan, G. (2012, March). Fine-grained private matching for proximity-based mobile social networking. In INFOCOM, 2012 Proceedings IEEE (pp. 1969-1977). IEEE. doi:10.1109/INFCOM.2012.6195574

Chapter 2
The Urbis Prototype Development:
A Touristic Guide Application

Ivaldir Honório de Farias Junior
UFPE University, Brazil

Nelson Galvão de Sá Leitão Júnior
UFPE University, Brazil

Marcelo Mendonça Teixeira
UFRPE University, Brazil

Jarbas Espíndola Agra Junior
UNICAP University, Brazil

ABSTRACT

Currently, tourists spend a lot of time planning their trips because they need to make the most of every moment. In this sense, technology has been a great ally, especially when performing and adapting this planning in the event of some unforeseen event during the journey. And the emergence of distinct types of mobile devices was presented as an opportunity to improve the experience of tourism significantly. In this context, this chapter aims to identify the main computing needs to a mobile application to support the promotion of tourist sites for the traveler. The authors adopted a literature review as the research methodology. The main result of this chapter is the proposal of the Urbis prototype, an application that aims to help tourists to know better the cities they are visiting, even in the absence of local information or a specialized tour guide.

DOI: 10.4018/978-1-5225-5270-3.ch002

INTRODUCTION

The constant development of information technology has driven major changes in modern society, resulting in changes in economic, political and social levels (de Andrade, 2006) (Anjos, Paula Souza, & Vieira Ramos, 2006). It is common to consider that technology is one of the progress engines, thereby providing the development of human knowledge (Pinheiro, Silveira, & Bazzo, 2007). Therefore, information technology turns out to be one of the main tools that companies have to differentiate themselves from their competitors in the market, ensuring their competitive advantage (Morgan, Translate, & Summers, 2008) (Neves, Semprebom, & Lima, 2011). One of the productive sectors starting to feel the changes brought about by the popularization of the Internet and the modernization of information systems is tourism. Vicentin and Hoppen define the impact of the information age in tourism businesses and their customers as follows:

For the customer this type of business, the tourism, when is still not experienced, it can only be perceived as a set of available information. Thus, coupled with the fact that the Internet is an information technology that enables widely available information quickly and easily, allowing the emergence of numerous websites specializing in tourism marketing on the Internet. These sites (companies) may be causing changes in the business models of those involved in tourism in Brazil. (Vicentin & Hoppen, 2002)

The information technology revolution is causing a profound impact on the way that trips are marketed, distributed, sold and delivered, simply because the business covering the travel is formed by information (Vassos, 1998). It is worth mentioning that the information is essential for tourism in this new information age. In this context, Bissoli states:

Tourist activity generates a great amount of information that has importance and strategic value in the tourist business, meaning that information should be treated as an element of organizational strategy. (Bissoli, 2001)

At this moment, the customer that was once exclusively bound to information coming from travel agencies, now has more power, interactivity, and flexibility, accessing the information and the product more directly and clearly (Anjos et al., 2006). It can be observed that in the last decade, the growing trend of the new traveler of being always connected. The emergence of different types of mobile devices is presented as an opportunity to improve the tourist's experience (Buhalis & Law, 2008). This work aims to contribute to improving the promotion of sights through the development of an application that helps tourists to better know the cities, even in the absence of a specialized tourist guide. Taking advantage of this form of the demand for products and services that take place in Brazil in the coming years, due to increased tourism projections.

The habit of traveling was always present in human history. The travel and the ways of realizing them diversified in every age and every civilization, acquiring different meanings by the people's material, acquired knowledge and cultural beliefs (Miranda, 2002). The tourism we know today is a peculiar phenomenon of the twentieth century, many researchers define tourism as "A way of life institutionalized for most of the world's middle-class population." Tourism has a significant importance in the world economy, and in 2001, it was considered the fastest growing industry in the world, reflecting the generation of jobs, income distribution and raising the conditions of life of host communities (Pearce, 2001).

With the development of the globalization, the advent of technological innovations and the expansion of the economy, tourism had an amazing breakthrough, reaching in 2003 the amount of U$ 3.5 trillion in financial movements, as stated by the World Tourism Organization (UNWTO). In this sense, many communities that had tourism potential and that were in financial crisis, resorted to this activity to boost their economies (Neves et al., 2011). According to EMBRATOUR (www.embratur.gov.br), only during the World Cup in Brazil in 2014, with games in 12 host cities, the Brazilian Ministry of Tourism received the admission of 1,736,645 foreign visitors, which according to the Brazilian Central Bank, generated the amount of U$ 1.578 billion. Only in three of the major events that Brazil received in recent years, World Youth Day (WYD), FIFA Confederations Cup and FIFA World Cup, directly engaged the amount BR$ 8 billion in the Brazilian economy. The rest of the chapter is presented as follows: a background section with the theoretical reference from tourism and android literature, related work, a section describing the development of the Urbis prototype, a usability analysis on the developed prototype, final remarks, and references.

BACKGROUND

Technology and Tourism

Tourists spend considerable time planning their trips, but they need to adapt in the event of any unforeseen, and these trips become not very structured (Neves et al., 2011) (Pareschi, Riboni, & Bettini, 2008). Brown and Chalmers (Brown & Chalmers, 2003) cite many points that exemplify the challenges that tourists usually find during their trips, one of the points mentioned by the authors seems quite simple, but it is quite common for people who are not familiar with the place, this point is: what can you do here? This decision is based on several factors such as the prevailing conditions, the purpose of tourist on site, the time of arrival, the main attractions and even the tourist's predisposition in spending money and time in commuting between places for seeking nearby attractions. Until some time ago the main sources of information for a tourist were maps, tourist publications, and guides, which end up being used together, and still, finding the location of a tourist attraction that is not on the route of the most visited attractions, could be a challenge. Often tourists had a vague notion of the location they visited (Tapscott, 1997).

With the Internet, the tourist can plan and fully program the trip to anywhere in the world without the need of leaving home. Being able to make reservations for hotels and airline tickets, make payments, rent vehicles, check the weather conditions at the destination, currency quotations, learn about the fate of culture, verify documentation needed, that is, almost everything that is essential to a trip. At this point, the use of technology can be of great assistance to tourists. According to Werthner et al. (Werthner, Klein, & others, 1999), the information technology brings some benefits for tourists, such as process automation, acceleration of routine processes, real-time information of destinations and greater and improved analytical tourist capacity by distributing information on services and products. According to Carvalho (Carvalho, 2004) and Guimarães (Guimarães & Johnson, 2007), the greatest impact of IT in customer relationships was the advent of systems known as CRM (Customer Relationship Management). Carvalho defines CRM as: "The most comprehensive set of processes, strategies, and technologies for the management of relationships with customers and prospects" but Guimarães describes CRM as something more than a set of strategies and technology, to this author, the CRM can be defined as:

A business philosophy, which is founded on the understanding of the client, trying to understand and influence their behavior, maintaining meaningful communication, improving the shopping, retention, loyalty, and also the return that favor to the company. (Guimarães & Johnson, 2007)

The consumer has been able to use technology to their advantage comparing prices, quality and service, in this context, the CRM emerged as a response to consumer behavior change and companies must be prepared to use these systems to create a relationship with the customer, offering the right product at the right time, creating an active customer loyalty.

Android's Architecture

The Android operating system architecture can be divided into four layers: Linux Kernel, Libraries, Application Framework and Applications (Lecheta, 2012) (Lecheta, 2013). The first layer is its Linux Kernel. To its development, it was used for version 2.6 of the Linux operating system. In it, we find the memory management programs, security settings, and various hardware drivers.

The second layer is the Libraries and Android Runtime layer. These Libraries are set of instructions that tell the device how to handle different data types, also including a set of library C / C ++ used for various system components. The third layer is the Application Framework, i.e. programs that handle basic phone applications. Developers have full access to this layer, using it as a set of basic tools with which to build their applications. The fourth layer is the Applications layer. This is the layer of interaction between the user and the android mobile. All the routinely used apps can be found in this layer, such as email clients, calendar, and maps, in addition to the basic functions concerning the device.

- **Google Maps API:** One of the features that draw more attention on the Android platform is the ease we have to integrate an application to Google Maps. In December 2012 Google released its new Android platform API integration with Google Maps, called Google Maps Android API V2. The whole framework has been redesigned to support views in 2D and 3D, which enables a significant performance gain, making animations and more fluid interactions to the user (Glauber, 2015) (Lecheta, 2013).
- **Localization Based Services:** Find bars, restaurants or the nearest hotel in an unknown location has always been a small challenge, actively dependent on the search for information with the people who best knew the region. But nowadays the situation has to rule the exception, since with the possession of a mobile device connected to the Internet with just a few clicks you can get this information (Ribeiro & Zorzo, 2009). The Location Based Services or LBS (Location Based Services) are services that use the user's position on any device that can be geographically located with some specific purpose (Porter, 2001). One of the markets that have been more promising for the use of these services is the mobile technology market, as well as offering devices with high processing power. The LBS has provided an improvement in internet connection technologies and new location techniques (Martucci, Andersson, Schreurs, & Fischer-Hübner, 2006).

- **REST Protocol:** The Representational State Transfer Protocol (REST) is an architectural style that consists of a coordinated set of architectural constraints applied to components, connectors and data elements within a distributed system. The REST ignores the implementation details of the components focusing on the roles of the components, the restrictions on their interaction with other components and interpretation of significant data elements (Fielding, 2015) (Rabello, 2009) (Reckziegel, 2009). The REST is intended as an image of the application design, and it behaves as a website network (a virtual state) where the user progresses with an application by clicking on links (state transitions), resulting in the next page (which is the next state of the application) being transferred to the user and presented for your use (Fielding, 2015). Usually, the file format used by RESTful services are JSON, XML or YAML (Reckziegel, 2009).
- **Open Data From Recife City Prefecture:** Launched in July 2013, the open data portal of the city of Recife has emerged as an innovative initiative regarding the publication of municipal open data in Brazil. This initiative is being led by EMPREL (http://www.emprel.gov.br/) and has the support of the town hall, as well as many other organs of the municipal government, relying on the collaboration of the professors from the Informatics Center of the Federal University Pernambuco (CIn / UFPE). For the implementation of the open data portal, it was allocated a multidisciplinary team and strongly committed to the project, which contributed greatly to its success and completion in record time (Recife, 2017).

RELATED WORK

Many applications are designed to help the large number of national and international tourists who visit Brazil in the coming years, using technology in your favor to address this shortcoming in tourist information in most Brazilian cities, hosting or not major events. There is a common feature among these applications, which is the development of specific regions of states and cities. It can be assumed that many tourist applications will also be developed, even for the same regions or targeting unique areas to interact with the end user, such as applications targeted to people with disabilities, the elderly and children. The user can use several applications simultaneously on your smartphone, each with its purposes and specific features, thus allowing freedom of choice. The following list of topics presents the projects related to this work.

- **Reviva Manaus:** The "Reviva Manaus" (in Portuguese) is an application that works as a tourist guide to the city of Manaus, created by the designer Gisely Mendonça, as a response to the problem of lack of tourist information and bilingual information signs in the city. According to the creator, the greatest difficulty faced during the application development was the collection of information of the city's sights. The system allows the user to find a brief history of the monument and can share it on the major social networks, emails, and taking pictures, checking in, rank as favorites and view the comments of other users. Therefore, working as a free dissemination for the town. Gisely Mendonça used concepts such as optimization of information, simplified visual language, usability, and interaction (Melo, 2012).
- **CURTA Curitiba:** Created by a partnership between the Brazilian Association of Bars and Restaurants of Paraná - Brazil (Abrasel - PR), CCVB - Curitiba, Region and Coast Convention & Visitors Bureau, Municipal Tourism Institute, Sebrae / PR and enabled by the local commerce, the free application "CURTA Curitiba" (in Portuguese) brings not only information on tourism but also on services, lodging, dining, directions on how to get around in the city, etc. Without wasting time and without being connected to the internet, it can search over the city, choose from tourist options, find hotel contacts, restaurants or where the tourist wants to go. The application is available on ITunes. More information on the site: http://www.guiaturismocuritiba.com/2011/03/aplicativo-curta-curitiba-para-tablets.html.
- **CURITOUR:** Created by systems analyst and UFTFR student, Rodrigo Fagundes, the CURITOUR is an application to facilitate the meeting of sights in the Brazilian city of Curitiba, serving as a virtual tour guide for the city. It allows us to find all the sights through a map within the application, once chosen identified the tourist spot on the map, it can be traced a route from our current point to the site of interest using the modes of "using a car" and "using a bus". The application also gives us the opportunity to create a personalized itinerary according to our program preferences for the day in the city.
- **Enguia:** Created by post-graduation students of the Institute of Higher Studies of the Amazon, the "Enguia" (in Portuguese) is an application with the goal of acclimatizing and guide the tourists in the Brazilian city of Belém. The application brings a new vision of tourist interaction vs. city, with important resources for providing a good impression of the sites to the tourists. Information such as transportation, sightseeing, and local dictionary, increases the overall convenience for the visitors (Carvalho, 2004). Table 1 shows a comparative view of the applications that were mentioned in this chapter.

Table 1. Tourism applications comparison

	Reviva Manaus	CURTA Curitiba	CURITOUR	Enguia
Touristic sites	V[1]	V	V	V
Access to maps	X[2]	X	V	V
Bus routes	X	V	V	X
Bike lanes	X	X	X	X
Hotels	X	V	X	X
Check-in	V	X	X	X
Favorite sites	V	X	X	X
Bilingual	V	X	X	X
Social networks	V	X	X	X
Collaborative	V	V	V	V

[1]The feature is available.
[2]The feature is not available.

It can be noticed that only some of the applications adds real value to the user experience, most of them focus on showing the basics of touristic site information, but "Reviva Manaus" stood out as a more user interaction oriented application, that's because it provides more integration features such as logging into social networks and sites check ins. It's worth also mentions that only the "CURTA Curitiba" and "CURIOTOUR" applications have the information of the bus routes in the city.

DEVELOPMENT OF THE URBIS PROTOTYPE

For the development of the Urbis application prototype, the first activity was the elicitation of its requisites, followed by all necessary UML modeling, an investigation of the sites geographic coordinates, the creation of the sights icons and finally, the implementation of the application for the Android platform.

Requirements Elicitation

Requirements elicitation is of foremost importance for the project's success, and the attention given to requirements will always be essential to projects of any sizes. Functional requirements are requirements that express functions or services that software should or may be able to perform or provide. Non-functional requirements are requirements that state restrictions or quality attribute in the software development process (Cysneiros, 1997). The analysis of related applications was the source of the Urbis requirements; this investigation was based on the identification of good practices and features that were proposed by these works. The identified requirements are listed in Table 2, Functional Requirements (FRs) and in Table 3, Non-Functional Requirements (NFRs).

Modeling

For the application modeling, the UML standard was used. The UML provides standards for the preparation of system architectural design plans, including conceptual aspects, such as business processes and system functions as well as concrete elements, such as classes written in a particular programming language, database schema and reusable software components. The AstahUML tool (Available at http://astah.net/) was used to create the diagrams. Figures 1 and 2 illustrate the class and use case diagrams for the proposed application.

Product Description

The Urbis application prototype was designed to work on smartphones with the Android 4.0 (Ice Cream Sandwich) or later. The application requires a data connection

Table 2. Functional requirements

FR001	The user can plot routes to desired sights.
FR002	The user can save the tourist site as favorite.
FR003	The user can add new sights.
FR004	The user can configure a sightseeing tour.
FR005	The user can view information about a particular tourist site.
FR006	The user can perform check in at the tourist site and share on social networks.
FR007	The user can register new sites of interest. (Hotels, bars, etc.).
FR008	The user can give notes to visited sites of interest and tourist sites visited.

Table 3. Non-functional requirements

NFR001	The application should be simple and intuitive for the user.
NFR 002	The application should respond quickly to commands on the screen.
NFR 003	The application should handle its inherent errors.
NFR 004	The application must be functional on smartphones.
NFR 005	The application must run on the Android operating system.

Figure 1. Class diagram

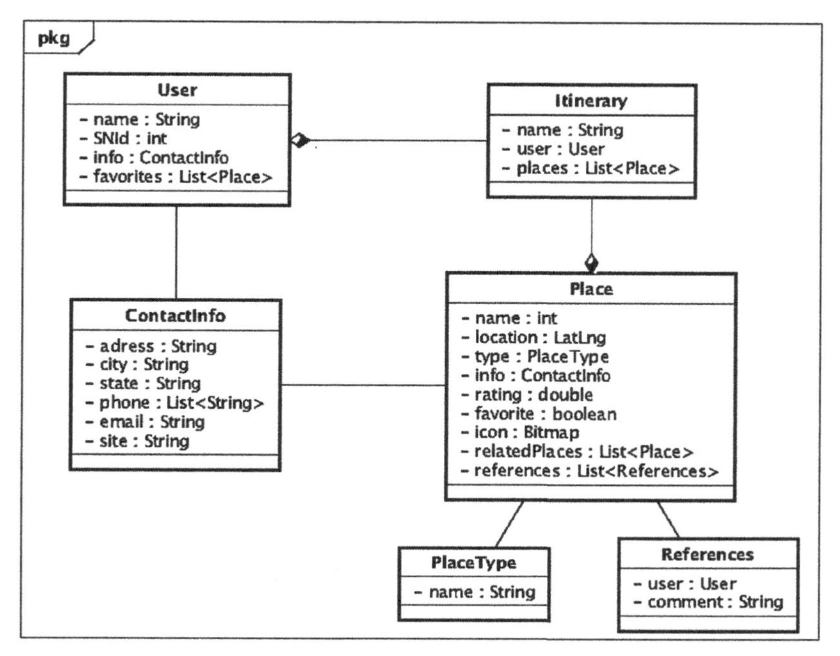

to the Internet to use the maps and to authenticate and receive data. The development of the application was made considering the client-server model, where multiple clients can simultaneously access the server for seeking information in a database, for retrieving and sending information on the sites of interest. This model allows greater flexibility with the previously persisted and new user data handling. This application is intended to be simple, interactive, easy and at the same time useful to the tourist who is interested in learning more about the sights of the city of Recife without the need of a tour guide.

The application displays a menu with the main categories of attractions available in the city, by clicking on one of the categories all the related information is displayed. The user will also have access to more detailed information about the site and have many actions, e.g., call the tourist spot, navigate to it or know a little more of its history. When the navigation feature is selected, the application will redirect the user to the desired navigation application, such as Waze or Google Maps. The user can also have the possibility to save one tourist spot as one of their favorites, with this action the tourist spot is moved to an exclusive list that will be easily accessible later. In Figures 3 and 4 are displayed the main application, sites of interest and

Figure 2. Use cases diagram

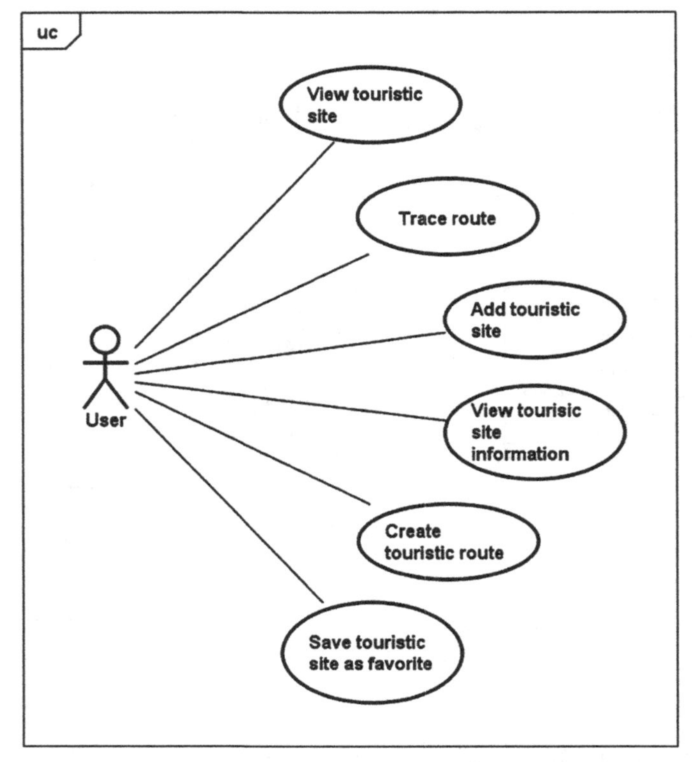

details screens respectively. All the displayed data were retrieved through the Open Data Portal of Recife Brazilian city.

USABILITY ANALYSIS

Interaction and User Experience

Preece, Rogers and Sharp (Preece, Rogers, & Sharp, 2005) states that, among the concerns for the development of an interactive product, the following elements should be considered: ease of learning, the effectiveness of use and the ability to provide a rewarding experience. It is up to the field of Interaction Design to meet these and other criteria that involve the relationships between humans and artifacts/services.

Figure 3. Urbis main screen

Interaction design means the creation of "experiences that improve and extend the way people work, communicate and interact." (Preece et al., 2005). The quality of these interactions with the artifact will impact on what is meant by user experience, which can be explained as:

Every aspect of the user's interaction with a product, service, or company that make up the user's perceptions of the whole. User experience design as a discipline is concerned with all the elements that make up that interface, including layout, visual design, text, brand, sound, and interaction. EU works to coordinate these elements to allow for the best possible interaction by users. (UXPA, 2017)

To guarantee the minimum qualities associated with a rewarding experience, the fields of design, software engineering and others involved in information technology development have set standards, guidelines, and principles that guide the creation of so-called friendly interfaces. These methods of improving ease of use during the design process are called usability.

Figure 4. Sites of interest and details screen

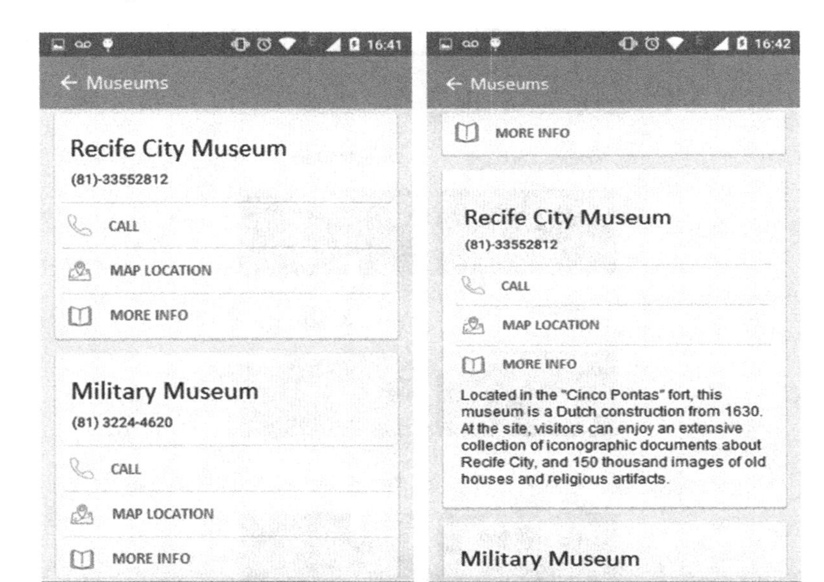

Usability

Usability is a term defined by ISO 9126 (ISO, 1991) as "the set of attributes that bear on the effort needed for use, and on the individual assessment of such use, by a stated or implied set of users." For Nielsen (Nielsen, 1994), usability is defined by 5 components:

1. **Learnability:** How easy is it for users to accomplish basic tasks the first time they encounter the design?
2. **Efficiency:** Once you have learned the design, how quickly can they perform tasks?
3. **Memorability:** When users return to the design after a period of not using it, how easily can they reestablish proficiency?
4. **Errors:** How many errors do users make, how severe are these errors, and how easily can they recover from the errors?

Heuristics

Based on these criteria and the evaluation of a series of usability problems, Nielsen and Molich (Nielsen & Molich, 1990) propose a series of heuristic principles, defined as "the method for finding the usability problems in the user interface design so that they can be attended to Part of an iterative design process ". From these principles, the author (Nielsen, 1994) started to a refinement of 10 criteria with the maximum explanatory power to evaluate the interfaces, they are:

1. **Visibility of System Status:** The system should always keep users informed about what is going on, through appropriate feedback within reasonable time.
2. **Match Between System and the Real World:** The system should speak the users' language, with words, phrases and concepts familiar to the user, rather than system-oriented terms. Follow real-world conventions, making information appear in a natural and logical order.
3. **User Control and Freedom:** Users often choose system functions by mistake and will need to clearly mark "emergency exit" to leave the unwanted state without having to go through an extended dialogue. Support undo and redo.
4. **Consistency and Standards:** Users should not have to wonder whether different words, situations, or actions mean the same thing.
5. **Error Prevention:** Even better than good error messages is a careful design which prevents a problem from occurring in the first place. Either eliminate error-prone conditions or check for them and present users with a confirmation option before they commit to the action.
6. **Recognition Rather Than Recall:** Minimize the user's memory load by making visible objects, actions, and options. The user should not have to remember information from one part of the dialogue to another. Instructions for the use of the system should be visible or easily retrievable whenever appropriate.
7. **Flexibility and Efficiency of Use:** Accelerators - unseen by the novice user - may often speed up the interaction for the expert user such that the system can cater to both inexperienced and experienced users. Allow users to tailor frequent actions.
8. **Aesthetic and Minimalist Design:** Dialogues should not contain information which is irrelevant or rarely needed. Every extra unit of information in a dialogue competes with the relevant units of information and decreases their relative visibility.
9. **Help Users Recognize, Diagnose, and Recover From Errors:** Errors messages should be expressed in plain language (no codes), precisely indicate the problem, and constructively suggest a solution.
10. **Help and Documentation:** Even though it is better if the system can be used without documentation, it may be necessary to provide help and documentation. Any such information should be easy to search, focused on the user's task, concrete steps to be carried out, and not be too large.

The Heuristic Evaluation

To identify initial problems in the Urbis application interface, this chapter proposes a heuristic evaluation based on 6 of the 9 criteria mentioned above, due to the prototyped nature of this application. For this study, some representative screens were selected (main screen, general menu tab, site of interest screen and detail screen of the site of interest), since the product is still in the prototyping phase. To categorize the identified problems, the work follows the criterion of the degree of severity defined by Nielsen (Nielsen, 1994), which ranges from 0 to 4 and is defined as follows:

Category 0: I do not agree that this is a usability problem.
Category 1: Cosmetic problem.
Category 2: Less usability problem.
Category 3: Greater usability problem.
Category 4: Usability catastrophe.

The analysis followed the order of items in Nielsen's heuristics:

- **Visibility of System Status:** The system displays an acceptable level of visual feedback. When selected, the items present in the upper area the titles of the sections, for example, placing the user within the navigation. In the main menu, the selected items differ by color. Improvement suggestions: Sound feedback could be included. Category 2.
- **Match Between System and the Real World:** The system is objective and clear in its language also provides easy interpretation and readability icons that serve as visual support for textual information of the search categories. Improvement Suggestions: The Main Menu Bar icons could be more intuitive and more representative of the associated functions. Category 1.
- **Consistency and Standards:** The application present satisfactory solution of color patterns, typography, makes appropriate use of font sizes hierarchies. The placement of the buttons, as well as the types of interactions, remain predictable, allowing the user to avoid unnecessary errors and to provide intuitive interactions. Suggestions for improvements: None.
- **Flexibility and Efficiency of Use:** For a prototype stage, the application meets this criterion very well, allowing for easy navigation. It is possible to observe a linear navigation through the screens, which allows the indiscriminate use by distinct types of users. Suggestions for improvements: If you want to consider the more experienced user, it would be possible to study the screens flow map and think about the various navigation possibilities to meet both the novice user and the experienced. Category 2.

- **Aesthetic and Minimalist Design:** The application serves more than satisfactorily this requirement, especially when dealing with a prototype. The elements are all in the right places from the point of view of the diagram, the typography presents size and contrast that allow good readability, the icons are quite simple, as well as the use of colors without extravagance, allowing the user to focus on what matters most without loss Information or distraction during use. Suggestions for improvements: None.
- **Help and Documentation:** This version does not provide help or documentation. Suggestions for improvements: create an initial tutorial to use or add a help section with general information of recurring questions. Category 2.

Considerations on the Analysis

After the evaluation, it is possible to observe that the application minimally meets some of the criteria established by Nielsen. In most cases, it is more than satisfactory. The problems that arise, when they arise, are mostly low-impact for user interaction at this prototype level (although very important for product development in its subsequent phases). All problems can be solved with improvements, with few cases where there is a complete violation of the Heuristic criteria, as in the case of aid documentation. Although partial and preliminary, this analysis reveals the points that need to be improved in the product. For this, it would be important to obey the suggestions of improvements indicated and move to the second phase of heuristics associated with a direct test with users to submit the product to a direct contact with possible users of the application.

CONCLUSION

The Recife city has several attractions, e.g., parks, public squares, and museums. Also has a special bus line that runs on the main touristic sites, with the accompaniment of a tour guide. However, the demand for these professionals is very high, and in some cases, the guides are not sufficient for the whole group. There are even some programs and sights without guidance and even information boards. During the usability analysis on the developed prototype, we could already apply six of the Nielsen's and Molich's heuristics. And although we analyzed this application in its preliminary state, the Urbis Prototype achieved acceptable results on its overall usability. Concerning the prototype's development weaknesses and limitations, we may state its dependency on external services, that usually, could not be replaced by alternative ones, the lack of its availability to iOS, and the lack of a deeper

Table 4. Comparison between tourism applications and Urbis

	Urbis	Reviva Manaus	CURTA Curitiba	CURITOUR	Enguia
Touristic sites	V[1]	V	V	V	V
Access to maps	V	X	X	V	V
Bus routes	X[2]	X	V	V	X
Bike lanes	V	X	X	X	X
Hotels	V	X	V	X	X
Check-in	V	V	X	X	X
Favorite sites	V	V	X	X	X
Bilingual	V	V	X	X	X
Social networks	V	V	X	X	X
Collaborative	V	V	V	V	V
Touristic sites	V	X	X	V	V

[1]The feature is available.
[2]The feature is not available.

architectural specification on its main core components. Future works include further testing on the prototype's functional and non-functional requirements, and improving its usability and stability. Table 4 presents the results of the comparison between Urbis and the tourism applications identified at Related Work Section. We proposed this comparison according to the specifications of the functional and non-functional requirements from all applications. The comparison shows that the Urbis offers the basic touristic functionalities (e.g., viewing the sights), and offers features for an aggregated user experience, such as viewing sites of interest (e.g., bars, restaurants, hotels). The possibility of tracing routes with the user-selected external application and the check-in and touristic route features. And just like the CURITOUR and "Enguia" applications, the Urbis has been developed with the focus on Brazilian tourism industry, to meet its national and international demands.

REFERENCES

Bissoli, M. Â. M. A. (2001). *Planejamento turístico municipal com suporte em sistemas de informação*. Futura.

Brown, B., & Chalmers, M. (2003). *Tourism and mobile technology*. ECSCW.

Buhalis, D., & Law, R. (2008). Progress in information technology and tourism management: 20 years on and 10 years after the Internet - The state of eTourism research. *Tourism Management, 29*(4), 609–623. doi:10.1016/j.tourman.2008.01.005

Cysneiros, L. M. (1997). Integrando requisitos nao funcionais ao processo de desenvolvimento de software. Rio de Janeiro: Academic Press.

de Andrade, J. V. (2006). *Turismo: fundamentos e dimensões*. Ática.

de Carvalho, R. L. (2004). *Gestão de relacionamento com o cliente via Internet para grupos de pesquisa*. Goiânia, Goiás, Brasil: Universidade Federal de Goiás.

dos Anjos, E. S., de Paula Souza, F., & Vieira Ramos, K. (2006). Novas tecnologias e turismo: Um estudo do site Vai Brasil. *Caderno Virtual de Turismo, 6*(4).

Fielding, R. (2015). *Representational State Transfer (REST)*. Retrieved June 23, 2017, from https://www.ics.uci.edu/~fielding/pubs/dissertation/rest_arch_style.htm

Glauber, N. (2015). *Dominando o Android Studio, do básico ao avançado*. São Paulo: Novatec.

Guimarães, A. S., & Johnson, G. F. (2007). *Sistemas de informações: administração em tempo real*. Qualitymark Editora Ltda.

ISO. (1991). *ISO/IEC 9126: Information Technology-Software Product Evaluation-Quality Characteristics and Guidelines for Their Use*. ISO.

Lecheta, R. R. (2012). *Google Android para Tablets*. São Paulo: Novatec.

Lecheta, R. R. (2013). *Google Android-3ª Edição: Aprenda a criar aplicações para dispositivos móveis com o Android SDK*. Novatec Editora.

Martucci, L. A., Andersson, C., Schreurs, W., & Fischer-Hübner, S. (2006). Trusted Server Model for Privacy-Enhanced Location Based Services. In *Proceedings of the 11th Nordic Workshop on Secure IT Systems* (pp. 19–20). Academic Press.

Melo, T. (2012). *Designer projeta aplicativo como guia de pontos turísticos, em Manaus*. Retrieved June 23, 2017, from http://g1.globo.com/am/amazonas/noticia/2012/12/designer-projeta-aplicativo-como-guia-de-pontos-turisticos-em-manaus.html

Miranda, A. L. (2002). *Da natureza da tecnologia: uma análise filosófica sobre as dimensões ontológica, epistemológica e axiológica da tecnologia moderna*. Curitiba, Brazil: Biblioteca do Centro Federal de Educação Tecnológica do Paraná.

Morgan, M. J., Translate, V., & Summers, J. (2008). *Marketing esportivo*. Thomson Learning.

Neves, I. A., Semprebom, E., & Lima, A. A. (2011). Copa 2014: Expectativa e Receptividade os Setores Hoteleiro, Gastronômico e Turístico na Cidade de Curitiba. In Simpósio de Administração da Produção, Logística e Operações Internacionais. (SIMPOI). São Paulo, Brazil: Academic Press.

Nielsen, J. (1994). *Usability Engineering*. Cambridge, MA: Morgan Kaufmann Publishers.

Nielsen, J., & Molich, R. (1990). Heuristic evaluation of user interfaces. In *Proceedings of the SIGCHI conference on Human factors in computing systems* (pp. 249–256). Seattle, WA: ACM.

Pareschi, L., Riboni, D., & Bettini, C. (2008). Protecting users' anonymity in pervasive computing environments. In *Pervasive Computing and Communications, 2008. PerCom 2008. Sixth Annual IEEE International Conference on* (pp. 11–19). IEEE.

Pearce, P. L. (2001). *A relação entre residentes e turistas: literatura sobre pesquisas e diretrizes de gestão. In Turismo Global* (pp. 145–164). São Paulo: Editora SENAC.

Pinheiro, N. A. M., Silveira, R. M. C. F., & Bazzo, W. A. (2007). Ciência, tecnologia e sociedade: A relevância do enfoque CTS para o contexto do ensino médio. *Ciência & Educação (Bauru)*, *13*(1), 71–84. doi:10.1590/S1516-73132007000100005

Porter, M. E. (2001). *Estratégia competitiva: técnicas para análise de indústrias e da concorrência* (Vol. 2). Rio de Janeiro, Brazil: Campus.

Preece, J., Rogers, Y., & Sharp, H. (2005). *Design de interação: Além da interação homem-computador*. Porto Alegre, Brazil: Bookman.

Rabello, R. R. (2009). Android: um novo paradigma de desenvolvimento móvel. *Revista WebMobile, 18*.

Recife. (2017). *Portal de dados abertos da Prefeitura do Recife*. Retrieved June 23, 2017, from http://dados.recife.pe.gov.br/

Reckziegel, M. (2009). *Entendendo os WebServices*. Academic Press.

Ribeiro, F. N., & Zorzo, S. D. (2009). A Quantitative Evaluation of Privacy in Location Based Services. In *Systems, Signals and Image Processing, 2009. IWSSIP 2009. 16th International Conference on* (pp. 1–4). Academic Press. 10.1109/IWSSIP.2009.5367735

Tapscott, D. (1997). *Economia digital: promessa e perigo na era da intelig{ê}ncia em rede*. São Paulo: Makron Books.

UXPA. (2017). *Usability Body of Knowledge*. UXPA.

Vassos, T. (1998). *Marketing estratégico na Internet*. Makron Books.

Vicentin, I. C., & Hoppen, N. (2002). Tecnologia da Informação aplicada aos negócios de Turismo no Brasil. *Turismo-Visão E Ação, 4*(11), 61–78.

Werthner, H., Klein, S., & ... (1999). *Information technology and tourism: a challenging relationship*. Springer-Verlag Wien. doi:10.1007/978-3-7091-6363-4

Chapter 3
Mobile Apps for Acting on the Physical Space

Sara Eloy
Instituto Universitário de Lisboa, Portugal

Pedro Faria Lopes
Instituto Universitário de Lisboa, Portugal

Tiago Miguel Pedro
Instituto Universitário de Lisboa, Portugal

Lázaro Ourique
Instituto Universitário de Lisboa, Portugal

Luis Santos Dias
Instituto Universitário de Lisboa, Portugal

ABSTRACT

This chapter focuses on the development of mobile applications in a research strategy that combines computer sciences and architecture and urbanism. The main goal of the research is to develop mobile applications that help specific target people in daily life situations and that clearly contribute for advances in the fields of computer sciences and social responses. The authors discuss a group of mobile apps that were developed for smartphones and tablets and that respond to the following broad goals: 1) mapping of the physical space in order to adapt it to respond better to the users' needs, 2) adaptation of the physical space to the users' needs, and finally, 3) give the users a better knowledge about the physical space they are in. For each app developed, the authors describe the research problem involved, the goals, the development process, and the developed solution as well as the tests conducted to measure their performance. Usability and satisfaction tests revealed that the developed apps have a good acceptance by the target users.

DOI: 10.4018/978-1-5225-5270-3.ch003

INTRODUCTION

With the generalization of the use of mobile devices as smartphones and tablets, mobile apps have had an enormous increase in number, variety, and processing power. The interesting illustration by Fling (2009) regarding the history of mobile shows how fast and powerful was the evolution of devices that allowed mobility and created all different possibilities by the new technologies introduced. We still live the transition from the desktop paradigm to the mobile / wearable paradigm using what, a few times ago, was an unimaginable digital processing power in the palms of our hands. We have this digital power when walking, strolling, jogging around by using digital processors' equipped smartphones as day by day inseparable companion devices. Although originally mobile apps were mainly offered for general productivity and information retrieval, nowadays we can find apps for almost all possible tasks. Using mobile devices became therefore a tool to reach information about almost all aspects of life. In this context developers drove rapidly into other categories that enable citizens to use their mobile devices as their powerful personal assistant. The development of such tools is therefore aiming at a large audience and supersedes nowadays the web tools designed for desktop computers. For several authors as Fling (2009) and Castledine et al (2011) mobile is (still) the next big thing, and a big reason is that it represents a new medium as well as a new business model.

This chapter focus on the development of mobile applications in the scope of work developed in a research strategy that combines computer sciences and architecture and urbanism. One of our broad goals in architecture design is to assess what are the real needs and requirements of users so that we can design buildings and public spaces that respond to them. This situation is especially relevant when we consider groups of population that have more specific needs like elderly, people with disabilities, children and people with low incomes and underprivileged (socially, economically and culturally), lacking functional reading skills and people digitally info excluded.

For this we developed mobile apps that help to observe people's behaviour when using the physical space by mapping it, and apps that directly help people to make choices about their living conditions. For mapping people's behaviour, we developed mobile apps that are in line with state of the art techniques for behaviour observation allowing counting and tracking people when they are using the spaces that are under analysis.

With the goal of adapting the living conditions to the users' needs and obtain the relevant information for achieving that goal we developed three mobile apps, two as an help to deliver inhabitants a description of the recommended characteristics of an appropriate functional programme for their house and a third app that empowers elderly to assess their living conditions regarding the safety of their living space.

The last group of mobile apps developed empower users to obtain a better knowledge about the physical space they are in. The two apps developed with this aim use augmented reality to add information that increases the available information existing on cities both during a normal walk around the city and during the experience of using a physical map to obtain information about the city.

To all our mobile apps we used a user centred-design methodology following usability engineering methods (Nielsen, 1992; Nielsen, 1995; Coorevits et al., 2016). During and at the end of the development we performed usability and user satisfaction tests and evaluations, in order to assess their interest and usefulness to the targeted final users. Results were very positive and in line with the goals of our research.

Context

The work described in this chapter was developed in the scope of a fruitful collaboration between the scientific fields of Architecture and Computer Sciences. Since some years ago the close collaboration of these two fields under the same school –School of Information Technology and Architecture – at ISCTE-IUL has proven to create a laboratory of experiments and collaboration between professors and students. The mobile apps here described were then developed under the scope of master thesis from Architecture and Computer Sciences, master course assignments, from the Human-Computer Interaction area courses, and a European research project (Organizational Life Assistant - OLA). In all of them, computer scientists and architects worked close together. The research problems regarding the use of the developed apps are brought by Architects and are close connected to the architects' activity and the will to obtain more information about their clients and provide them with tools that give them more autonomy.

Method

The method used for the development of each mobile application described in this chapter was an user centred-design methodology following usability engineering methods (Nielsen, 1992; Coorevits et al., 2016) as well as the 10 Nielsen heuristics (Nielsen, 1995). The methodology used for each app development included the following stages: i) definition of personas, ii) definition of scenarios of use, iii) definition of features to implement, iv) development of the app, v) performance of usability and satisfaction tests. During the development a constant interaction between potential users and developers was performed so that constant feedback could be integrated and developers could work on solutions to the problems discovered.

In the group of seven mobile apps described in this chapter, six were developed as part of education assignments, both semester and annual, and one was developed in the scope of a funded European research project. For this reason, there were different time periods to perform all the stages of the apps' development.

The first stage of definition of personas was conducted in two different ways: a) for the first six education assignments apps, this stage included first the definition of the target users and secondly the description of each persona that represented each target group; b) for the app developed in the context of an European research project the process for definition of personas was preceded by quantitative surveys by questionnaires in Poland, Portugal and Sweden as well as, in depth-interview in Sweden. The definition of scenarios was done based on the personas defined. For all apps these scenarios were then used to design mock-ups of the envisioned apps. From these scenarios the features for implementing the apps were extracted. The development stage included alpha and beta testing for all the apps in order to improve the quality of the apps and ensure good satisfaction results. Usability and satisfaction tests were performed under a protocol that ensured that every subject got the same introduction and followed the same tasks. The protocols used stressed out the test was meant to evaluate the graphical interface and interaction and the performance of the apps as well as the application's relevance rather than the subject's performance.

Structure of the Manuscript

This chapter is divided in five sections. In the Introduction we set the target of the paper, detail the context of our research and the method used. In the three following sections we describe the development of each of the three groups of mobile applications developed by detailing the different technical implementation aspects as well as the different aims. In the last section we discuss and give final considerations on the work developed.

MAPPING THE PHYSICAL SPACE

For mapping people's behaviour, we developed two main apps that are in line with state of the art techniques for behaviour observation. These apps allow to monitor people in a non-intrusive way and enable designers of social researchers to collect data on people behaviour so that it can be analysed in order to improve the quality of the built environment. Two mobile apps were developed, a first one for counting people regarding customizable criteria – The Counting app (Lopes et al., 2014) – for smartphone and a second one for tracking motion flows in a built area – Manual Tracking app – for tablet.

Counting App

Observation techniques include actions like counting people when performing some specific action. The traditional counting method is based on manual annotations on paper sheets, a slow process, and error prone during the annotation phase and during the input stage for later use and analysis. Counting apps exist (Kelly, 2015) but they do not allow the customization of criteria to be counted neither the aggregation of data. The aim of the Counting app is counting specific groups of elements doing specific actions when using observation technics. For monitoring people's behaviour non intrusively, to gather data to be used and imported for Space Syntax studies and applications, we developed the Counting app in the context of the projects *Lisbon Pedestrian Network* (Guerreiro et al., 2012) and *Close to cities and closer to people* project (Moural et al., 2014). This app was developed as part of a semester academic course.

The app development used a participatory and iterative design strategy, with a basic specification established by knowledgeable technical practitioners; followed by end users' testing, implementation and successive users' testing and feedback input to implementation, done in short time loops. This process enabled to add to the basic needed features some important simple and effective details suggested by testers from the target users group, resulting in a final app that exceeded the initial goals. Examples of added features, for instance, are the counting vibration feedback that makes the app faster to use, and pre-set fast counting specifications that speed the app immediate use.

The app was developed in 2013 in Java using the Eclipse IDE (Integrated Development Environment) with the Android Development Tools ADT Plugin for Eclipse.

The user starts by defining a virtual gate (or group of virtual gates for simultaneous observations) where items are to be observed. Items can be people, groups of people, or any other user definable items (cars, buses, ...) that can traverse the gate(s) to be studied: a set of basic items are predefined (Man, Woman, Child, Youngster, Adult, Old person, ...) and any other items can be added or edited by the user in order to expand the items to be studied. The counting is done in the user specified time frame. The app automatically produces a CSV formatted report, or an aggregated report from multiple counts, directly usable in Space Syntax contexts.

Single Count enables the user to start counting in a very fast way after defining the Local Name and Gate Name, followed by the Time interval and the items to be counted. User predefined lists of items can be instantly loaded or newly created: a study project might impose that all observations follow a predefine items subject study and loading the predefined items list guarantees consistency along the project's different observations (Figure 1).

Figure 1. Single Count screens example, from left to right: starting screen, defining the gate, choosing the items to count

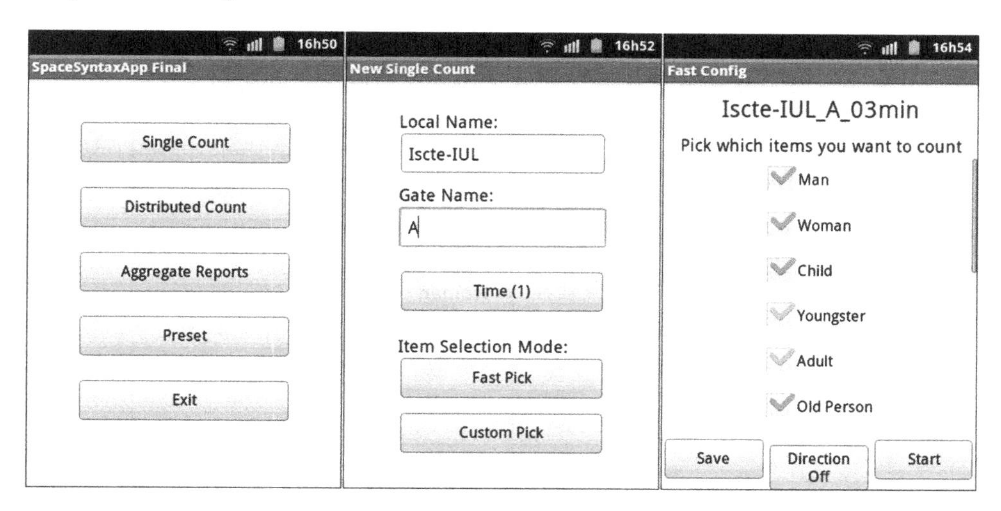

The actual items counting layout can be defined automatically choosing Fast Pick, where the screen area for each item is automatically attributed, or it can be user defined with a Custom Pick layout for space input aggregation that best suits the user's input practice or organization: different items categories may be placed where the user wants them, easing the counting input observation task (Figure 2).

The Counting screen shows rectangles with the Items' names inside: this is the "+" area, where one touch of the user increments by 1 the item's counting, the total counted so far shown under the item's name. To the side of each item the "-" area is available for any needed counting correction, decrementing by 1 the current item value. Initially the "+" and the "-" areas in the interface where geometrically equally the same. Applying the direct observation technique (Rettig 1994) of the testers in action showed:

1. The "+" counting input was rarely mistaken;
2. The rare accounts of "+" input mistakes where due to the equal geometrical space attributed to the "+" and "-" functions;
3. Ease of use and input error minimisation where obtained unevenly attributing less area to the "-" function and a bigger area to the "+" function;
4. Placing the "+" areas in the center and the "-" areas in the screen borders enabled easier and faster "+" input, the main function, since the finger that touches the "+" travels the least between input actions;
5. In practice it was observed that in the end the "-" function was rarely necessary.

Figure 2. A Single Count screens example on Custom Pick (left) and a report listed from a previous counting (right)

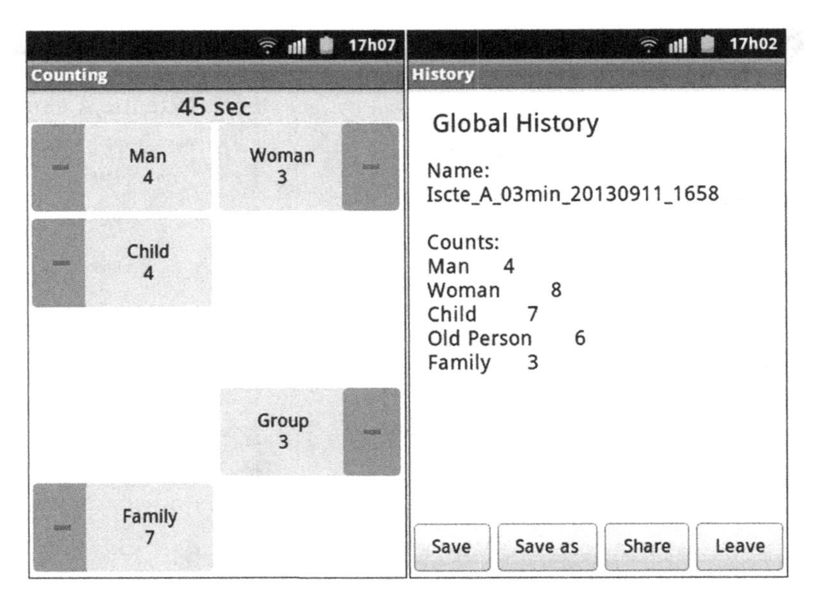

One user testimony stated: "The app is very useful, easy to use and much better than using a sheet of paper. The reports also eased the work, enabling me to create graphics with no need to introduce the counts one by one". Without publicity the app became somehow popular by word of mouth between end target users.

Initially developed for Architects and Architecture students as the final target users and audience, the finalized Counting app – for its interface simplicity – found itself being used by other researchers interested in the gathering of information for Space Syntax studies, for instance by social sciences researchers.

Manual Tracking App

Observation techniques also include actions like tracking people when they are walking on a space. Traditional tracking methods is usually done on a paper plan resorting to handwritten notes which is very time consuming and less accurate during the analysis of results. Several tracking apps based on GPS exist, but this implies that the subject knows that he/her is under observation. *Manual Tracking* is an app designed for tablet computers to ease the task of tracking the flow of people walking on cities. *People Watcher* is a similar app developed for iPad (Dalton et al., 2015).

The Manual Tracking app was developed in Java using the Eclipse IDE (Integrated Development Environment) with deployment to Windows 8 for the Surface tablet. This app was developed as part of a semester academic course.

This app provides a touchable interface where the user can pick points and build the path that a specific person (subject) is following. *Manual Tracking* provides several tools to add information to the track such as writing text or recording audio notes, add unusual events (i.e. a car crash that would affect the subject's path choice) or register interesting viewpoints that the subject has taken. Thus, a story of the track can be built so that it can be revisited later. Figure 3 shows a path that has been recorded anchored to track points that can be corrected by dragging them on edit mode.

The app lets the user sign up or login to his account and view his projects associated with floor plans/maps and subjects. This is a way of keeping everything organized. The home screen of the user account management is presented in Figure 4. On the left side, there is a list of shortcuts for the projects, plans and subjects associated with the logged in account, as well as an option for shortcut or to immediately start a tracking task.

Quick Mode is an option to start a tracking task without associating it to a project or a test subject. In this mode the user only needs a floor plan. This is meant for unexpected situations where the user spots a subject and wants to start tracking him immediately.

Figure 3. Manual tracking interface with a track in red

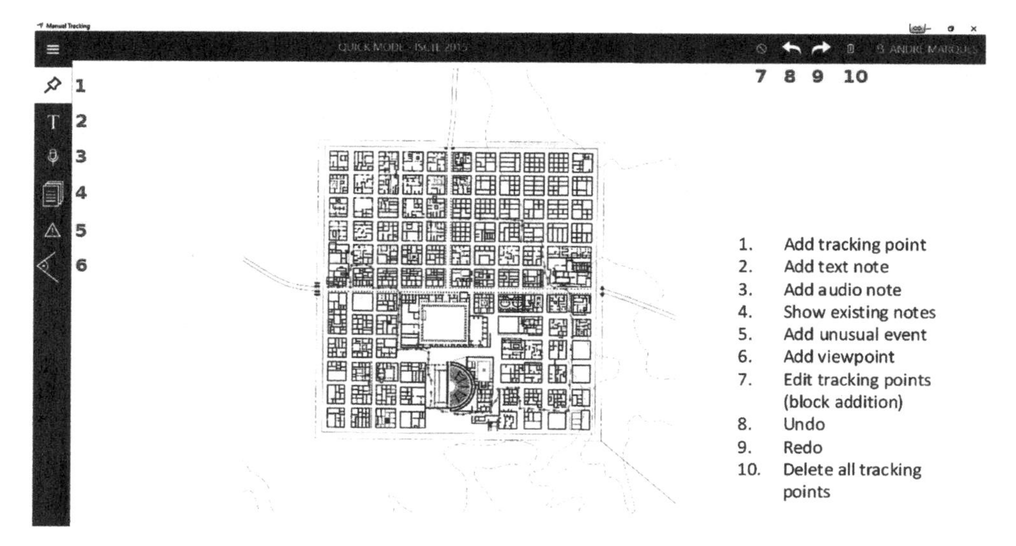

**For a more accurate representation see the electronic version.*

Figure 4. Home screen of account management

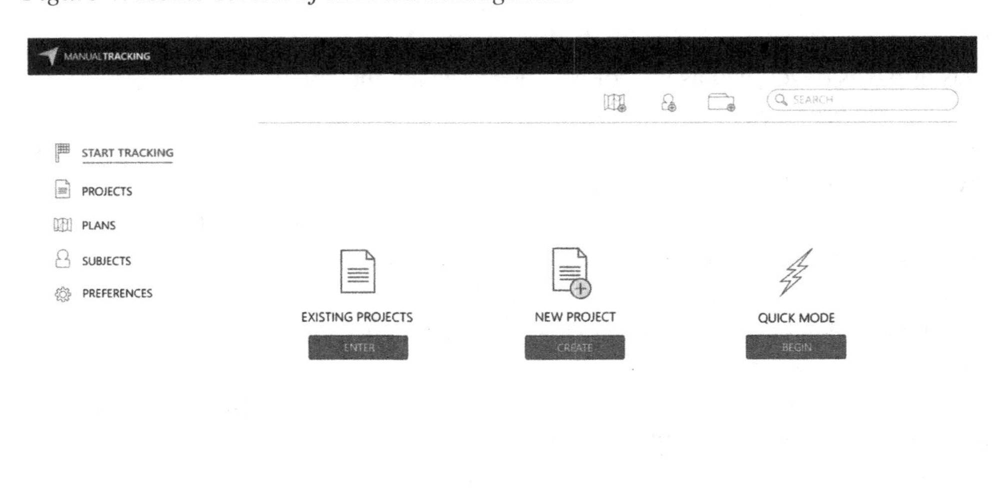

We performed usability tests for both described parts: the user account management and the tracking interface. An alpha testing for a first assessment and the possibility to include improvements in the app (1st series of tests) and a beta testing (2nd series of tests) were performed. We tested the flow of the user management interface, the whole iconography for both parts and asked more subjective questions to assess the acceptance and relevance of the application. In the beginning of the tests participants were introduced to the context in which this application is meant to be used to provide a frame for their evaluation on the application when it came to its relevance. Participants were asked to think aloud. Researchers iterated from *lo-fi* to *hi-fi* prototypes (Rettig 1994) and present the results of the two series of the *hi-fi* stage in the charts below.

Both series collected the feedback of a total of 27 subjects (19 for the 1st series, 8 for the 2nd series), 15 women and 12 men, with a minimum age of 21 years old, a maximum of 54 and an average of 30 for the 1st series and 25,38 for the 2nd series. The subjects were required to be related to Architectural Design, Urbanism, Civil Engineering or any other space use related topic. Almost all (26 out of 27) of them were architects and architecture students. 62,96% of them hold a Master, 11,11% hold a Bachelor and 25,92% hold a PhD.

Each test was divided in two stages: the first one was about the account management flow and the second about the tracking interface. Figure 5 and Figure 9 are relative to the first stage and the rest of the figures are relative to the second stage.

Subjects were asked what they thought each icon meant. Figure 5, tests on the readability of the account management iconography, shows results above 5,75 in a 1-7 scale for almost all icons, exception made for the subject icon, that still showed an improvement as it was changed from the 1st to the 2nd iterations.

Once again subjects were asked what they thought each icon meant and Figure 6 shows the results regarding the readability level on the iconography tests of the tracking interface. Some of the icons were already very readable from the start, i.e. the icons of "Add tracking point", "Delete all tracking points" or "Redo", so there was no need to change them. Others were changed from the 1st to the 2nd series and registered an improvement on readability such as the "Add text note", "Add viewpoint" or "Undo". "Add audio note", "Add unusual event" and "Edit tracking points" registered meaningful fluctuations that were not totally expected since they weren't changed between both series. This may be due to the different test subjects and to the overall difference on the iconography that present a slightly different visual context. Lastly, the "Show existing notes" registered no differences even though its icon was changed.

Figure 7 and Figure 8 present the results of the usability of touch and gesture when manipulating the tracking interface. Adding a trackpoint was easy as it only required to touch wherever the user wanted to add it. Editing a trackpoint required fine tuning to adjust the touching area around the trackpoint where the user could drag it, resulting in an overall improvement of the easiness of editing trackpoints. As for the sensitivity, the values were optimal and adequate as they represent the usual way of panning, zooming and rotating, exception made to the "Edit sensitivity" that is affected by the abovementioned fine tuning of the touching area.

Figure 5. Results of the account management iconography tests

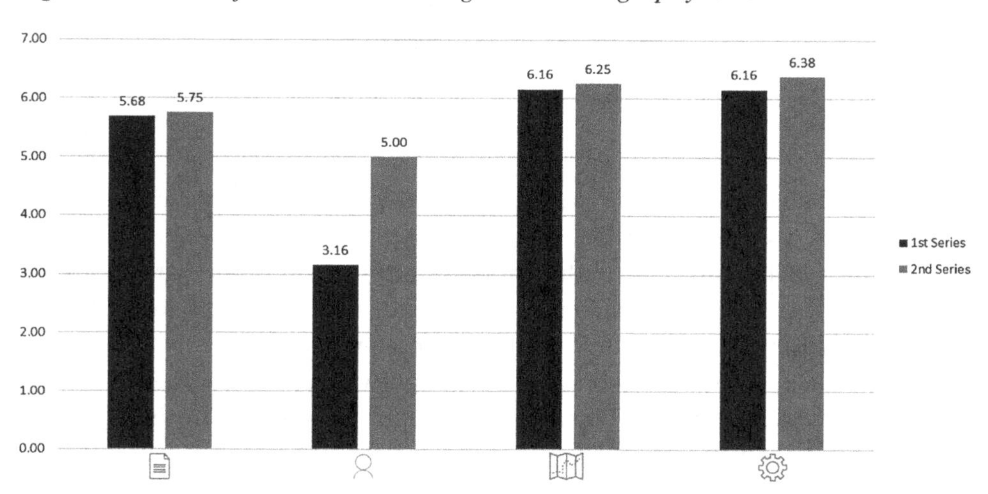

Figure 6. Results of the tracking interface iconography tests. The "add text note", "show existing notes", "add viewpoint" and "undo" icons were changed from the 1st to the 2nd series of tests.

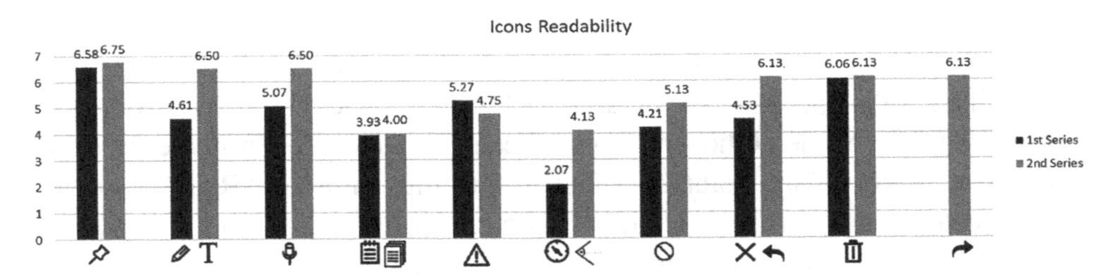

Figure 7. Results of the usability of touch and gestures of the tracking interface (1 is very hard and 7 is very easy)

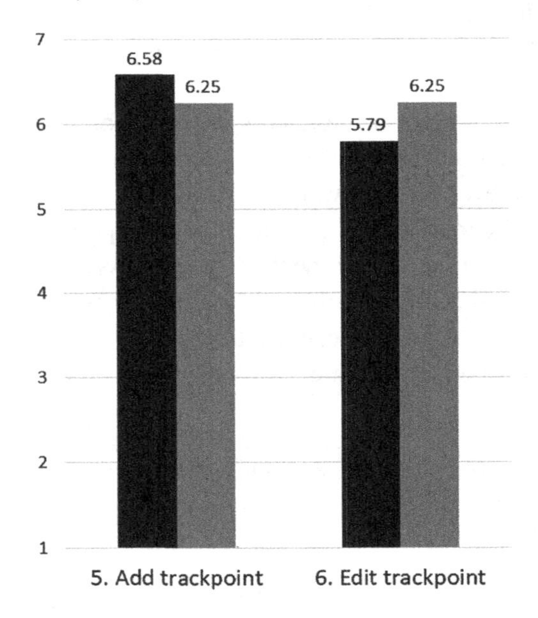

Figure 9 shows the results of the general questions assessing the relevance of the application, its usability and its *Quick Mode* feature. Users gave high value to *Quick Mode* which holds the core functionality of the application that is tracking someone. Furthermore, they found the application useful and intuitive which is a good evidence that this approach is of use for these professionals.

With the tests performed over this application we can conclude that there is a need for what it provides. The two tests iterations allowed us to improve the application flow as well as its iconography that resulted in an overall improvement

Figure 8. Adequacy of the edit trackpoint interaction area for dragging and transforming sensitivity (1 is too insensible, 4 is adequate and 7 is too sensible)

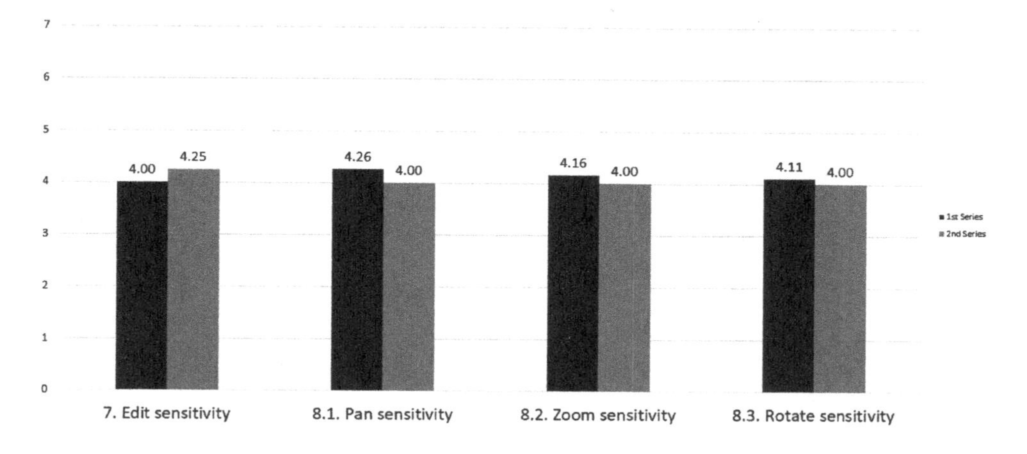

Figure 9. Relevance of the application, its quick mode and usability

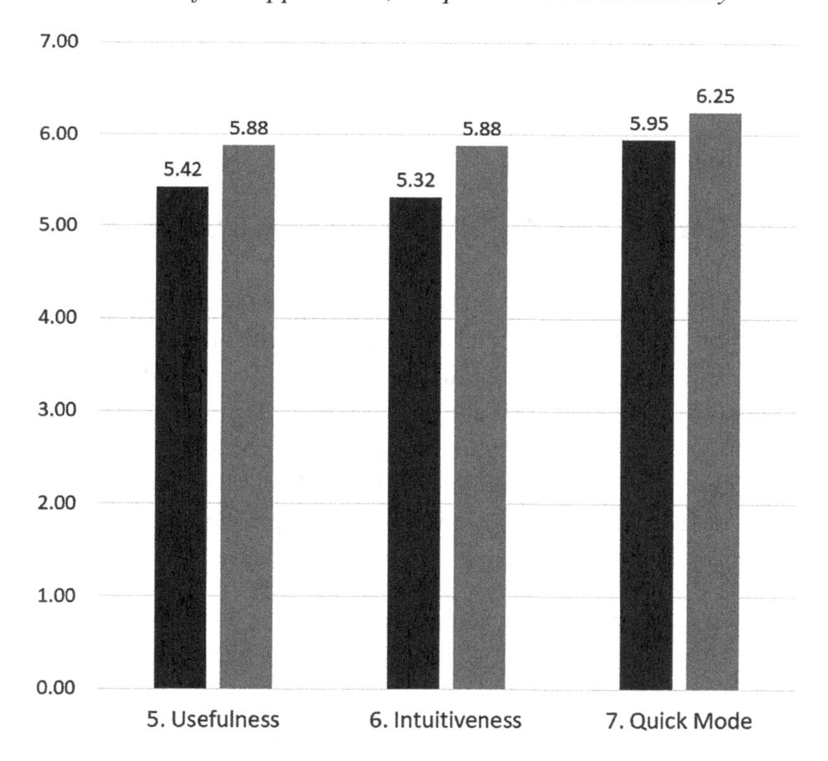

Figure 10. Results for "Would you use this application?"

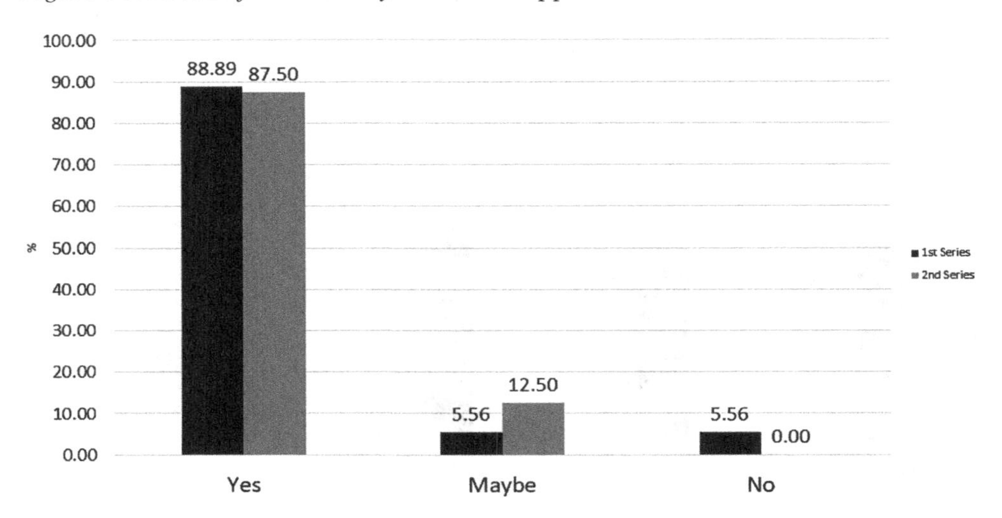

of its acceptance. For future work we aim at developing an export format compliant with GIS supported file formats, perform more usability tests to improve the icons that received a lower score on readability, as well as add other features that allows the integration of more information in this context.

ADAPTING THE PHYSICAL SPACE TO THE USERS' NEEDS

With the aim of assisting people to make choices about their living conditions and adapt them to their needs we developed three mobile apps for different context all of them related to the physical living conditions. House Definer 1 and House Definer 2 are mobile apps for smartphones that allows a person or a family to obtain a functional programme for their future house by inputting the household characteristics. In the scope of the research and innovation project OLA (Organisational Life Assistant), which aims at developing ICT solutions for elderly people, we are developing a mobile app that will be used in a tablet for environment analysis.

House Functional Programme Definer

Although architects are trained to observe and propose housing functional programmes to their clients, it often happens that clients don't really know what they want. In large design projects the client is asked to deliver a detailed design brief where all his/her requirements are included and specified. Although of major relevance, this design brief is difficult to do by the regular person not used to order new buildings to be built, e.g. someone that wants to build their own house. For those clients we envisaged an app that would guide them through the regular housing functional requirements which include type, number and size of rooms as well as the connections between them. This app will therefore help clients to define their goal. Correia et al (2010) developed a similiar software, PROGRAMA, for desktop, that uses the user and site data to generate a design brief for an adequate house design for the context of Malagueira mass customization housing grammar. Our apps are broader in portability.

With House Definer 1 and House Definer 2 apps, families can input their characteristics and obtain functional programme recommendations for their houses. Housing Definer 1 focus especially on the development of the interface that allows a clear and unambiguous identification of the household and House Definer 2 focus especially on the output of the app regarding different types of households. Several iterations are possible so that families can play with the apps and search for the better option for them. House Definer 1 was designed for a context of non-technological savvy users or very low digitally skilled users; and House Definer 2 for a European context for users with technological skills.

House Definer 1

With House Definer 1 we wanted to solve the problem of letting (near) illiterate and non-digitally savvy people to use an app that would enable them to define their family members composition in order to later define their house structure and needs. Worst case scenario would be an intermediary info gathering technician that, with the help of the family members, would use the app while the family members would follow every single step of the input process, being able to understand all the elements at stake being edited. The context: defining the house needs for underprivileged people from under development countries where populations living in poverty areas lack the very basic tech skills. Examples of such populations can be found living in slums, like in Brazil ("favelas"), Angola ("musseques") and Mozambique ("caniços"). Previous house solutions applied to those impoverished populations adopted the "one size/plan fits all" approach, applying an European house structure to the local populations, regardless of local housing needs/habits and also not paying attention to the nucleus family structure and environment that influences the house basic structure. Non-digitally savvy people and local family structures, that can be quite different from European patterns and thus are not part of the European app developers' background, these elements pose interesting challenges to be solved in the context of an app development.

From the interface point of view, the major problem was considering the population and the family unit in some examples of under development countries, namely Angola and Mozambique, characterized by attributes common in those regions, to which European app developers are not familiar with:

- Although in the European countries we live in the digital paradigm where actions like drag & drop, scroll, pop-up windows and menus, pull-down menus and drop-down list menus are pervasive, invasive and daily; for the population in low income countries with low or no digital user expertise, these actions are barriers. How do we develop an interface away from the de facto standard, usable by non-digitally savvy people?
- The extended family, where a group of people is living together under the same roof linked by family ties different from the European ones; a family can be made of parents, children, cousins, aunts, nephews, nieces, grandparents and others (friends, friends of friends, may be still relatives or not).
- There is a strong possibility of having amputees in the household since Angola and Mozambique are included in the list of countries with the most land mines still active placed in the fields, the frequently walking passages, the common and popular accesses to water sources, due to long duration civil wars that used population terror tactics thru land mines, very cheap to place, strong terror inducers, devastating in human kills and in body damage to the survivors, creating a high population of amputees. From a 2011 report, in Mozambique there are approximately 20 people injured / killed by land mines each month (Owsley, 2011).

As a consequence, the family definition should include the possibility of many members, including amputees. The interface should be readable for the non-tech savvy person.

At the time of development (2012) the app was developed in Java using the Eclipse IDE (Integrated Development Environment) with the Android Development Tools ADT Plugin for Eclipse. The app was developed as part of a semester academic course.

The icons study evolved in three phases, first with a more artistic approach (Figure 11). Tests with target users representatives showed these icons did not ease immediate readability. Subsequent iteration studies (Figure 12) stabilized the icons interface in a more standard approach to include man, woman, old man, old woman, boy, girl, man amputee, woman amputee, baby, pregnant woman (Figure 13).

As for the input approach targeting users with no digital user expertise, the 10 Nielsen heuristics (Nielsen, 1995) were applied to the extreme, taking all the time into account that "You are not your user" (quote). The devised approach was labelled the "You See What You Input" (YSWYI) approach. The YSWYI approach features:

Figure 11. First icons study for the app icons: artistically oriented, this first study served to show readability problems from the target audience point of view; a more direct and main stream family members' graphic icons was developed.

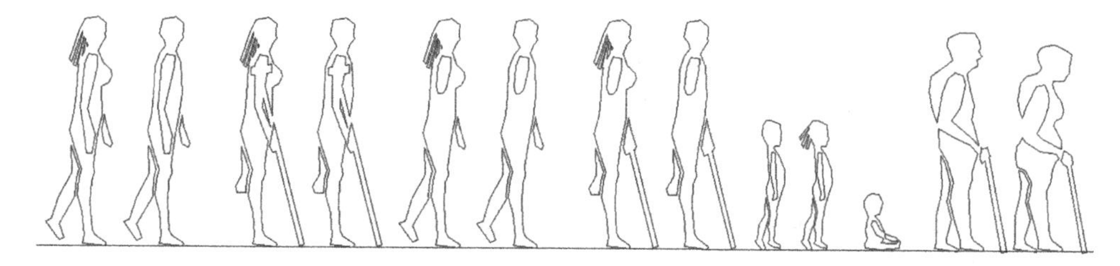

Figure 12. Second study for the app icons: Since the app will be mainly based on icons and their signification, tests with target users representatives showed the children's icons to be not proportionate for the intended age representation.

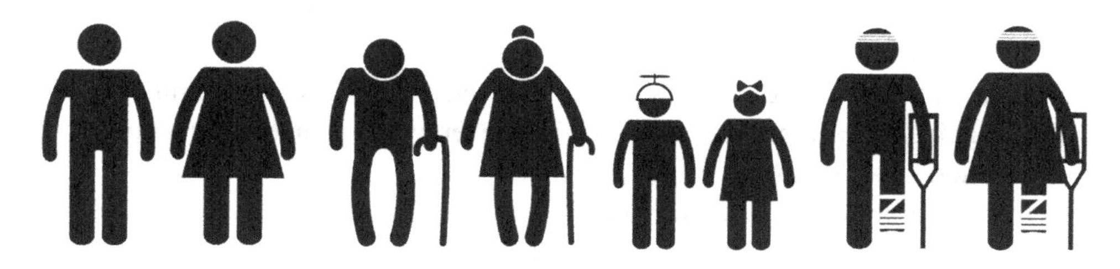

Figure 13. Third study for the app icons, final icons: funny hat removed and youngsters' height reduced. Added baby and pregnant woman.

1. **No Hidden Information:** There is no hidden information, all that is needed is visible;
2. **The Basic Unit Is the Full Screen:** There are no scrolls or other hidden forms of information to be fetched;
3. **Input Information Visually Present:** You see what's being introduced;
4. **Minimal Information Per Basic Unit to Avoid Confusion:** A unit being an element to be counted (woman, man, …) the associated information should be

kept as minimum as possible, what might be information to us, might represent distraction and noise to the target user;

5. **Complete Information Input:** Successive screens, step by step, screen by screen;
6. **YSWYI:** Simplicity and usability in the real world, the equivalent to Nielsen's "Match between system and the real world".

The development approach followed, again, Nielsen's (1995) heuristics in a very tight loop: each step, each decision, would be confronted with "You are not your user", using the heuristics to validate and reason as in much depth as possible on the value, importance and consequences at stake. Going to Mozambique or Angola for tests with local people was being considered. For lack of financial backup, a local test strategy was used:

1. Target testers were selected among expert testers used to apply the Nielson heuristics;
2. Random testers were given the app to play with, to test if it was immediately recognizable no explanations at all whatsoever were given, their reactions and interaction were observed and annotated;
3. Local old people with no smartphone experience were given the app with a simple question: could you play with this and, with it, tell us about your current / past family?

Initial informal and preliminary tests were done by 4 people. Iterative tests were done, and in total they included 5 heuristic experts, 32 undifferentiated testers (22 students and 10 office workers) and 6 old aged people, non-smartphone savvy.

When there is no input (0 item) a "shadow" item is shown in the screen: it induces the user to touch it and immediately grasp the interaction process, adding or removing family members is a one touch screen action. Touch once and the item adds and becomes full dark, another touch and the item is removed. Touch once and away from the initial item icon and the distance in-between is automatically filled with as much items as the area in-between olds. Or touching on the icon on the navigation section, one item is added at a time. Scaling is done automatically to fit the screen. At any given time a Report can be saved locally or sent by email (Figure 14).

The tests done showed a fast and natural adaption of every tester to the interface and interaction. Even technophobic people used the app for family definition with no constrains and very naturally, addressing it as a "non computer", more like a challenge or a game. The tests showed the concept worked, as far as the local testers were concerned. But in the end, the hopefully predictable success of this solution

Figure 14. Except for the final report (R), the interface and interaction follows the principle of YSWYI by touching a screen area; touching the icon area several items can be inputted with one touch; touching the navigation section inputs one item at a time; at any time the current report can be saved locally or sent by email.

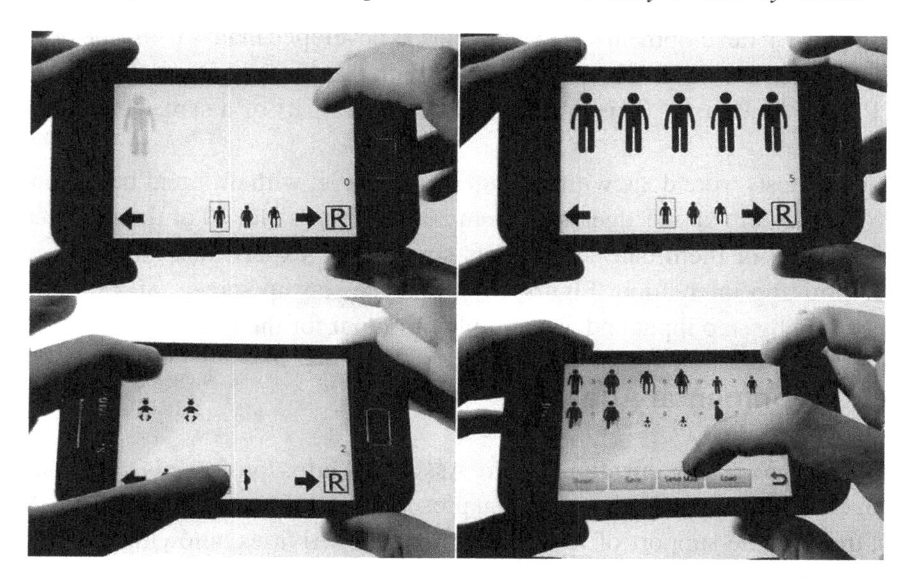

has to be tested and validated in the real world with the families to whom it was designed to.

House Definer 2

The House Definer 2 app is the frontend for the generic definition of customized housing layouts. The user inputs information on the cohabitation group (a single person, a family or any combination), the members' ages, gender, family relationship, if there is/are disable person(s) in the group, and the app automatically defines the housing architecture functional programme for that cohabitation group. The app's output can then be feed to a backend process where a shape grammar approach is applied to generate several solutions, either in a house rehabilitation process or for a new house. For the functional programme, including information on number and type of rooms, and connections between rooms, Portuguese housing legislation and related literature was used (Eloy, 2012).

The app development followed a step by step approach, from the specification iteration process involving architects that supplied the needs to be addressed, followed by a "low fidelity" (lo-fi) implementation (Rettig 1994), used to test the concepts and debug the interface and interaction early on. Heuristic evaluation and users' evaluation / feedback / observation were used along the lo-fi phase and in the app coding phase, done for Android smartphones, enabling the final users' interaction refinement. Development was done for and with the end user, with regular iterative tests.

At the time of development (2013) the app was developed in Java using the Eclipse IDE (Integrated Development Environment) with the Android Development Tools ADT Plugin for Eclipse. This app was developed as part of a semester academic course.

The final tests were done with a group of 12 people, with different backgrounds and no architecture knowledge, aged from 20 to 49 years old, 10 of them had basic level studies, 7 of them had no previous smartphone experience. All testers had a successful app interaction. Figure 15 shows the startup screen, an example of a guided step by step input and an example of output for the house specification.

OLA Environmental Analysis

The OLA project, Organizational Life Assistant (http://project-ola.eu/), tackles societal challenges by providing a virtual presence that addresses older adults' daily needs, through the support of related instrumental activities, allowing the elderly to be more independent and self-assured. Additionally, they can have a healthier,

Figure 15. Left: The app enables the user to choose between a Typical standard cohabitation group house or a Custom personalized one; Load Data makes it possible to recover previously saved work. Middle: input is introduced in a step by step guided way. Right: Example of output list with the final house specification.

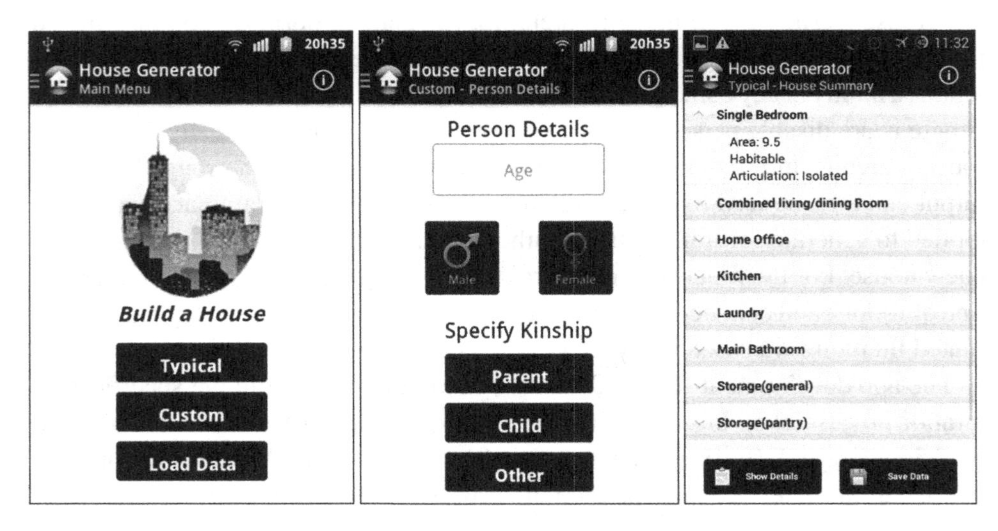

safer and more organized life by facilitating high quality assistance provided by caregivers. This is achieved by mediating interaction between elderly and its formal or informal caregivers through technological devices, like standard computers, tablets and home automation modules, using innovative multimodal software, adapted to the physical/healthy and cognitive characteristics of older adults. OLA, also provides personalized well-being and safety advices to seniors in order to handle unwanted age-related health and safety situations at their home.

In the scope of OLA ambient assistive living solutions, the OLA Environment Analysis (OLA-EA) and Safety Advisor systems envisages to identify architectural barriers and hazarding situations that may interfere in seniors' home security and quality of living. We envisage that such a system will be used via a tablet that, when pointed at the rooms of the senior's house, will give a feedback using Augmented Reality about possible hazards. Up to our knowledge there is no existing similar app developed. For our system algorithms for real-time object recognition and scene understanding are being developed to analyse and decide which action to be taken in order to support the elderly by suggesting environment changes and providing hints/advices for safety and accessible environments (Ourique et al., n.d.).

The basis to avoid architectural barriers and hazarding situations depends on the use of Universal design strategies. Considering that existing houses may not follow all the good practices of universal design and may hide some potential hazards and not accessible areas for people with mobility impairs, the OLA-EA and Safety Advisor systems will help to advise them about their own house.

Using the concept behind this design several guidelines, standards and norms have been introduced both at the European level, by the European Commission and other European organizations (CEN-CENELEC, 2014; Aragall, 2003; Aragall et al., 2013) and at the national level (HM Government, 2014; Anon, 2014; Diário da República 1ª série, 2006; Anon, 2010). In this research we have used these regulations and recommendations. One of the norms used as a proof of concept concerns the width of interior passage ways. This norm is related to the architectural barrier caused by a narrow passage way due to a distance between two walls or two elements, e.g. two pieces of furniture, being under a certain threshold value.

In more detail, one of the mock-up scenarios for OLA-EA is as follows:

- On the bottom left corner of the tablet screen, a floor plan is presented showing the architectural barriers that have been detected so far;
- The user (typically the caregiver) walks around the house holding the tablet, and red dots (warnings) appear superimposed on the environmental barrier that was found (Figure 16);
- When the user taps on the red dot the associated barrier is highlighted and a corresponding popup message is presented (Figure 17);

Figure 16. A red dot is shown on the display of OLA-EA indicating an hazard.

Figure 17. The user taps on the red dot and the app OLA-EA shows what is the hazard (highlights the place and give a text message). In this case the hazard is a loose cable on the floor.

- When the user taps on the warning area, the recommendation for that specific situation is exhibited as a new popup message (Figure 18);
- The user can close the previous popup messages and restart the navigation on the house to find new hazards.

To develop the OLA-EA and Safety Advisor systems the following stages need to be considered: i) Object Recognition at the Category level so that the system can recognize different pieces of furniture, ii) identification of the bounding box of each piece of furniture, iii) determination of the distance between neighbouring pieces of furniture, based on the bounding boxes, iv) definition and implementation of action rules when the encountered distances are not according to the regulations.

The first stage predicts the category of never-before-seen objects using as input on a-priori knowledge of a set of objects (i.e. the dataset) (Proença, 2013). The dataset comprises eight categories of miniatures of furniture commonly found inside the house (i.e. arm chair, bed, chair, medium height furniture, sofa, table, tall pieces of furniture) with a total of 76 instances of objects. The data collection capture setup is described in Pascoal et al. (2015). The second stage is a preliminary stage that allows to know the limits of each piece of furniture on the three-main coordinate axis. For the third stage, our implementation determines the distance between the bounding boxes of neighbouring objects - there is no limit on the number of objects present on the scene. The distances are computed over the X axis. In the final stage,

Figure 18. The display shows the recommendation for a specific warning, in this case the recommendation will be to remove the cable.

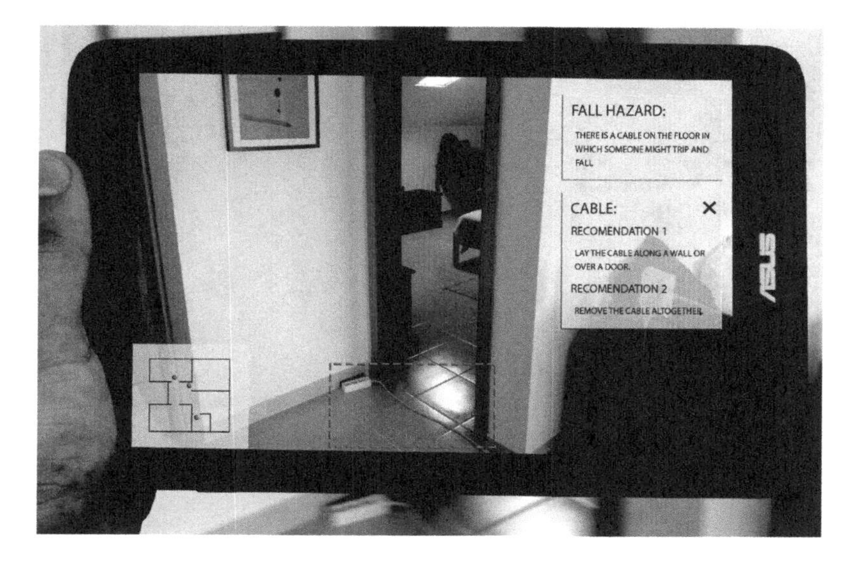

a colour code is used to signal the violation of the norm of the interior passage way: if the distance between two objects is higher than a certain value, determined by the applicable regulation, its representation is shown in red, otherwise it is presented in green (Figure 19).

INCREASING THE KNOWLEDGE ABOUT THE PHYSICAL SPACE

In todays' society, the availability and presence of digital technologies in our day to day life makes possible to augment users knowledge on their surrounding environment with customized information in real-time. The second broad goal of our research in architecture is to provide the users with a better and customized knowledge about the physical space they inhabit. With such aim in mind we develop Augmented Reality based technological solutions that customize the information that users received according to their own interests. SeeARch and Arch4maps are both mobile solutions of that kind for which the target audience includes people ranging from the city inhabitant up to tourists, interested in cultural aspects of the city. Both these solutions were developed to be used in a tablet device and both use Augmented Reality as the approach to increase the information people may obtain for a specific topic of interest.

Figure 19. Distances between neighbouring objects

SeeARch

SeeARch (Raposo et al., 2017; Raposo, 2016) is an Augmented Reality mobile tool for outdoor use that is designed to increase the knowledge about the city. With such an app, the users will acquire, in real time, relevant and customized information about several aspects of cities which augments their real experience of the place and therefore increase the engagement and interest towards the city, both by citizens and visitors. Apps like Street Museum (http://streetmuseum-ar.com/), Urbasee Project and Urbasee Future (https://urbasee.com/en/) are already existing and enable tourists to get a broader knowledge about several cities. SeeARch advances by using uses Augmented Reality to recognize the volumetry of buildings and superimpose on them relevant information, namely 3D models, while visiting the city. The goals of the app are to provide customized information to the users/visitors according to their own interests and time to visit the city.

SeeARch prototype was developed as a Windows App that allows users to follow a pre-defined path in the city. One of the aims of this app is to give extra information on the regular city fabric and not only on the monumental and historic city fabric that all the city guides include. Therefore, the prototype developed uses the neighbourhood of Alvalade in Lisbon which is one of the most known by its innovative urban planning dating from the 50s. The app includes information about several regular buildings of the neighbourhood that are special in the city because they constitute an architectonic type, usually called *rabo-de-bacalhau,* repeated several times along the city (Eloy, 2014).

The app SeeARch was developed for a Windows tablet with the APIs (*Application Programming Interfaces*) NUTTS (*Natural Ubiquitous Texture Tracking System*) (Bastos & Dias, 2008) and OSG (*OpenSceneGraph*) (Wang & Qian, 2010). The programming language used was C++, common to the used APIs. To capture the real world the tablet's camera was used.

The information available for each building is: i) a 3D model of the original architectural project categorized into various layers according to the various design specialities: architecture, structural engineering, infrastructures (water, sewage, electricity network, air conditioning, etc.), interior architecture, etc., ii) photos, ii) detailed descriptive text; iii) technical drawings, and iv) videos.

We present some screen shots and functionalities of the app as a use scenario:

- After starting the app, a menu will appear with the various Points of Interest (POI) – e.g. relevant buildings and public artworks (Figure 20);
- The user will be prompted to choose a POI that he/she intends to visit;

- After arriving at the POI, in this case a building, the user changes to the camera mode and points the tablet towards the façade of the building;
- After the app recognises the building façade, the 3D model of the building will be superimposed (Figure 21);
- After the identification of the building the user can navigate the various functions of the app;
- The user can consult the additional information made available by the app:
    - A text description of the history of the building;
    - Photographs of the interior of the building, facade in its original state or historical photos;
    - The user can also superimpose the original facade or the facade of buildings that previously occupied the same lot (Figure 22);
    - Technical drawings of the building;
- The user can also interact with the 3D model by:
    - Performing live sections, horizontal or vertical (Figure 24);
    - Scaling the model (Figure 25);
    - Rotating the model;
    - Explore the various components of the building – e.g. structure or infrastructure;
- Finally, the user can exit the building menu and explore other POIs or close the app.

Figure 20. Initial display of the SeeARch app showing the POI

Figure 21. SeeARch display after the identification of the façade of the building to be viewed

Figure 22. SeeARch displaying a superimposed augmented photo

In the current version of the app, AR registration is achieved by using the facade as a visual marker, adopting Computer Vision techniques. SeeARch uses the well-known SIFT (Scale Invariant Feature Transform) algorithm (Lowe, 1999), which allows the identification of key features in an image based on its grey level contrasts - in this

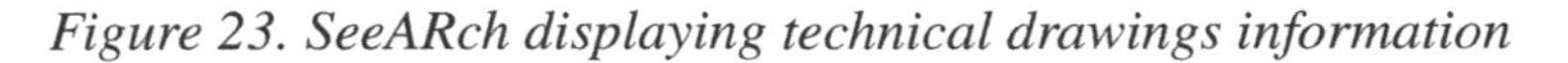

Figure 23. SeeARch displaying technical drawings information

Figure 24. SeeARch performing a vertical section on the registered 3D model of the building

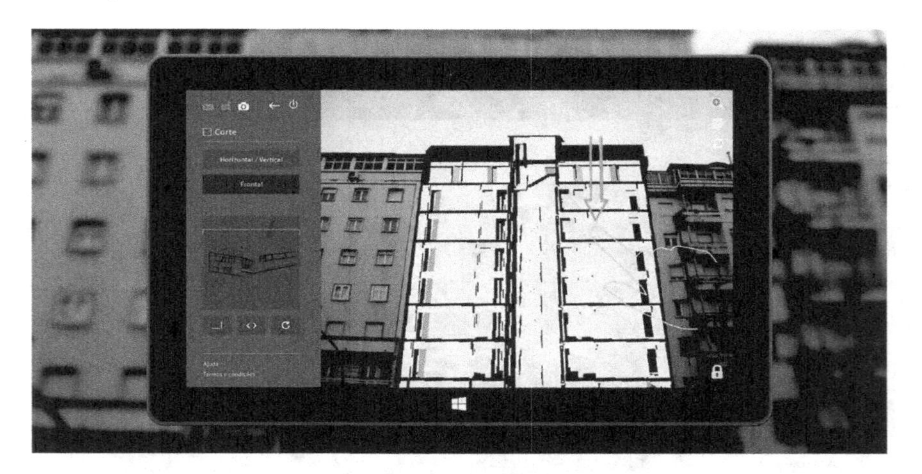

case the features found in the facade of a building. These key features, called Startup Features and Tracking Features allows the system to first recognize and then follow the facade through our in-house developed NUTTS (Natural Ubiquitous Texture Tracking System) technology. When the system recognizes a sufficient number of SIFT features in the facade, the registration of the digital content - which consists of a 3D model – occurs, and the model is superimposed over the image captured by the camera of the tablet (Raposo et al., 2017).

Figure 25. SeeARch performing a scaling of the model

The SeeARch prototype was tested with a total of 10 users to collect preliminary feedback about the acceptance of the proposed app (Raposo et al., 2017; Raposo, 2016). The sample was divided into 6 men and 4 women, with ages ranging from 15 to 65 years old and was constituted by four frequent users of the Alvalade quartier, five architects or architecture students, and one tourist. In this satisfaction test the users were asked to navigate the various functions of the app, followed by a questionnaire and finally a brief open question interview. From the questionnaires, we concluded that the participants considered the app easy to use, that it helped to understand the POI in question, and that the app offered more information than traditional city guides. Finally, from the follow-up interviews we determined that the information currently contained in the app – e.g. 3D models, technical drawings and the content of the descriptions – were mainly designed towards architects and urban designers and that a diversification of the content available would increase the offered value of the app.

Arch4maps

In the continued search to explore how technological solutions can expand upon traditional mediums of information the app ARch4maps explores the limitations of traditional city paper maps and guides and takes advantage of augmented reality. By using augmented reality, ARch4maps provides an hybrid paper-digital response to the problem by enabling users to expand on the already available information on paper and add additional information that paper do not support - e.g. interactive 3D models and a larger amount of information. Although, up to our knowledge, no similar app exist that combines traditional paper maps to AR features, other apps exist that augmented information on a bi-dimensional support as books as *Wonderbook: Book of Spells* (https://www.playstation.com/en-us/games/wonderbook-book-of-spells-ps3/) and *Moby Dick* (https://vimeo.com/42273996). The application was envisioned as a tool to help its users exploring the city heritage before and after their visit to a city and to ease the access to knowledge on various points of interest of a city.

Arch4maps (Gaspar et al. 2016; Gomes 2015) runs on a Windows tablet and recognizes features on a city traditional paper map and overlays, in real-time, digital content related to buildings which are relevant in the city. A short demo video of the application is available on the following link: https://vimeo.com/154310723After. The app SeeARch was developed for a Windows tablet, Microsoft Surface Pro 2, with the APIs (*Application Programming Interfaces*) NUTTS (*Natural Ubiquitous Texture Tracking System*) (Bastos & Dias, 2008) and OSG (*OpenSceneGraph*) (Wang & Qian, 2010). The programming language used was C++, common to the used APIs. To capture the real world the tablet's camera was used. The system registers the location of those buildings in the traditional map, enriching the user experience through Augmented Reality, with several levels of multimedia information for each building. Displayed buildings can be queried and filtered by associated meta-data. The main target users of this system are people interested in architecture and history that want to have a more inclusive and customized experience/solution where they can select the information that is of interest to them.

The Valmor and Lisbon Council prizes map (1902-2002) was used as a proof of concept (Figure 22). With a dimension of approximately 60 by 100 cm, on the front side it shows a map of the city with the buildings locations identified by a number, letter and color code and, on the right side the basic information of the 115 works highlighted in the map. On the flip-side of the map there is more information on the awards and building descriptions and images, organized by decades.

We present some screen shots and functionalities of the app as a use scenario:

- After starting the app, a menu appears with the feed from the tablet's camera;
- The user points the tablet towards the map (Figure 26);
- All the buildings that have received the Valmor award will be superimposed to the map represented in by a red or white sphere (in the proof of concept only the red dots had all the information included) (Figure 27);
- The user has now access to the Search Menu with options to search by architect name, by decade of construction, by the type of award, by conservation state and if it's open to public visits;

- The user then chooses a particular building by the query's results or by clicking in the superimposed sphere;
- The 3D model of the selected building appears superimposed in the camera feed in substitution of the building spheres (Figure 28);
- In the Model menu, the user can access additional information to visualize like the description of the building and its history (text), technical drawings and photos (images);
- The user can scale and rotate the model, perform vertical and horizontal sections (Figure 29), as well as, hide and highlight various layers.
- The user can then leave the model menu and return to the map menu or close de app.

The Arch4maps prototype was tested with a total of 11 users to collect preliminary feedback about the acceptance of the proposed app. The sample ranged from 21 to 35 years old people and was constituted by four practicing architects, four students and three persons with several different occupations. 90% of the users reported that the app makes the maps more and much more informative. 90% of the sample reported that Arch4maps facilitates the understanding of the information contained in the map. How clear is the application on displaying the buildings through the 3D model, 82% responded that it's clear. Around 63% of the users reported that they would certainly or most likely use the application.

Figure 26. Arch4maps app recognizing the map

Figure 27. App print screen of augmented map

Figure 28. Arch4maps showing a 3D model of a Valmor awarded building registered on the map

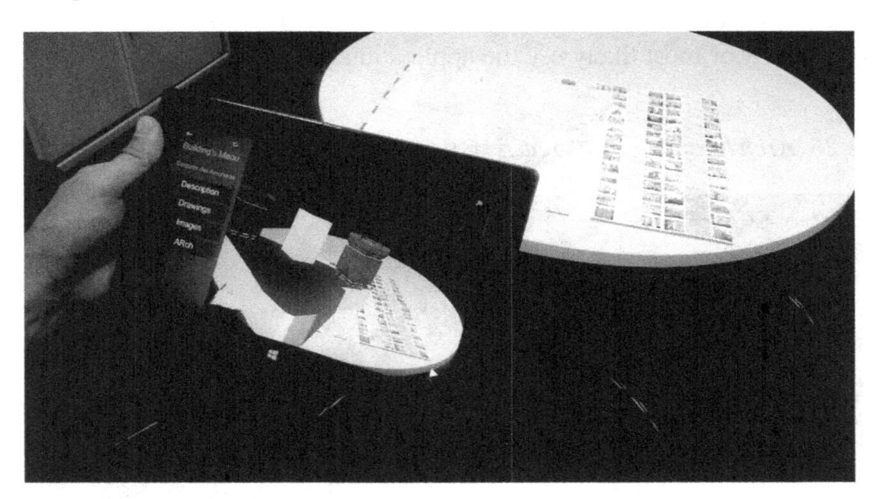

Figure 29. Photo of the model viewing experience

FINAL CONSIDERATIONS

The apps that are described in this chapter were designed in the scope of a close collaboration between architecture and computer sciences with the aim of simultaneously develop solutions for problems raised in architecture (mapping the space and increasing knowledge) and develop programming skills and new solutions in computer sciences. The approach taken was then a close collaboration between these two fields and between several levels of research from master students to senior researchers and from master course level to funded European research projects.

The main goal of this app development is to act on the physical space and improve it by making it more adaptable to its users. For this the methodology for the development of all apps included the definition of personas and scenarios. At the end tests were performed with a sample that represented these personas so that the solutions developed could be improved with the feedback of the real target users and therefore the potential of satisfaction of users would be high.

REFERENCES

Anon. (2010). *Plan- och bygglag (2010:900)*. Available at: http://www.notisum.se/rnp/sls/lag/20100900.HTM

Anon. (2014). *Construction Law*, Poland. Available at: http://www.architektura.info/index.php/prawo

Aragall, F. (2003). *Technical Assistance Manual*. Available at: http://www.eca.lu/index.php/documents/eucan-documents/13-2003-european-concept-for-accessibility-2003/file

Aragall, F., Neumann, P., & Sagramola, S. (2013). *Design for All in progress, from theory to practice*. Available at: http://www.eca.lu/index.php/documents/eucan-documents/29-eca-2013-design-for-all-in-progress-from-theory-to-practice/file

Bastos, R., & Dias, M. S. (2008). Automatic Camera Pose Initialization, using Scale, Rotation and Luminance Invariant Natural Feature Tracking. *Journal of WSCG*, *16*(1–3), 97–104.

Castledine, E., Eftos, M., & Wheeler, M. (2011). *Build Mobile Websites and Apps for Smart Devices*. SitePoint.

CEN-CENELEC. (2014). *Guide for addressing accessibility in standards*. Available at: www.cen.eu

Coorevits, L., Schuurman, D., Oelbrandt, K., & Logghe, S. (2016). Bringing Personas to Life: User experience Design through Interactive Coupled Open Innovation. *Persona Studies*, *2*(1), 97–114. doi:10.21153/ps2016vol2no1art534

Correia, R. C., Duarte, J. P., & Leitão, A. M. (2010). MALAG: a discursive grammar interpreter for the online generation of mass customized housing. *Proceedings of the workshop in 4th Conference Design Computing and Cognition*.

Dalton, N., Dalton, R., & Höelscher, C. (2015). People Watcher: an app to record and analyzing spatial behavior of ubiquitous interaction technologies. *Proceedings of the 4th International Symposium on Pervasive Displays (PerDis '15)*, 1–6. Available at: http://oro.open.ac.uk/42643/

Diário da República 1ª série. (2006). *Regime da acessibilidade aos edifícios e estabelecimentos que recebem público, via pública e edifícios habitacionais*. Available at: https://dre.pt/application/dir/pdf1s/2006/08/15200/56705689.pdf

Eloy, S. (2014). "Rabo-de-bacalhau" Building Type Morphology : Data to a Transformation Grammar-based Methodology for Housing Rehabilitation. *Arq. Urb*, *11*, 31–47.

Fling, B. (2009). *Mobile Design and Development: Practical concepts and techniques for creating mobile sites and web apps (Animal Guide)*. O'Reilly Media.

Gaspar, F. (2016). ARch4maps : a mobile augmented reality tool to enrich paper maps. In *Proceedings of the Symposium on Simulation for Architecture and Urban Design* (pp. 179–182). London: SIMAUD. Available at http://www.scs.org/simaud

Gomes, S. (2015). *Requalificação do Bairro S. Nicolau e Realidade Aumenta Aplicada nos Mapas de Arquitetura*. Instituto Universitário de Lisboa.

Government, H. M. (2014). *The Building Regulations 2010. Access to and use of Buildings. Document M*. Available at: http://webarchive.nationalarchives.gov.uk/20151113141044/http://www.planningportal.gov.uk/uploads/br/br_pdf_ad_m_2013.pdf

Guerreiro, R., Eloy, S., & Lopes, P. F. (2012). *Lisbon Pedestrian Network. 2ª Conferência do PNUM Morfologia Urbana nos Países Lusófonos*.

Kelly, K. M. (2015). *Apps for Behavior Analysts*. Available at: https://batechsig.com/2015/03/09/apps-for-behavior-analysts/

Lopes, P. F. (2014). Counting App for local observations and space syntax. In A. Moural, P. F. Lopes, & S. Eloy (Eds.), *CLOSE CLOSER: Close do Cities, Closer to People* (p. 24). ISCTE-IUL.

Lowe, D. G. (1999). Object Recognition from Local Scale-Invariant Features. *Proceedings of the International Conference on Computer Vision*, 1150–1157. Available at: http://www.cs.ubc.ca/~lowe/papers/iccv99.pdf

Moural, A., Lopes, P. F., & Eloy, S. (Eds.). (2014). *CLOSE CLOSER: Close do Cities, Closer to People*. Lisboa: ISCTE-IUL.

Nielsen, J. (1992). Usability engineering lifecycle. *IEEE Computer*, *25*(3), 12–22.

Nielsen, J. (1995). *10 Usability Heuristics for User Interface Design*. Available at: https://www.nngroup.com/articles/ten-usability-heuristics/

Ourique, L. (n.d.). *Home mobility hazards detected via object recognition in augmented reality*. (forthcoming)

Owsley, S. (2011). Soldiers Take First Step in Combating Mozambique's Landmines. *U.S. Africa Command*. Available at: http://www.africom.mil/getArticle.asp?art=6854

Pascoal, P. (2015). Retrieval of Objects Captured with Kinect One Camera. In *SHREC'15 Track: Eurographics Workshop on 3D Object Retrieval*. Eurographics Workshop on 3D Object Retriev.

Proença, P. (2013). *Object Category Recognition through RGB-D Data*. ISCTE-IUL.

Raposo, M. (2016). *Ver a Arquitectura através das tecnologias digitais*. Instituto Universitário de Lisboa.

Raposo, M., Eloy, S., & Dias, M. S. (2017). Revisiting the city, augmented with digital technologies : the SeeARch tool. In *Proceedings of the REHAB 2017 conference*. Available at: http://rehab.greenlines-institute.org/

Rettig, M. (1994). Prototyping for Tiny Fingers. *Communications of the ACM*, *37*(4), 21–27. doi:10.1145/175276.175288

Wang, R., & Qian, X. (2010). *OpenSceneGraph 3.0: Beginner's guide*. Packt Publishing Ltd.

Chapter 4
Development of a Framework for Technological Embedding in Private Social Solidarity Institutions:
Technology for Inclusion in the Daily Activities of Third Sector Institutions – The Portuguese Case

Luis Barreto
Polytechnic Institute of Viana do Castelo, Portugal & University of Aveiro, Portugal

António Amaral
Polytechnic Institute of Viana do Castelo, Portugal

Teresa Pereira
Polytechnic Institute of Viana do Castelo, Portugal

Filipe Carvalho
Polytechnic Institute of Viana do Castelo, Portugal

ABSTRACT

The economic and social challenges felt in recent years because of the financial crisis impact wave were somehow attenuated by the silent work provided by the third sector institutions. Therefore, the incessant instability of the markets, as well as the population life-expectancy increasing and the implications thereof require new approaches towards pointing strategies to mitigate these problematic situations.

DOI: 10.4018/978-1-5225-5270-3.ch004

For that reason, the development of technological solutions and applications for the private social solidarity institutions is an utmost challenge towards guaranteeing their sustainability and efficiency over time. The adoption of such solutions should be properly conceived to enhance their efficiency of the daily routines and to fulfill and inclusion of all users, while trying to reduce the technological literacy. The development of a technological framework to support the adoption of the practices, selection of technical requirements, and functionalities is seen as a great contribution for setting the roadmap that should be followed. This chapter explores the development of a framework for technological embedding in private social solidarity institutions.

INTRODUCTION

In recent years the society, in general, has been affected by a profound economic and financial crisis with broad social impacts for families affected by unemployment, excessive debt, and low or none expectation of state support. Consequently, such crisis triggered the augment of social exclusion among citizens, especially those who were more fragile and helpless forcing them to transform, to adapt and to react to the adversities, which were imposed. Nevertheless, the difficulties faced by these groups of people were profound and the conditions to overcome them were weak and incipient.

Therefore, the Private Social Solidarity Institutions (PSSI) have become essential to attenuate the social problems and combine innovative solutions to increase the limits and scope of their action.

The emergence of organizations with objectives to promote the quality of life and social support highlighted the weight of a sector, usually referred to as the Third Sector or also as Social Economy. The role of Third Sector institutions has been enormous on trying to mitigate the social scourge, furthermore the amount of financial, human and technological resources is meager and deficit among most of these institutions. Therefore, the management of third sector organizations becomes even more important in the current socio-economic context, particularly due it to the increasing demand for effective social responses in the different existing institutions, whether governmental or private nature. Hence, the study of technological solutions for supporting the management practices can be pointed out as extremely relevant, as a way of increasing sustainability of these organizations as well as to enhance its performance in the normal course of its activities.

In the third sector, there is a considerable diversity of organizations. The study will focus on PSSI Institutions, which are usually associated with health and well-being and social action activities. Small Private Social Solidarity Institutions, which are the most dominant type of institutions in Portugal, lack of financial, human and capable resources. They are in partly supported by state reimbursement, which is not enough for their sustainability. However, such institutions are very important for the wellbeing of elderly people. With the population aging, such institutions clearly need their activities to be supported by technological solutions that will allow them to improve their operations, as well as to be able to take care more efficiently of their clients and resources.

In the literature review it was not found any technological frameworks for small PSSI. Nevertheless, their specific challenges need to be supported, in a daily basis, through technological solutions that can be based on cloud and mobile applications. Additionally, these institutions have several procedures, which are common and thus may become automatically handled, such as home support, medical attendance and meal delivery. However, and mainly because of the limited amount of resources, which might cause new challenges that can be addressed in terms of better and dynamic activities and thus respond more accurately to their client needs. Another important aspect that must be considered is the protection of those clients private information. Currently the increase use of technological gadgets and the huge number of sensitive information processed by these organizations, it is demanded an adequate management of information, to ensure the confidentiality, integrity and availability of information. In this context, it is proposed the use of security standards such as the family standard ISO/IEC 27000 to manage information security, as well as the implementation of adequate security policies followed by awareness initiatives to the security risks, to promote a safer security practices regarding sensitive information management.

The main aim of this work is to define a set of needs and requirements, which were supported by structured interviews to some PSSI technical directors, as well as some visits to the institutions towards validating some of the intended perceptions, and propose a technological framework to enhance a daily operations and management activities. The interviews and visits will also be important to understand the relevancy of each requirement and its technological challenges, and at the same time, to understand how new and crossed needs and requirements can be added.

RESEARCH METHODOLOGY

The proposed methodology is composed of two main steps. It began with a literature review about the topics under research, mainly regarding technology for social inclusion, technical aspects about security and global demography. Secondly, it was decided to make an exploratory study to obtain the professional's opinion about the importance of technology into the daily routines of Private Social Solidarity

Institutions (PSSI), based in their experience. The interview was considered the most appropriate procedure to collect information about the usability benefits that technologies might have in this type of organizations.

Towards knowing better their reality and guarantee the systematization of the information, a structured interview procedure was adopted, in which all interviewees were asked to answer the same questions. This allowed us to explore and compare different answers given to the same questions.

The questions were divided in three main groups. The first group of questions was dedicated to the institution' characterization, namely the identification of the services provided, their procedure to collect funds, the academic background of their employees. The second group of questions was dedicated to understanding how institutions develops/manages their operations. The emphasis was on the activities that might have a direct support from technological solutions, such as the implemented mechanisms/procedures adopted by the institution to control and report important information, among other situations. Finally, in the third group of questions we've explored their personal views and perceptions about how the information technologies might be used to help the PSSI in their daily routines. It is also important to verify if there are conditions to implement such technological solutions and in which dimensions they are most suitable and requested.

BACKGROUND

World Population Continuous Aging Tendency

During the last century, a great number of people around the world is living longer and healthier, which constitutes one of the greatest achievements of mankind. According to the World Population Prospects - The 2017 Revision, the world's global population is nearly 7.6 billion, which means that during the last twelve years the population increased approximately one billion inhabitants (UN-DESA, 2017).

As presented in Table 1, the rhythm of the population growth projected to the world and by region will increased almost 50% from 2017 to 2100. It is notoriously that the Africa region will the major contributor for this. It is also important to be aware that some regions like Asia and Europe will shrink over the years, which means that the birth rate will continue to decrease and the overall percentage of elderly population will continue to rise.

According to the National Institute on Aging from the United States Department of Health and Human Services (NIOA, 2007) there are obvious challenges and opportunities that clearly displays why population aging cannot be avoided. Namely:

Table 1. Projections of the world and regions population

Region	Population (Millions)							
	2017	%	2030	%	2050	%	2100	%
World	7 550	100%	8 551	100%	9 772	100%	11 184	100%
Africa	1 256	16,64%	1 704	19,93%	2 528	25,87%	4 468	39,95%
Asia	4 504	59,66%	4 947	57,85%	5 257	53,80%	4 780	42,74%
Europe	742	9,83%	739	8,64%	716	7,33%	653	5,84%
Latin America and the Caribbean	646	8,56%	718	8,40%	780	7,98%	712	6,37%
Northern America	361	4,78%	395	4,62%	435	4,45%	499	4,46%
Oceania	41	0,54%	48	0,56%	57	0,58%	72	0,64%

Source: Adapted from United Nations, Department of Economic and Social Affairs, Population Division (2017). World Population Prospects: The 2017 Revision. New York: United Nations.

- For the first time in history, the amount of people aging 65 or over outnumbers children under the age of 5;
- For the same reason previously pointed, the number of people at the age of 85 and over are now the fastest growing age-interval of many national populations.
- The major cause of death among older people in developed and less developed countries is related to chronic no communicable diseases, which is a natural consequence of the life expectancy increasing over the years;
- While world population is aging at an unprecedented rhythm, the total population in some countries is simultaneously diminishing;
- The patterns of work and retirement are shifting because of the life-expectancy augment, people are spending a larger portion of their lives in retirement which progressively strains existing health and pension systems;
- The traditional family structures are changing and as people live longer and have less children, older people have fewer options for care and support in the end of their life;
- Social insurance expenditures are escalating in an increasing number of countries, which creates structural and sustainability issues on these current systems;
- New economic challenges are emerging through the enhancement of the elderly people worldwide. Therefore, this situation will contribute to drastic effects on social entitlement programs, labor supply, trade, migrations

patterns and savings policies around the globe which will require new fiscal approaches to proper balance all these different challenges.

These previously mentioned aspects will oblige governments worldwide to plan for the long term towards guarantying the proper countermeasures to mitigate the effects of the accelerated rhythm of the population aging. Currently, Europe has an average of four people at working age for every older person, but in 2050 it is estimated only to have two workers per older person. This scenario will, in most of the countries, arouse the amount of public expenditure devoted to the social insurance to the elders. For that reason, aside the public effort towards creating a sustainable environment for bearing these demographic effects the social responses will have to increase, creating suitable strategies to become efficient towards responding to the problems of the age-related chronic diseases and conditions.

The Portuguese demographic profile follows the patterns previously mentioned. According to the data available (World Bank, 2017; INE, 2015) the reduction of the Portuguese population is increasing and the ratio between elders and youngers is pointed to deeply aggravate in the following years.

The social problems caused by the financial crisis experienced and the demographic tendency for the following years will request the reinforcement of the nonprofit institutions from the third sector. These institutions are doing an upmost job on the inclusion of people with limited conditions and which need special care and support on an important period of their life - that is constantly increasing. The introduction of some technological solutions, in the operational procedures with the elderly and staff as employees and management, will be critical for enhancing the overall efficiency and sustainability of the PSSI.

ICT: The Path Towards Social Inclusion?

It the recent years an unprecedented technological evolution and increasing trend to connect devices have been evolving. Additionally, it is noticed an accelerated global aging of the population, becoming essential and strategic to take advantage of the Information and Communication Technologies (ICT), to provide technological solutions for the elderly living assistance. Although, the availability and usage of ICT in nonprofit institutions is nowadays very limited. Especially, because of the investment required to assure the identification of infrastructures and staff skills to properly operate.

In practice, the ICT use in healthcare is not new and has been widely used in healthcare institutions for monitoring vital signs and triggering alerts when emergency situations occur, also enabling the collection and record accurate patients' vital data, turning this data available to the medical staff or caretakers, in a timely manner.

Moreover, and with the advent of the Internet of Things (IoT) trend, significant changes in everyday life will certainly occur, specially promising new challenges to the current health solutions in order to provide personalized and collaborative ways of care. This evolvement may promote a more personalized, preventive and collaborative form of care, enabling patients with chronic diseases, for example, being able to manage their own health and share their health data with medical staff. Systems with sensors capabilities and mobile notifications have been developed and provide a secure and cost-effective option for independent living. These solutions can be particularly helpful in rural areas, where the number and availability of emergency teams with adequate responsive means is most of the times poor and very limited.

In this context, it is undoubtedly the advantages and the innovative services the technological evolvement may bring to the elderly, to the medical staff and caretakers. Nevertheless, the technological development and the sophistication of new devices usually reveal unexpected and underestimated security vulnerabilities.

Nevertheless, the increase reliance on the technology requires strategic efforts to protect the technological resources and the IT infrastructures, and ensure the safe, uninterrupted operations and safeguarding the computers, programs and data files. The amount of the daily collected data, the integration of new systems and applications, the potential access by unauthorized users, the lack of computer skills may contribute to the occurrence of information security incidents. In this context, it is very important to have adequate security procedures and good-practices to manage the collected data and the patients' information storage. It is essential to implement adequate mechanisms to ensure the information security properties named as Confidentiality, Integrity and Availability (CIA) of the patients' healthcare information.

Following, it is briefly presented the concept of each of these properties:

- **Confidentiality:** It is an essential requirement to ensure that only authorized parties can access information. This means that only authorized parties can read, print or simply knowing that a specific asset exists (Pfleeger & Shari, 2007). Similarly, the National Institute of Standards and Technology (NIST) Computer Security Handbook, (SP 800-12) defines confidentiality as being "a requirement that private or confidential information cannot be disclosed to unauthorized individuals". This property is essential to avoid inadvertently or intentionally unauthorized (non-legitimate) disclosure of information. Unintentional disclosure may be difficult to envisage. For example, do caretakers know what information are they allowed to share with their immediate organizational colleagues? Does this differ from the information they may share with the public? (NIST, 1995)

- **Integrity:** It is a security property that according to (Pfleeger & Shari, 2007) has different meanings in different contexts. This author points out that the integrity of an item is preserved when the item is: precise, accurate, unmodified, modified only by acceptable ways, modified only by authorized people, modified only by authorized processes, consistent, internally consistent and meaningful and usable (Pfleeger & Shari, 2007). Similarly, NIST (SP 800-12) (NIST, 1995) refer there is integrity of information when it is timely, accurate, complete and consistent. The NIST definition of integrity (SP 800-12) is very limited when compared with the (Pfleeger & Shari, 2007) assumption.
- **Availability:** It applies both to data and services. Data and services need to be available to legitimate entities at agreed times, and at acceptable levels, as specified in the quality of service agreed in a service level agreement or similar commitment. However, those assets can often become unavailable due to faults, attacks or errors in a service or supporting devices. According to (Pfleeger & Shari, 2007) and other standards, an object or a service is considered available if it is present in a usable form, it has capacity enough to meet the service's needs, it is making clear progress, and, in wait mode, it has a bounded waiting time. Lastly the service should be completed in an acceptable period of time.

These CIA properties have been at the root of information security since the start of digital area and the growing need for information security. One of the challenges is to achieve the right balance between the information security goals and the desired flexible level. In a general way, any intention to protect a specific resource must take into account the flexibility required to ensure its primary functional purpose. An adequate strategy involves the definition of the information security goals of the institution, this means the identification of the information resources that need to be protected, followed by the implementation of the adequate security procedures/ controls. In addition, the controls performance should be monitored and assessed on a regular basis, and just in case a specific control is not working as expected, it should be followed by the implementation of the necessary adjustments or updates to the control, in order to ensure the efficacy and efficiency of the control. The principle should follow a continual improvement process. In practice, organizations need to be agile, responsive, flexible and dynamic when they are establishing their security objectives and especially when they are conducted by an overall need for continuous improvement.

In this context, the control access of patients' private information is a security objective that must be ensured. The use of ICT promotes the large collection of sensitive data, thus it is demanded an efficient management of information in order to ensure their privacy and its misuse by third parties. Notwithstanding it is important to achieve a balance, since the confidentiality protection of patient's information should not compromise its availability. In practice, the private records of a patients or the medical prescription should be available to specific authorized caretakers. The information manager should clearly define which patient's information caretakers should access and ensure that this information is not shared to third and non-authorized parties. In fact, the elderly people are very vulnerable to frauds and burglary and in some cases inside people of the institution perform these incidents. So, this means that the patient's private information should be properly managed in order to not promote this kind of incidents, which may result into serious impacts. A properly security management strategy demands for a rigorous process, where every person interacting with critical information resources need to be aware and participate in security management, both adopting secure behaviors and continuous evaluating security control's performance.

The security risk environment is continuously changing and the attacks are getting more and more sophisticated. Institutions must consider how they are going to succeed to these risks, since the technical controls alone are no longer guaranteed, but mainly dependent on other security requirements such as legislation, culture or the environment (Hawthorn, 2015; Rouse, 2008).

International Security Standard ISO 27002: Information Technology, Security Techniques, Code of Practice for Information Security Management

Nowadays there are security standards and methodologies available to assist organizations to manage information security. To effectively create a solid structure for an information security program, organizations need to analyze and implement their methodology in alignment with either of the following international standards and methodologies.

ISO 27002 is part of the ISO/IEC 27000 family of standards. It is an information security standard published by ISO/IEC, entitled Information technology – Security techniques – Code of practice for information security management. ISO/IEC 27002 is an advisory document, to define the requirements for initiating, implementing, maintaining and improving information security management practice, in order to promote confidence in inter-organizational information collaboration. The ISO 27001 replaced the BS7799-2 standard, which was primarily published as a code of practices. As this matured, a second part emerged to formally cover security management system. The ISO 27001 is harmonized with other standards. In fact, ISO 27001 is a formal specification that uses ISO/IEC 27002 to indicate suitable information security controls within the ISMS (Information Security Management

System). However, since ISO/IEC 27002 is a code of practice or guideline, rather than a certification standard, organizations are free to select and implement the controls, according to their security control objectives, which should reflect the organizational security requirements. Implementing ISO/IEC 27002 involves a cost-effective plan that includes appropriate security controls for mitigating identified risks and protecting the confidentiality, integrity and availability of an organization's information resources. It also involves ongoing monitoring to ensure that these controls remain effective (Saint-Germain, 2005).

The standard comprises eleven information security controls and seeks to address security compliance at all levels: managerial, organizational, legal, operational and technical. It specifies 35 control objectives, consisting of general statements of security goals for each eleven domains. The standard also includes 114 controls that identify specific means for meeting the control objectives. The ISO/IEC 27002 security domains are (ISO/IEC, 2009):

1. **Security Policy:** Demonstrates management commitment to, and support for information security.
2. **Organization of Information Security:** Develop a structure for the coordination and management of information security in the organization. Assign information security responsibility.
3. **Asset Management:** Perform an inventory and classification of all critical or sensitive information assets.
4. **Human Resources Security:** Manage the security aspects related to employees, in order to reduce the risk of error, theft, fraud, or misuse of computer resources by promoting user training and awareness regarding the risks and threats to information.
5. **Physical and Environment Security:** Protect the information processing facilities, in order to prevent unauthorized access.
6. **Communications and Operations Management:** Manage technical security controls in systems and networks, in order to reduce the risk of failure and its consequences and develop incident response procedures.
7. **Access Control:** Restriction of access rights to networks, systems, applications, functions and data in order to detect unauthorized activities.
8. **Information Systems Acquisition, Development, and Maintenance:** Prevent the loss, modification, or misuse of information in operating systems and application software.
9. **Information Security Incident Management:** Responding appropriately to information a security breach that was exploited by an attacker.

10. **Business Continuity Management:** Develop the organization's capacity to react rapidly to the interruption of critical activities resulting from failures, incidents, natural disasters or catastrophes.
11. **Compliance:** Ensure that all laws and regulations are in conformance with information security policies, standards, laws and regulations.

The security standards help organizations align adequate security practices, according to their activity requirements. ISO/IEC 27001 has put forth guidelines intended to provide a level of protection for information resources. A large part of effective security practice is reaching a common level of proficiency, since patching systems in a timely way and configuring them in a secure manner, increases the likelihood that an organization will remain secure. The adoption of security standards, enables institutions to demonstrate their commitment to secure information resources and to ensure confidentiality, integrity and availability of patient's information. They also provide their partners and patients with greater confidence in their capacity to prevent and rapidly recover from any interruptions to the provided services.

INTERVIEWS ANALYSIS

Private Social Solidarity Institutions in Portugal

The design and adoption of technological solutions into the PSSI daily routines requires, as previously mention in the research methodology, to know with rigour the type of activities and resources available in the daily routines. This turned out to be essential for us to be keen on developing the technological framework. Therefore, we were able to perform sixteen structured interviews to managers from the different PISSs. The structured interviews were accompanied by a visit to the services and activities performed towards seeing in loco the conditions, thus allowing to gain a better picture of the environment and of the possible requirements/barriers that should be properly addressed.

Over time, the PSSI have developed a wide range of services to respond to the necessities of the Portuguese population. As presented in Figure 1, the nursing home and the home support are clearly the most frequent services, available in twelve and thirteen institutions, respectively.

Regarding the academic qualifications of the managers, it is possible to notice that the most common degree among our sample is the bachelor, as it can be confirmed in Figure 2. From all the institutions interviewed for this study, only one institution has managers with unfinished elementary school, three institutions have managers with elementary school and six have managers with high school. The characterization

Figure 1. Types of services available in PSSI

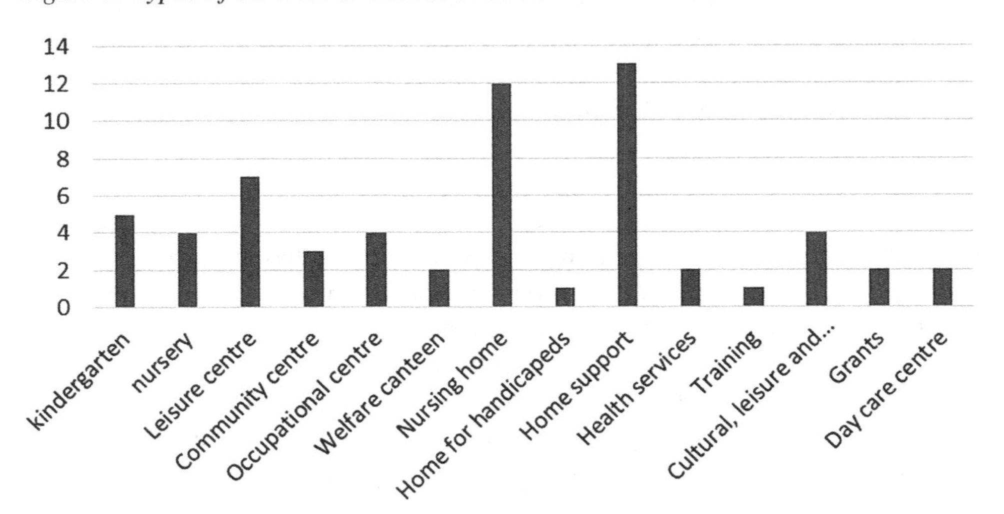

Figure 2. Academic qualifications of the managers interviewed

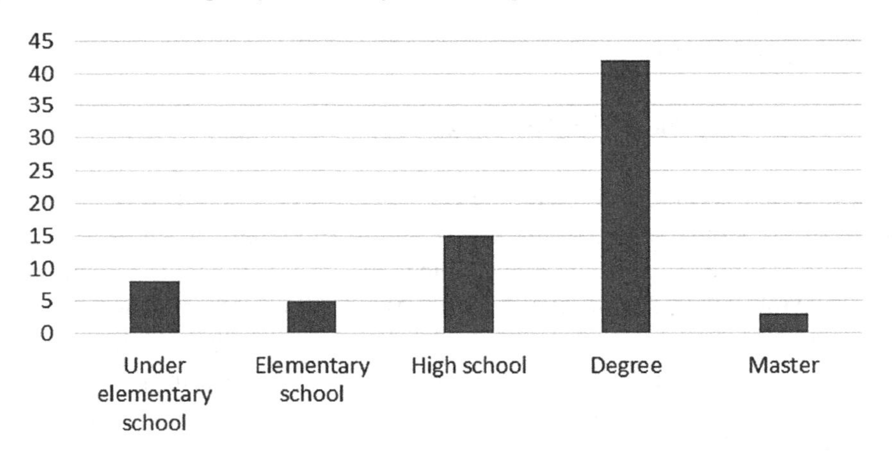

of the managers' level of qualification was important because it could work as a predictor of the level of awareness about new approaches, involving technologies, into their operations. In our opinion, a leader with a clear view of the possibilities could work out as a facilitator, triggering the embedding of the technological solutions.

Regarding the staff level of qualification, the situation is quietly different from the previous situation. As displayed in the Figure 3, the number of collaborators with lower academic qualifications is substantially higher. This could be an issue, because a great number of these collaborators are older and less experienced on using technological solutions. Therefore, it could limit the design of some technological applications towards guaranteeing the universal usage of the prescribed solutions. It is important to be aware of these manners to restrict the technological literacy issues of the staff.

The number and type of PSSI' staff is important is terms of dimensioning the technological structure, as well as to consider the typology and costs of solutions designed. As pointed by Figure 4, the great majority of the staff is permanent with full time contracts.

The number of clients that benefit from the services provided by these institutions is very divergent as we may conclude by the standard deviation of over 12000, as pointed in Table 2. This situation may be caused by the different services performed by the institutions. In a future work extension, it will be necessary to distinguish the number of clients for each type of service to be able to perform a better statistical analysis.

In Table 3 it is shown the collected information about how institutions control their documents. It is important to point that that the institutions that use both paper and digital control are only considered in the digital option. Some of the institutions have already implemented the ISO 9001 standard, which obliges to establish some procedures towards managing the information flow.

As it may be realized there are still a significant number of activities that don't have any kind of registration or that have only a paper registration. The only activities whose register is not made by all institutions are visitor identification, visitor registration and medical appointments. On the other hand, the only activity that has digital support in all institutions, as expected, is the invoice. We asked managers to describe how they define who has access to the information of their clients. All

Figure 3. Academic qualifications of PSSI employees

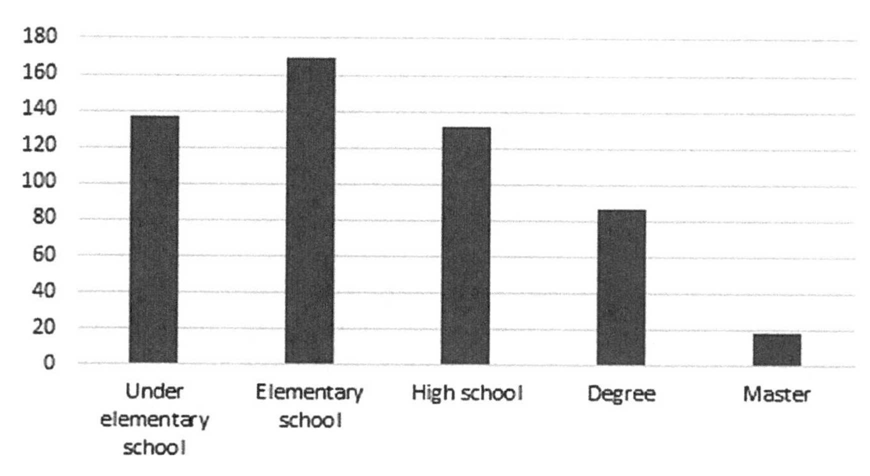

Figure 4. Type of PSSI employees

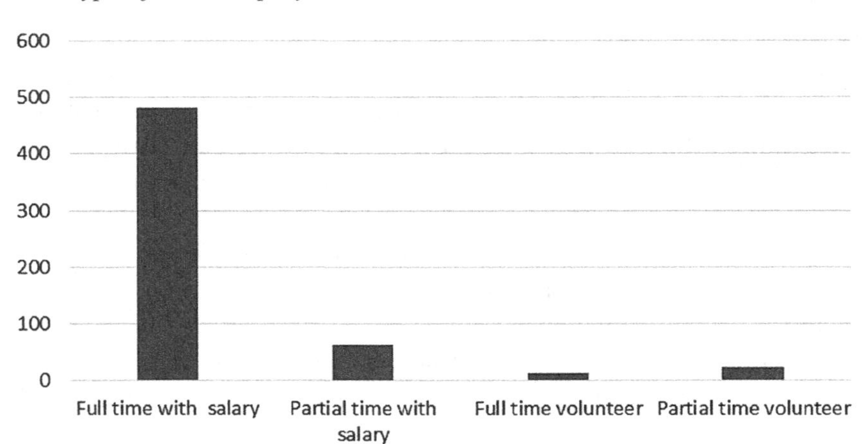

Table 2. Sample descriptive statistics

Minimum	Maximum	Average	Standard Deviation
9	48000	3920	12119

Table 3. Type of information control performed by each PSSI

Situation	No Register	Paper	Digital
Clients documents	0	7	7
Visitors identification	7	5	2
Invoice	0	0	13
Visitors registration	7	4	1
Contact of family people	0	6	8
Medical appointments	2	6	8
List of clients' necessities	0	5	7
Medical prescriptions	0	3	10

managers referred that not all the collaborators have access to all information. In some institutions, there are a specific group of collaborators that have access to all the information and in other institutions the information is given to people according to their competencies. However, the control of this information and the people that have access to it is only performed by some institutions.

Some institutions recognized the importance of access control to the information but still do not do it. The two most obtained responses: "we intend to improve this control" and "in the control quality system we described the procedure of controlling the information, but in practice we do not do it", show this concern. When we asked if any employees could access, change or even destroy documents with information about the clients, more than 50% of the institutions responded affirmatively. This fact emphasizes the importance to control the access to the information.

Regarding internal communication, we noticed that most of the PSSI uses oral communication, e-mails and customized documents. Those PSSI customized documents are used to answer to different requirements. The existence of such tools, designed by the institutions, suggests that they did not find a proper tool in the market that satisfies their needs.

With respect to external communication, besides the oral communication and the e-mail, institutions also use flyers and social networks. In general, institutions choose the people they communicate regarding their interests. Nevertheless, there are still a large number of institutions that do not do it, even though they recognize its importance as we may notice from responses such as: "We do not always have careful to direct the information. We are trying to improve."

The way that institutions deal with the information that might be used to improve their every-day tasks is very different from one institution to another. There are institutions that do not register any information and others that use methodologies of constant improvement. Nevertheless, in general, the information technologies are not yet implemented or its implementation is still limited.

As it may be observed from Figure 5, most institutions (63%) consider that the ICT have an important role in their institution and 12% wish to use them more.

Figure 5. Role of ICT in PSSI

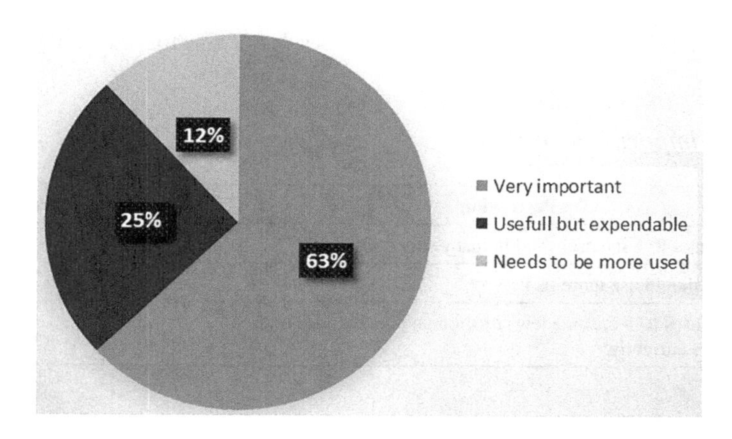

The Figure 6 shows in which situations ICT are currently being used.

As it may be observed, ICT are used more frequently in external communication. Nevertheless, when confronted with the competencies of collaborators regarding the usage of ITC, the managers, in general, said that not all collaborators have the required competencies to use this type of technology. The implementation of a digital tool would require some training.

We asked the managers how they feel about the intensity and the importance of ICT in their institutions and the answers are presented in Table 4.

As we may realise, from Table 4, most managers consider that their institution uses ICT intensely and in many situations, and only three managers consider that ICT does not have, or needs to have, an important role in the institution. When confronted with the possibility to introduce ICT in other activities, as support tools, managers presented a diverse range of options. This, in some sense, contradicts the previous answer and emphasises the importance of developing tools specifically designed for this type of institutions. Among other activities suggested by the

Figure 6. Usage of ICT for type of activity

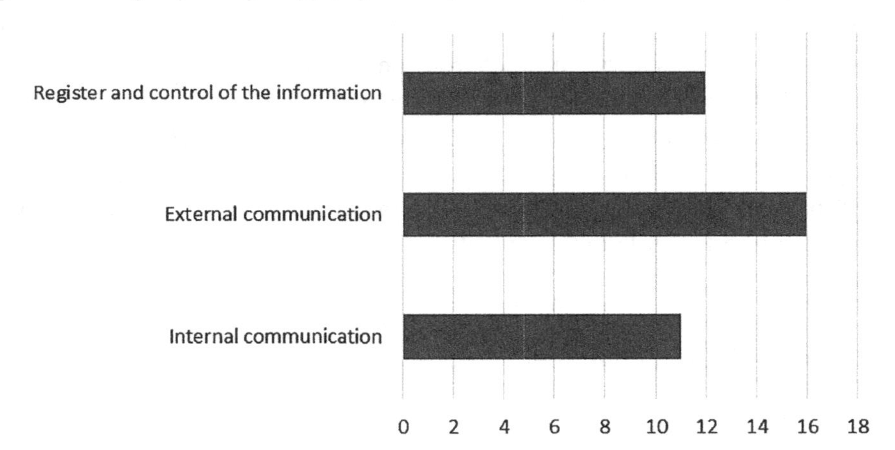

Table 4. ICT importance in PSSI

Perception	Number of Institutions
The institution uses ICT intensely and in many situations	8
The institutions should use more ICT	5
The institutions uses ICT in some few situation and do not need to use it more than it does currently.	3

managers, we highlight the following: "registration of activities"; "internal and external communication"; "control of medicines taking"; "visits control"; "drugs management"; "economical activities control"; "searching for more efficient routing plans".

Finally, we intended to know what are the main issues to implement the ICT as support tools. Only two institutions claim that they do not need to use more ICT. The other institutions mentioned the low budget (11 institutions), unprepared collaborators (7 institutions) and inexistent tools adapted to their needs (5 institutions).

In this context, various typologies of activities may be addressed. These PSSI normally have, regarding care for the elderly, two main action areas: older people whose support is provided on the premises of the institution; and older people whose support is provided in their homes (designated by home support). A new dimension area is also important, in our point of view, the integrated management support for the institution. For the home support dimension area, the proposed framework, with different users' profiles, such as PSSI employees, the medical professionals and the institutional manager. The home support framework will also include a wearable allowing mobile tracking and monitoring and generating alerts in real time that will help the institution to enhance the elderly support. The on premises support will include wearables, indoor navigation systems and monitoring and tracking systems. The other dimension of the framework, which will be more concerned with the integrated management of the institution, predicts a route optimization system, a medicines elderly profile and a logistics support application. These framework dimensions, as they handle with more sensitive and private information, must be fully compliant with information security management standards and definitions. The definition and use of such standards will be an important aspect for the fully acceptance of the technological frameworks by the PSSI.

The development of such framework is of capital importance for establishing new technological solutions, properly adopted to the social institutions reality. These solutions, will be sensible to the user-friendliness design because most of the staff in these institutions has a lack of technological knowledge and competencies. Therefore, the framework frontend clearly must be conceived according to these limitations and to allow a fast learning and usability. The PSSI, with limited dimension which are the most common institutions in Portugal, do not have enough resources to improve their work, as well as to overcome some of the common problems and requirements for increasing the service quality and satisfaction levels of users and relatives with the institution and their staff.

TECHNOLOGICAL FRAMEWORK

With the objective to improve quality of service on a daily basis, and also to help to improve resources of the Small Private Social Solidarity Institutions, which are the most dominant type of institutions in Portugal, and normally lack of financial, human and capable resources, we have defined a support technological framework – that we call Advanced and Virtual Technological Framework for Small PSSI (PSSI AdVirTec). This technological framework, that will be described next, pretends to be a cost-effective solution for most of those institution's needs, and somewhat validated with the survey previously presented.

These solutions, will be sensible to the user-friendliness design because most of the staff in these institutions has a lack of technological knowledge. Therefore, the framework frontend clearly must be conceived according to these limitations and to allow a fast learning and usability. The Figure 7 shows the proposed framework. According to Figure 7, the blue arrows present the information flow, while the white arrows evidence the data and communication flow.

Architecture

This PSSI AdVirTec will allow the collection, integration and availability of important and on real time information, allowing the Private Social Solidarity Institutions (PSSI) to be able to take actions to be more efficient and to support their clients. This framework will be a unified information management platform that will deliver capability across application domains critical to the PSSI. These Private Social Solidarity Institutions normally have, regarding care for the elderly, two main action areas: older people whose support is provided on the premises of the institution; and older people whose support is provided in their homes (designated

Figure 7. Technological framework for PSSI

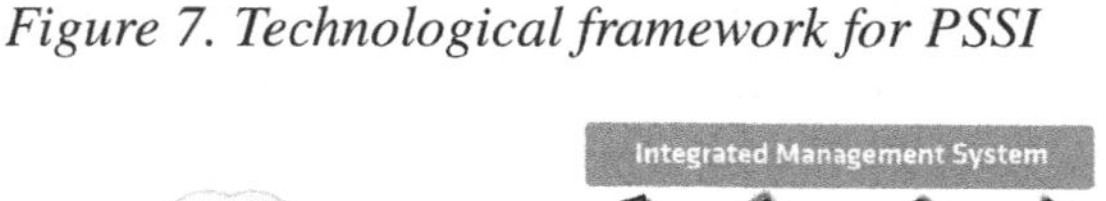

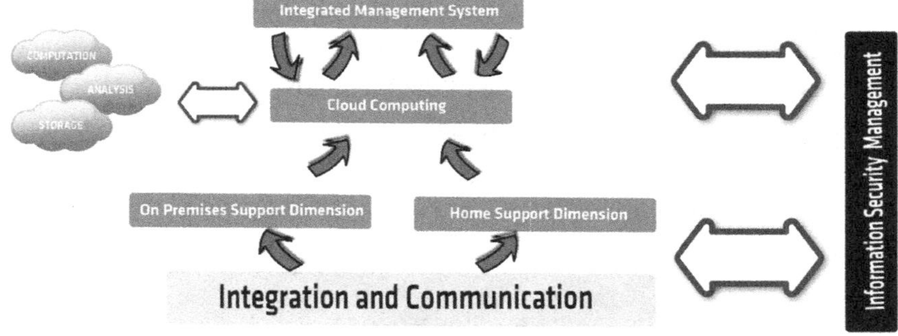

by home support). A new dimension area is also important, in our point of view, the integrated management support for the institution. Each one of these areas have specific requirements but also common requirements.

This technological framework is an information framework enabled by the Internet of Things (IoT) (Greengard, 2015) and the use of wearables (Lukowicz, Kirstein, & Tröster, 2004). This provides a means for consolidating tasks, monitoring and sharing of data between the previous referred action areas. Such institutions can benefit from IoT capability, which includes the effect on citizens (health and well-being). Clear the Internet of Things can bring huge benefits in these Small Private Social Solidarity Institutions. However, IoT is also a big game changer for security. Everything is moving faster than we thought, and because of the ubiquitous interconnectivity between devices, users, and distributed networks, defending a single place in the network is becoming increasingly ineffective. The practical reality is that most IoT devices are not designed with security in mind. Thus, the proposed framework also takes into consideration the security needs when using IoT and cloud based systems.

Home Support Dimension

The proposed framework will be used by different users, namely the PSSI employees, the medical professionals to record the health patient records and the institutional manager. In order to ensure the information's confidentiality, integrity and availability, as well as to ensure the accurate patient information, each user will have different access privileges in the system.

In the scope of the home support dimension area, the proposed framework, with different users' profiles, will include a wearable allowing mobile tracking movements and monitoring of some important vital information of the elderly people, like, for example, monitoring the user heart. Tracking those movements will be of special importance, as it will allow to know if an elderly is still for a very long time. If the elderly is still for a pre-defined threshold, and if the heart rate is not normal, the wearable will send an alert to the integrated management system. The integrated management system will then forward an action alert to an employee, and the employee will check if there is any problem with the elderly. The alerts are generated in real time, allowing to take actions more efficiently and more targeted, helping the institution to enhance the users support.

On-Premises Support Dimension

In terms of the framework for the on-premises support of the elderly, wearables will also be the main pieces of it. The wearables of the elderly will use indoor navigation systems and monitoring and tracking systems. The employees' wearables will be simpler, and they will just be used for receiving alerts and action information. One action information, can be that one of the elderly must take his medicine. Tracking systems are also important in the on-premises support, as they will alert when an elderly is moving to a certain distance of the institution facilities.

Integrated Management System

The other dimension of the framework, which will be more concerned with the integrated management of the institution, a route optimization system will be taken into consideration, a medicines elderly profile and a logistics support application will be part of it. A route optimization system will be important when one elderly person of the home support does not need, during a specific period, support. The route optimization system will allow to check if the route plan defined for the home support vehicle is actually the best or if there is a better one. The route optimization system will allow to reduce costs in combustible and to maximize the time of the home support employees. The medicines profile of the elderly, will be important in managing the medicines intakes. The system will send an alert to a specific employee informing that a specific elderly must take its medicine. When the medicine is taken the employee must inform the system. Then, the system updates its medicine database and prepares the information alert accordingly to the elderly needs.

This framework, as it handles with more sensitive and private information, must be fully compliant with information security management standards and definitions. The definition and use of such standards will be an important aspect for the fully acceptance of the technological framework by the Private Social Solidarity Institutions. This framework and its integrated management system will be cloud based, allowing to reduce costs in terms of on-premises applications and hardware.

Integration and Communication

The Figure 7 module integration and communication, uses as the main key technological enabler the Internet of Things (IoT). For this support framework we divide it in three different domains: network-centric IoT, Cloud-centric IoT, and data-centric IoT, which correspond to the framework requirements (see Figure 7).

Network-Centric IoT

The vision of IoT can be interpreted in two ways: 1) “Internet” based and 2) “Object” based. The Internet-based architecture main focus will be the Internet services while data are contributed by the objects. For the collection, processing, analysis and dissemination of valuable information gathered in the on-premises dimension and the home support dimension, especially in this framework through wearable devices, Wireless Sensor Networks (WSN) will have the main role. With the availability of sensors that are smaller, cheaper, more intelligent, and widespread (e.g., vital monitoring), WSN plays an important role in social and health sensing applications.

Cloud-Centric IoT

To integrate the ubiquitous on-premises support and home support information with the integrated management system, as well as to maximize the full potential of Cloud Computing, the framework presented uses the Cloud as an aggregation point. This allows, that new sensing services can be fully added without too many changes, developers can provide their health software tools; and computational intelligence experts can provide their data mining and machine learning tools useful in converting information to knowledge. Cloud computing is able to offer these services as infrastructures, platforms, or software. Many new applications focused in IoT and social life can be developed and introduced into the framework background. The Cloud integrates all dimensions of ubiquitous computing delivering scalable storage and computational resources to build new services. Moreover, Cloud offers a unique opportunity to create efficient IoT monitoring and sensoring models.

Data-Centric IoT

The use of WSNs, together with wearables and other IoT services will generate a massive amount of data. Data-centric IoT concentrates in all aspects of data flow, including collection, processing, storage, and visualization. Efficient heterogeneous sensing of the elderly information needs to simultaneously meet competing demands of multiple sensing paradigms. This will also have huge implications on network traffic, data storage, and energy utilization.

Extraction of meaningful information from raw data is nontrivial. It usually involves pre-processing and event detection. Events need to be detected in long multivariate time-series data. For the purposed framework, adaptability and robustness of algorithms to compare data at large scales are essential. To further make sense of the information and convert it into knowledge, it is important to make use of genetic algorithms, evolutionary algorithms and, as the more developments are done, neural networks are also necessary (Ranjan & Om, 2017). These algorithms will improve automated decision making and provide useful to reinforce the policies.

Security and Monitoring

PSSI AdVirTec presents many security challenges namely the need to provide protection to the elderly's sensitive information. The implementation of security mechanisms has to be defined from the beginning of the framework's development and implementation, allowing for its better integration and acceptability (Kim & Solomon, 2012). This subject needs to be tackled according to the three main security characteristics: integrity, confidentiality and availability. One of the main security issues relies in the information collected in the stored elderly profiles. Digital signatures are security mechanisms that provide the integrity of those profiles by enabling the detection of unauthorized modifications. If the digital signature does not match the information contents then the profile must be marked as not valid. Confidentiality relates mainly to the access to sensitive information by authorized individuals. It is obtained by controlling access to information and by protecting it while in transit along network communications. Access control policies need to be defined by the PSSI administration. These policies must be implemented using Role-Based Access Control (RBAC) (Ferraiolo, Sandhu, Gavrila, Kuhn, & Chandramouli, 2001). As the equipment will mostly be wireless/mobile, it is necessary to protect the network infrastructure from eavesdropping, so it is necessary to encrypt all information whilst in transit. Availability focuses on the means to provide continuous access to information by authorized users. Equipment and power redundancy, backups and system monitoring need to be put in place to guarantee availability of the system at all times. Monitoring sensors also need to be installed in order to detect problems in any of the three security characteristics, as well as for instance systems' malfunctions, errors, services that are not working and even improper behaviour.

FUTURE RESEARCH DIRECTIONS

As future developments, there is an ambition, and a real need, to apply this framework into some PSSI with different sizes and complexities, as well as to compile and identify a set of metrics adjusted to each service typology towards helping to potentiate the service quality level and the overall efficiency of the available activities. The aim is also to provide a significant number of technological applications/tools to support the implementation and the embedding of the framework through a roadmap, thus allowing to support the technology adoption in a step by step logic as well as to enhance the technological skills of the PSSI' staff, in a suitable and efficient manner.

The usage of technological applications in mobile formats is spreading every day, which constitutes an undeniable tendency through the universalization of the smartphones through all population spectrums. Furthermore, the underlying concepts regarding the smart cities and the Internet of things (IoT) are sharpening the presence of technological systems with reinforced influence and a wide set of applications and usages in which the third sector will be an avid consumer towards responding to the increase demand of their services and activities.

CONCLUSION

The number of studies under development in the third sector has increased in recent years. This could be easily explained but the enhancement of its importance to society and general as well as its substantial growing in terms of economic value in relation to the gross domestic value (GDV). The amount and extend of services and activities has grown considerably during the financial crisis period. Nevertheless, the existence of these institutions had been created to respond to the necessity of some society groups without any kind of support. The actual conditions are still favorable to the growth of these institutions, especially among the elderly. The unstoppable tendency of the population aging worldwide, and the continuous increasing of the population life-expectancy is creating new necessities related to the support services to deal with the chronic diseases which are affecting the senior group of the population.

The appearance of technology solutions, platforms and applications for the inclusion with especial emphasis in particular groups like people with special needs and institutions of the third sector, allows to extend and improve their operations. In this context, various typologies of activities may be addressed. These PSSI normally have, regarding care for the elderly, two main action areas: older people whose support is provided on the premises of the institution; and older people whose support is provided at their homes (designated by home support). In our point of view, the integration of a new dimension area is also important, the integrated management support for the institution. Each one of these areas have specific requirements but also common requirements. For the home support dimension area, the proposed framework, with different users' profiles, will include a wearable allowing mobile tracking and monitoring and generating alerts in real time that will help the institution to enhance the elderly support. In terms of the proposed framework for the on premises support of the elderly, it will also use a wearable, it will use indoor navigation systems and monitoring and tracking systems. The other dimension of the framework, which will be more concerned with the integrated management of the institution, will facilitate a route optimization system, a medicines management profile for each elderly and a logistics support application will also be an important part of it. As the defined framework dimensions handles with sensitive and private information, must be fully compliant with information security management standards and definitions. The definition and use of such standards will be an important aspect for the fully acceptance of the technological frameworks by the PSSI.

The development of such framework is of capital importance for establishing new technological solutions, properly adopted to the social institutions reality. These solutions, will be sensible to the user-friendliness design because most of the staff in these institutions has a significant lack of technological skills. Therefore, the framework frontend clearly must be conceived according to these limitations and to allow a fast learning and usability. The PSSI, with limited dimension, which are the most common institutions in Portugal, do not have enough resources to improve their work, as well as to overcome some of the common problems and requirements for increasing the service quality and satisfaction levels of users and relatives with the institution and their staff, thus it is clearly important to use a new and simple technological framework that can take to another lever their services and satisfaction levels.

REFERENCES

Ferraiolo, D. F., Sandhu, R., Gavrila, S., Kuhn, D. R., & Chandramouli, R. (2001). Proposed NIST standard for role-based access control. *ACM Transactions on Information and System Security*, *4*(3), 224–274. doi:10.1145/501978.501980

Greengard, S. (2015). *The internet of things*. Cambridge, MA: MIT Press.

Hawthorn, N. (2015). 10 things you need to know about the new EU data protection regulation. *ComputerWorld UK Daily Digest*. Retrieved June 10, 2017, from: http://www.computerworlduk.com/security/10-things-you-need-know-about-new-eu-data-protection-regulation-3610851/4/

INE. (2015). *Envelhecimento da população residente em Portugal e na União Europeia*. Retrieved July 3, 2017, from https://www.ine.pt/xportal/xmain?xpid=INE&xpgid=ine_destaques&DESTAQUESdest_boui=224679354&DESTAQUESmodo=2&xlang=pt

ISO/IEC. (2009). *ISO/IEC 2nd WD 27002 (revision) - Information technology - Security techniques – Code of practice for information security management*. ISO Copyright Office.

Kim, D., & Solomon, M. (2012). *Fundamentals of information systems security*. Sudbury, MA: Jones and Bartlett Learning.

Lukowicz, P., Kirstein, T., & Tröster, G. (2004). *Wearable systems for health care applications*. Retrieved June 3, 2017, from https://www.researchgate.net/publication/8480044_Wearable_systems_for_health_care_applications

NIOA. (2007). *National Institute of Aging. Why population Aging Matters – A global Perspective*. Washington, DC: U.S. Department of State.

NIST. (1995). *SP 800-12: An Introduction to Computer Security: The NIST Handbook*. NIST. Retrieved March 3, 2017, from http://csrc.nist.gov/publications/nistpubs/800-12/handbook.pdf

Pfleeger, C., & Shari, L. (2007). *Security in Computing* (4th ed.). Prentice Hall PTR.

Ranjan, P., & Om, H. (2017). Computational Intelligence Based Security in Wireless Sensor Networks: Technologies and Design Challenges. *Studies in Computational Intelligence Computational Intelligence in Wireless Sensor Networks*, 131-151. doi:10.1007/978-3-319-47715-2_6

Rouse, M. (2008). *EU Data Protection Directive (Directive 95/46/EC)*. TechTarget. Retrieved June 10, 2017, from: http://whatis.techtarget.com/definition/EU-Data-Protection-Directive-Directive-95-46-EC

Saint-Germain, R. (2005). *Information Security Management Best Practice Based on ISO/IEC 17799*. The Information Management Journal.

UN. (2017). *World Population Prospects: The 2017 Revision*. New York: United Nations.

UN-DESA. (2017). *World Population Prospects: The 2017 Revision*. Department of Economic and Social Affairs, Population Division.

Word Bank. (2017). *Population ages 65 and above (% of total)*. Retrieved July 3, 2017, from http://data.worldbank.org/indicator/SP.POP.65UP.TO.ZS?locations=PT

KEY TERMS AND DEFINITIONS

Application: A program or piece of software designed to carry out a particular function.

CIA: Confidentiality, integrity, and availability, also known as the CIA triad, is a model designed to guide policies for information security within an organization.

Inclusion: The action or state of including or of being included within a group or structure.

Internet of Things: A system of connected computing devices, mechanical and digital machines, objects, animals, or people that has the ability to transfer data over a network without requiring human-to-human or human-to-computer interaction.

ISO 27000: An information security standard published by the International Organization for Standardization (ISO) and by the International Electrotechnical Commission (IEC).

Private Social Solidarity Institutions: Non-profit institutions, created by private initiative, with the purpose of giving organized expression to the moral duty of solidarity and justice between individuals.

Research Methodology: The process used to collect information and data for the purpose of making analysis and decisions.

Role-Based Access Control: Method of regulating access to computer or network resources based on the roles of individual users.

Technological Framework: An essential supporting structure for the development of a high-tech platform.

Third Sector: The activities carried out by public and private social solidarity institutions.

Wearable Technology: Electronics that can be worn on the body, either as an accessory or as part of material used in clothing, and has the ability to connect to the Internet, enabling data to be exchanged between a network and the device.

Wireless Sensor Network (WSN): A wireless network consisting of spatially distributed autonomous devices using sensors to monitor physical or environmental conditions.

Chapter 5
Use of Mobile Devices in Science Education in a Brazilian Public School Located in a Region of High Social Vulnerability:
A Case Study

Isabela Silva
Universidade Federal de Santa Catarina, Brazil

Loren Mattana Viegas
Universidade Federal de Santa Catarina, Brazil

Karen Schmidt Lotthammer
Universidade Federal de Santa Catarina, Brazil

Zeni Marcelino
Universidade Federal de Santa Catarina, Brazil

Karmel Silva
Universidade Federal de Santa Catarina, Brazil

Juarez B Silva
Universidade Federal de Santa Catarina, Brazil

Simone Bilessimo
Universidade Federal de Santa Catarina, Brazil

ABSTRACT

This chapter aims to present a case study about the use mobile devices as a tool to practice in science classes in subjects related to the parts of the human body and digestive system. This case study was carried out with 24 students of the fifth grade of a public school located in a region of high social vulnerability of the city

DOI: 10.4018/978-1-5225-5270-3.ch005

of Araranguá, Santa Catarina, Brazil, in partnership with the Laboratory of Remote Experimentation (RexLab). For the practice, the following applications were used: Human Body Systems 3D, Human Anatomy and Puzzle Anatomy. At the end, students were asked to respond to a questionnaire about the level of satisfaction related to the use of tablets in the classroom. Through the answers, a positive reaction related to the integration of digital technologies in the classroom was perceived, characterizing, thus, an assertive opportunity to continue the use of mobile devices in the school environment, giving an improvement in the quality of teaching of the sciences provided by the applications, considering the benefits of digital inclusion.

INTRODUCTION

The use of the New Information and Communication Technologies (NICT) is increasingly present in the routine of students of different educational levels and also of the most varied parts of the world. According to the author Coll (2013), the technological advance has brought many changes in a profound way in several areas; among these areas is the educational area, which works to bring the school environment new ways to improve learning and make teaching easier for both teachers and students – for Presby (2017), the use of ICT at classroom improve students' engagement and achievement.

There are many projects and programs currently in place to bring mobile devices into schools. According to the document "The future of mobile learning" (UNESCO, 2014), in formal education there are two popular models of mobile learning in schools, called One Computer per Student, in which each student receives his own device for free, and Bring Your Own Device (BYOD). There are also a number of university research and community projects to provide public school students with access to devices, since the infrastructure and budgets of these institutions are often precarious. In addition to the fact that most students have mobile devices, there are those who do not have them and will need to use the devices of third parties, such as students living in regions of high social vulnerability, thus promoting digital inclusion among these students.

Within this atmosphere, which is governed by new digital technologies ubiquitous in the routine of the individuals, it becomes of great necessity that institutions of education of basic education, level of education that in Brazil understands elementary, secondary and high school, seek adaptation to this reality. In this way, schools will provide their students with enhanced learning, linking the technological apparatus to the teaching of various subjects.

From the technological applications that supported the present study, it was possible to observe that the use of technological resources, such as mobile devices allied to content taught in the school environment, as well as mobile educational applications, can give students an even more complete experience of learning, also stimulating teachers of basic education to add this type of resource in their educational routine. Therefore, although there is a call for digital media integration in curriculum by current learning standards, rural schools continue to have access to fewer resources due to limited budgets, potentially preventing teachers from having access to the most current technology and science instructional materials (TIGHE, 2016).

Digital inclusion is a strengths-based approach that promotes technology equity for citizens through the creation of meaningful pathways to access skill building opportunities and resource innovation (COLFIELD-POOLE, 2016). According to Acioli (2014), digital inclusion is a type of social inclusion that allows to expand the use of technology as a field of realization of fundamental rights and duties of individuals.

Effectively integrating ICT in teaching has several benefits that research has found, such as increasing the quality of learning, providing easy and quick access to a very high volume of information and knowledge, reducing educational expenses, indirectly creating learning 2 experiences, increasing interest in learning, and increasing learning opportunities (FALATTA, 2016); so, the integration of ICT in public schools, specially rural schools, which suffer of a lack of resources, characterizes as digital inclusion, because it allows students who had no access to technology to use it to improve their education at school.

The present study aims to present a case study of digital inclusion, in which students from a school in a region of high social vulnerability could access mobile devices at classroom to complement their lessons. The case study was carried out in a school located in a rural region of Araranguá, a city located in the state of Santa Catarina, Brazil. In order to promote digital inclusion among students and teachers at the Otávio Manoel Anastácio School, located in a rural and remote area of the city, mobile devices were applied in a Science class, together with the teacher responsible for the subject, with which held meetings to integrate the use of mobile applications to strengthen the thematic of human body: digestive system.

The organization of this document is divided in some topics that support the research. Before Introduction, it will be presented the Methodology applied in the research; then, a section correlating Digital Inclusion and Integration of Technology in Education; a section explaining about Use of Mobile Devices in Science Teaching; a section presenting a brief historic about Integration of Technology in Education in other countries besides Brazil. Before these topics, the Case Study is presented, and then the Results and Conclusion.

METHODOLOGY

In order to carry out this work, a bibliographical research was initially carried out in order to recognize previous studies about the use of technology in the classroom, to identify some points: types of technological resources already used, methodologies for student evaluation and difficulties encountered by the teachers in carrying out the actions of technology integration in their classes.

For the bibliographic search, searches were done through the following databases: Eric, IEEE Xplore Digital Library and Scopus, searching for articles published in the last 4 (four) years, focused on technological practices carried out in the classroom with secondary and high school classes, use of mobile devices, integration of technologies into education, and tools for practice in science education.

The work was developed with the support of the Remote Experimentation Laboratory (RExLab), Federal University of Santa Catarina, Brazil. The projects of RExLab have as main focus the integration of digital technologies in several educational levels, but mainly the basic education, which in Brazil means the schools years from Kindergarten to High School. In this way, this research group has partnerships with several basic education schools in the public network. The school chosen for the application of the case study was the Otávio Manoel Anastácio School, located in the city of Araranguá/ SC.

The school was chosen because it was located in a region of high social vulnerability, therefore its students suffered from lack of opportunities and their education was even more precarious than that of students from other educational institutions. In this way, one of the objectives of the present work becomes the digital inclusion of these students in relation to digital technologies. The group chosen for the application of the research was a class of the fifth year of basic education, and the applications of technology were carried out during a science class, lasting an hour and thirty minutes in total. The group consisted of 24 (twenty-four) students. In addition, RExLab provided 24 tablets so all students could use mobile devices during the survey and one trainee to assist the teacher and students in handling the tablets.

In order to begin work organization, a meeting was held with the teacher responsible for teaching science classes to the fifth-grade class in the afternoon period of the school in which the case study was carried out. the theme proposed by the teacher during the class in which digital technologies would be applied, as well as it became possible to verify the contents that would be taught in that semester in order to identify what would be the most appropriate subject to be addressed by making use of technological resources in set.

Thus, after finding that the use of mobile devices would be of great value in the practice of subjects related to the parts of the human body and the digestive system, we looked for applications aimed at this subject that approached the contents presented in the classroom and that could be accessible and playful enough to be used in a class of fifth grade in elementary school, in which students usually have between nine (9) and ten (10) years of age. In this way, the following applications were achieved: Human Body Systems 3D, Human Anatomy and Puzzle Anatomy. These applications will be detailed in one of the next sections, called "Digital Technologies in Teaching Sciences".

Subsequently, the applications have been tested and installed in 24 (twenty-four) mobile devices (tablets) provided by RExLab to perform the practice among students in the classroom. In addition, an intern from RExLab, a student in an Information and Communication Technologies (ICT) undergraduate course, was provided to assist the teacher responsible for the class during the class in which the case study was carried out.

For the use of the mobile applications, students remained in the classroom, where each received a tablet with the applications installed, accessing them individually. To assist in these actions, in addition to the teacher responsible for administering the subject, the RExLab intern assisted the students in the use of the mobile applications and mobile device.

RELATED WORK

This section aims to present related work in relation to the present chapter; the following subsections, called "Digital Inclusion: Integration of Technology in Brazilian Basic Schools", "Use of Mobile Devices in Science Education" and "Integration of Technology in Teaching: International Cases" will demonstrate the context in which the research is based.

Digital Inclusion: Integration of Technology in Brazilian Basic Education

The Information and Communication Technologies (ICT) are increasingly integrated into the daily life of society. Today, it becomes difficult to imagine how society would work without tools and resources such as smartphones, tablets, desktops, laptops, softwares, the Internet, virtual games, among others, which are very familiar and act as facilitators of teaching-learning tasks is only one of the most diverse areas that can be covered by the use of digital technologies.

Howard and Thompson (2015) define technology as all kinds of Information and Communication Technologies (ICT) resources existent in notebooks, educational games or smartphones. These resources are each day more present in several actions executed in our daily life, since access to News to communication with other person through distance.

From this context, initiatives were created with a view to digital inclusion in society and education. According to Cogo, Brignol and Fragoso (2014), digital inclusion initiatives aim to give access to technology to groups of greater social vulnerability, which would find enormous difficulties in acquiring or making use of ICTs in another way.

In Brazil, actions of digital inclusion began in the 1970s with activities that included the acquisition of computers and their installation in public spaces of several municipalities of the country.

In 1984, through Law No. 7,232, the Special Secretariat for Informatics was created, whose function was to foster technological development in government sectors related to information technology, as well as to analyze and decide on the execution of projects.

Subsequently, several other projects were developed, for example: the National Program of Informatics in Education (Proinfo), between 1997 and 2006, whose main objective was to promote the educational use of computers in public schools, taking computers and resources to several municipalities in Brazil. According to data from the Ministry of Education of the Brazilian Federal Government, the Proinfo Program totaled 147,355 microcomputers acquired and 5,564 attended during the 9 years of validity (Ministry of Education, 2006).

In 2009, Decree No. 6,991 established the National Program to Support Digital Inclusion in Communities called Telecenters. Telecentres are public and community environments that allow free access to computers and the internet, available for multiple uses including free and assisted browsing.

More recently, in the year of 2014, administrative order number 2,662 created the Electronic Government- Citizen Assistance Service (Gesac), which provides for the provision of a broadband internet connection free of charge to communities in a state of socio-economic vulnerability, with the objective promote digital inclusion in the country.

Thus, despite several projects developed with the aim of promoting digital inclusion and technology integration in both society and education, it was seen that these actions did not effectively cover all municipalities in Brazil. In addition, due to the lack of maintenance of the equipment purchased and installed in the laboratories, these became scrapped, making it impossible to use both social and didactic. In relation to the laboratories installed in educational institutions for didactic use, data from the National Institute of Educational Studies and Research Anísio Teixeira (INEP) in 2015 indicate that 45% of Brazilian public schools have a computer lab, with an average of 7.3 computers per school.

Thus, although there are microcomputers available in schools for use by teachers and students, there is no integration of technologies in the teaching and learning process, as it is more important than learning how to use a particular program to find effective ways of integrating the teaching-learning process (Rodrigues, 2016). In order for ICTs to be integrated into teaching, the role of teachers is fundamental in orienting students to the conscious use of the tools and information made available.

According to Bonilla (2010) the training of teachers for use of ICT in the classroom emerged in schools from the implementation of Proinfo, occurring in the distance modality. However, great difficulties were encountered since teachers did not have familiarity with computers, making it difficult to relate them to virtual learning environments (Bonilla, 2010).

Data from Cetic.br (Regional Center for Studies for the Development of the Information Society) in 2016 state that only 11% of teachers received training courses through the school in which they work. However, even with the difficulties encountered, 79% of teachers state that the use of ICTs in the classroom stimulates collaboration among students in the accomplishment of tasks.

In view of this, the difficulties encountered by teachers of basic education in Brazilian public schools to integrate ICTs into their practices are visible, given the lack of infrastructure and orientation. However, the benefits of this integration to the teaching and learning process are also known, because regardless of the quality of the continuing education received, it is conditioned to the time and physical resources that teachers have to perform it (Maltempi & Mendes, 2016).

On the other hand, laws have been developed to determine the ban on the use of mobile devices in schools. According to Law No. 14.363, in Santa Catarina state (Santa Catarina, 2008), the use of cell phones in public and private schools in Santa Catarina is prohibited. If technology brings so many benefits into our daily lives, it becomes difficult to understand why laws like this are passed. New technologies have much to contribute to basic education through the various resources it offers, and avoiding them in schools only harms teaching.

In Brazil, there is a high and worrying rate of school dropout, and this rate tends to increase with the passing of school years. The traditional teaching methodology can be seen as a factor that discourages students from their studies. Faced with information and practices adverse to their daily lives, students may feel uncomfortable, something that culminates in the generation of difficulties, resulting in repetition and avoidance, as Neri (2009) says. So, it is necessary to search for solutions that motivate the students, with the objective of presenting the content addressed in the room in a more interesting way to them, using tools of their daily life to facilitate access to information and demonstrating how the content can be applied in real life.

In this way, a good solution to be considered is the use of technology in the classroom, because, in addition to using familiar objects in the students' daily life, it promotes a greater search for information and a greater socialization of knowledge, providing the development of classroom activities that broaden the concept of the classroom, as well as the expectations of the students involved in the context.

Thus, for the integration of ICT in the practices of teachers to be carried out effectively, it is necessary to collaborate the elements that make up an educational institution: students, teachers, parents, management and current legislation. In this way, the support of the institution as a whole promotes the necessary support so that the teacher can innovate in its practices and include technological resources to it.

Use of Mobile Devices in Science Education

A few years ago, the classroom was restricted to the physical walls of an educational institution, and teachers were the only transmitters of knowledge to which students had some kind of contact. Didactic material was written on blackboards through white chalks, and didactic material was addressed to students through printed books, and in Brazilian schools with lower financial resources, copied by mimeographs. The resources used by students to study outside of school were books borrowed from the library, the textbook, and exercises and activities addressed by teachers as homework assignments. Thus, from the popularization of the use of technology in society, it was observed the great use of mobile devices by young people and adults.

Mobile devices have become increasingly popular over the years, with place in the hands of more than half the Brazilian population. According to the Mobile Learning Policy Guidelines (UNESCO, 2014), there are more than 3.2 billion mobile phone subscribers worldwide, making the mobile phone the most widely used interactive ICT on the planet. The main factors that allow this popularization are the low cost, practicality, capability and high functionality that these devices offer. These factors also allow the devices to be used for educational purposes, since tools with such characteristics can function as innovative solutions to improve teaching.

According to Teleco data, in 2015 Brazil ranked 5th in the world in accesses to mobile devices, and according to data reported by the Brazilian Institute of Geography and Statistics - IBGE (2016), in 2014 it was pointed out that 77% of Brazilians whose age was over 10 years old, had a mobile phone for personal use; moreover, 91% of users accessed the Internet through mobile devices in a complementary way, while 18% accessed exclusively through smartphones and tablets. This information is positive, as it demonstrates the high applicability that mobile devices have in public education education (IBGE, 2016). In view of this, since these resources are widely used by a large part of the Brazilian population, the opportunity has also been found to make use of them also for the learning of students, an action called m-learning or mobile learning.

The term m-learning is defined by Fakomogbon and Bolaji (2017) as an electronic learning medium that makes use of devices such as laptops, smartphones and tablets as tools. M-learning has different approaches, focusing not only on technology alone, but on the changes caused in learning through the use of technology. Li & Wang (2017) categorize m-learning approaches in: focus on technology, concentration on changes in learning resulting from the use of technology, mixed type that unites technology and changes in the teaching and learning process, and finally, behavioral description in the use of technology. Therefore, for mobile devices to be used in teaching, planning is necessary so that the impacts of the use of these are satisfactory to the point of providing the student with autonomous learning.

Among the advantages of using m-learning in teaching, authors Ibrahim, Ahmad, and Shafe (2016) cite the sharing of teaching materials, as it enables students to carry out the educational work with their own resources from a mobile device to any time and place. Kurtz (2016) mentions that m-learning considers the diversity and individuality of students, respecting their learning styles, the time available for such action and especially their place of study, since it allows the student to devote himself to his studies of any place, making use of a device connected to the internet.

In this context, data from Cetic.br (Regional Center for Studies on the Development of the Information Society) for the year 2016 indicate that in Brazil there is a great use of mobile devices by young people and that these use for educational purposes. In an interview conducted, 93% of Brazilian students, aged between 11 (eleven) and 17 (seventeen), claimed access to the Internet through a mobile phone, and 82% of respondents use the internet as a tool for solving questions passed in the classroom by the teacher.

Since the mobile devices used for learning in no way replace the classroom (Kurtz, 2016), the role of teachers in this process is of paramount importance. Teachers, when equipped for the pedagogical use of technological resources, have the power to instruct students in the use of technology as a whole, in order to instigate them to research and constant learning. Mobile devices support not only learning, but also educational goals by optimizing interactions between teachers and students and between students themselves (Fakomogbon & Bolaji, 2017).

Among the difficulties encountered by teachers in the use of mobile devices in Brazilian classrooms are: the lack of continuous training, as already mentioned, and the legislation in force. In the state of Santa Catarina, located in the south of Brazil, Law No. 14,363 approved in 2008 prohibits the use of cell phones in classrooms of public and private schools in the State of Santa Catarina, even for pedagogical use. Thus, although all the advantages of using mobile devices in a pedagogical way in the classroom are proven, Brazilian state legislation becomes a barrier factor.

However, despite the difficulties, teachers seek alternatives for their students to make pedagogical use of their mobile devices, giving as a suggestion the use in extra-class hours.

This applicability also represents a solution to a problem found in most Brazilian public basic education schools. According to the School Census, made available online by the Lemann and Meritt Foundation (2016), in 2015 it was found that only 44% of the schools in the public network had computer labs for student use, and many of these labs often do not have the number of computers per student, which is often more than 40 (forty) students per class. In addition, it is common that the maintenance of these computers is precarious for several factors; among them are the lack of professionals or low-budget educational institutions for computer repair, and many laboratories are not used because they are not functional.

In this way, in addition to being used as a tool for information sharing, mobile devices are considered important tools to support student practice, since it allows access to several multimedia resources such as: question bank, simulations, remote experiments, 3D modeling, among many others. The use of these resources allows the students to understand and learn, because it makes them feel motivated and focused on what is being addressed (Ibrahim, Ahmad, & Shafe, 2016).

Aimed at teaching science, the use of mobile devices is considered to be great allies in practical classes, supplying the lack or lack of laboratories for such, given the importance of practice in understanding such content. According to Minussi and Wyse (2016) the teaching of science in schools is not very well regarded by students due to the difficulty encountered by the teachers in relating the theoretical content to the daily life of the students. Therefore, students feel discouraged to believe that it is not necessary to learn these subjects (Minussi & Wyse, 2016).

In view of this, mobile devices become valuable tools in this process, because through a resource that, in general, the students themselves already have is provided the practice of the contents covered in the classroom. A study conducted by Santos and Freitas (2017) corroborates this idea when, in a study carried out, it compares the performance of two high school classes in a school in the north of Brazil. In this study, the authors compare the classes, making use of traditional teaching methodology in one of the classes and in the other, uses the mobile devices as a tool to practice, accessing applications focused on biology. Through a questionnaire, it can be seen that the class in which the practice was carried out using the applications installed in the mobile devices, the students got better use of the subjects taught previously in the classroom.

The use of technological resources in education is also described by Bressler and Bodzin (2016) report a study developed with two groups of eighth grade students from the United States. To perform this research, two groups of students were analyzed: one of them was made up of 59 students who made use of games to perform the practice and another group, formed by 120 students who participated in the common class with laboratory practice. In both scenarios students worked in small groups to measure scientific practices and make observations in the classroom. Finally, through the research carried out with the two groups of students, it can be seen that the students who used games as a tool to practice the theoretical content addressed in the classroom, obtained higher scores when compared with the other group where the practice was performed in the traditional model.

In a study conducted by Steinert, Hardoim and Pinto (2016) it is possible to observe the importance of educational practices adapted for the use of ICTs. Held with a high school class from a public school in the city of Cuiabá-Centro Oeste Brasileiro, the research exposes access to the SAMBI-Biomedical Health application, and the CienTI blog for the study of infectious-contagious diseases, making use of hybrid methodology for this purpose. Hybrid teaching methodology is described by the authors as a mixture of traditional educational practices together with technology use, forming groups of students where each group will perform a distinct activity in relation to the content presented, and students should rotate between these stations of study. Through the practice carried out with the students, it was possible to observe that the students were able to assimilate the contents more easily as well as to relate them with others already discussed previously.

Thus, in addition to the applications and features already mentioned previously, there are now several other resources available for application in didactic use in both mobile devices and notebooks or desktops.

Integration of Technology in Teaching: International Cases

In addition to Brazil, several other countries are engaged in integrating technology into the educational process. Males, Bate, and Macnish (2017) report the use of mobile learning as a tool for the development of writing, grammar, reading and calculus of students in the five, seven and nine years of an Australian school. According to the authors, analyzing the use of mobile devices year after year, the students who made use of this tool stood out in relation to the other students in the country, due to their greater use in the contents approached digitally.

Ally, Balaji, Abdelbaki, and Cheng (2017) describe the use of tablets in schools in remote locations in Pakistan. Used by about seventy students from classes of eighth, ninth and tenth years, the tablets gave access to the Optus system that have several didactic contents without need of access to the internet to consult them. According to a survey conducted by the authors, after the use of tablets, students became more interested in studies and learned how to use technology for educational purposes.

In the United States, Clayton, and Murphy (2016) cite actions developed at a suburban Detroit school that encourage students to create tutorial videos about the use of certain tools and applications. To this end, students have learned how to make use of Instructional Design in order to create the video in order to achieve its intended purpose. Thus, students were able to realize that videos are also important educational tools.

Finally, among many cases of technology integration in the teaching and learning process carried out around the world, we have mentioned the case described by Liu and Li (2017) where iPads are used in secondary school education in England as a tool for learning methodology based on problem. In order to solve the problems proposed in the classroom, these are related to the contents addressed, iPads are consulted in order to consult relevant information that may be useful to solve the proposed questions. In a questionnaire carried out with teachers and students of the schools where iPads were used, most cited "independence" as a major advantage of using the tool, since students can guide their research and thus their learning.

CASE STUDY

This case study was applied on August 10, 2017 at the Otávio Manoel Anastácio School, located in the city of Araranguá, Santa Catarina. The educational institution attends pre-school, elementary and secondary classes. The participating class was a fifth-grade class in the afternoon, and the application of digital technologies was done during a science class lasting about one hour and thirty minutes.

As for the technologies used throughout the application, there were 24 (twenty-four) tablets, used individually by the students and free mobile applications, that were downloaded by PlayStore, belonging to Android systems.

For the accomplishment of this work the following mobile applications were used:

- **3D Human Body Systems:** This application, developed by Evobooks, allows the user to visualize the human anatomy and its main characteristics in a detailed way, mapping the digestive systems (esophagus, pharynx, larynx, liver, intestine and others) (heart, veins, capillaries, valves and others). Among the features and tools of the 3D Human Body Systems application are state-of-the-art 3D graphics and models, as well as free traffic between animated objects, with explanatory texts related to the graphics representing organs of the human body.
- **Human Anatomy:** This application was developed by Pome Games & Apps and features various human body organs and a description of each. The Human Anatomy mobile application is a set of questions related to the functions of organic systems, and can be used by students of different ages. Initially, the user should click on the organs to read about their functions and

learn how they work, then be introduced to the game, where they should fit each organ in its place in the human body system.

- **Puzzle Anatomy:** Developed by Gemini Software, the mobile app Puzzle Anatomy is a puzzle game where pieces are the human organs and students need to put them in the correct position. Like Human Anatomy, Puzzle Anatomy is also a puzzle-style game, aimed at initiating children and adults in relation to the knowledge of the human body and its organs. This game is designed in the form of an interactive puzzle, in which the main organs must be recomposed in the right position within the human body; each movement of the pieces of the puzzle will correspond to a reaction of the human body itself.

In relation to the agents involved in the case study, the following members participated: a pedagogue, who was a teacher responsible for the class; a student in an Information and Communication Technologies undergraduate course, a trainee at RExLab, responsible for preparing the mobile devices to be used and giving instructions for use to the students involved; 24 (twenty-four) class students of the fifth grade, who were aged between nine (9) and ten (10) years.

The objective of the lesson, which was planned by the teacher responsible for the group together with the trainee of RExLab, was to review the content related to the systems theme of the human body, with emphasis on the digestive system and the organs of the human body that compose it. The use of the mobile applications enabled an overview of the organs, allowing the visualization of their location and

Figure 1. Physiology: mouth
Source: 3D Human Body Systems Application.

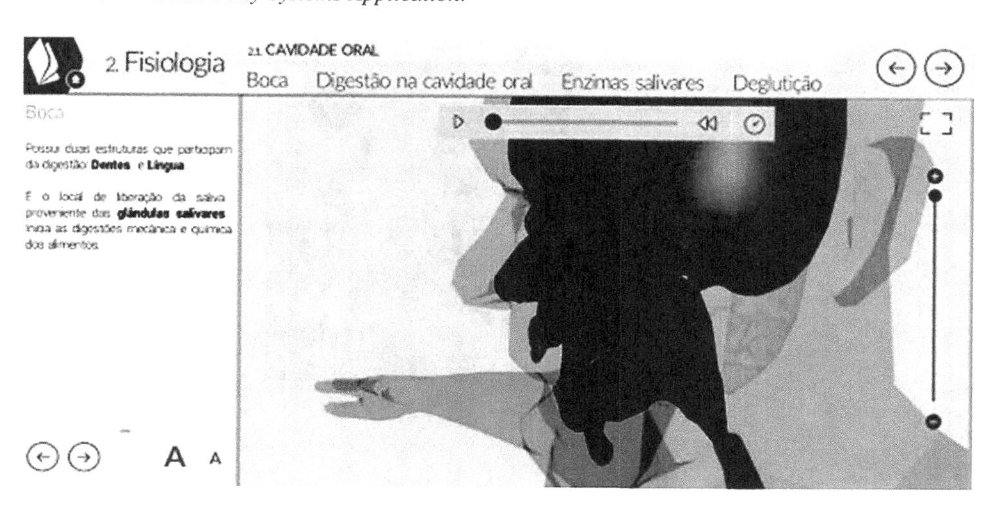

Figure 2. Organs of the digestive system
Source: Application Human Anatomy.

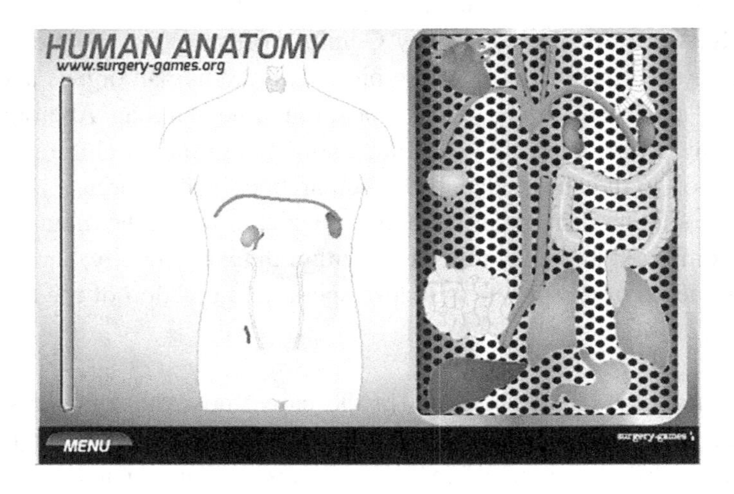

Figure 3. Organs of the human body
Source: Puzzle Anatomy application.

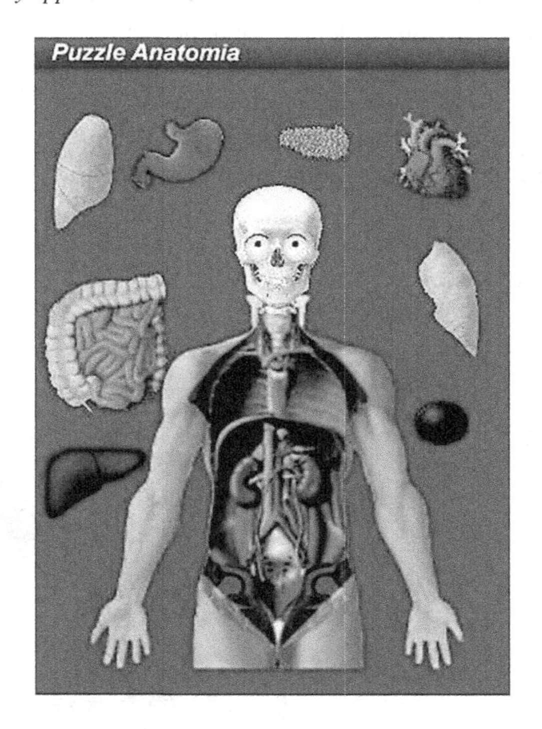

perception of their importance in the digestive system, as well as for the functioning of the human body in general, among other curiosities informed by the mobile applications used during the case study.

Figure 4-7 show some photographs of the case study.

In relation to the difficulties encountered and the way in which students reacted regarding the use of technology in the classroom, there was some difficulty in interpreting the descriptive texts present in the interface of the mobile applications used, however all the doubts obtained by the students were quickly remedied with the teacher responsible for the class. According to the teacher, the use of technology was satisfactory with respect to the learning and interaction of the students involved, who

Figure 4. Class of 5th grade students participating in the case study
Source: Authors' elaboration.

Figure 5. Students performing activities with the aid of mobile devices
Source: Authors' elaboration.

Figure 6. Student using mobile app
Source: Authors' elaboration.

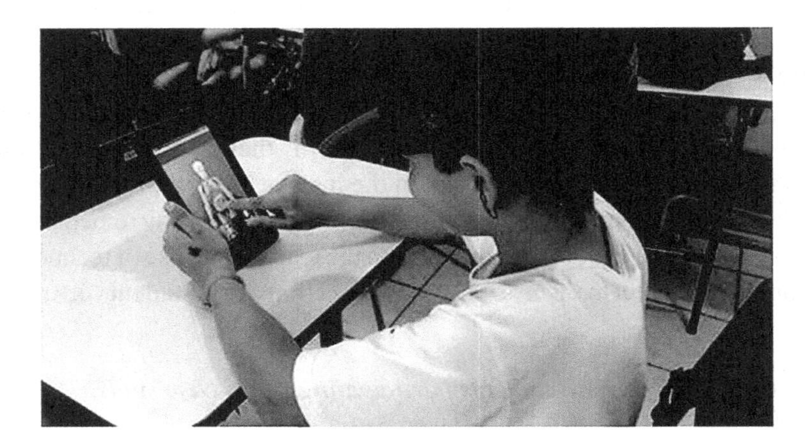

Figure 7. Student using mobile device
Source: Authors' elaboration.

demonstrated a positive behavior during the class, since they liked the differentiated didactics, as well as felt the will by digital technologies to be part of their daily.

After the application of technology in the classroom, the students participating in the activity were invited to respond to a questionnaire regarding their experience regarding the use of digital technologies in the classroom. This questionnaire was developed through the Google Forms platform and was answered by the students through the mobile devices used during the class activity. Student responses will be explained in the "Results" section, to be described below.

RESULTS

In this section the results will be presented regarding the answers given by the students involved in the case study in relation to the questionnaire applied at the end of the application of digital technologies in the science class, in order to verify students' satisfaction with technology, as well as verify their opinion about their learning and this method of differentiated teaching and the impact that the use of digital technologies caused in relation to their conception of the classroom.

This questionnaire is used for all applications carried out by RExLab and is called "Student Profile and Technological Profile Questionnaire", aiming to obtain data regarding the use of technology when accessed by children. This questionnaire also has other similar versions for teenagers and adults. The complete questionnaire with all the questions follows in appendix. Figure 8 shows screens of the questionnaire accessed by a mobile device.

The most relevant questions for presentation in this document were selected, and the analysis of their answers is presented in the following paragraphs. The eleventh question was intended to ascertain the satisfaction of the students involved regarding the use of tablets in class. The statement presented the question "How do you feel about using tablets in class?". According to the answers, the use of tablets was well accepted, because all the students felt good with the use of the mobile devices in class, with 95.8% chose the option "I loved" and 4.2% chose "I liked". The scale for response has 5 (five) levels, ranging from Adorei to Poor (possible levels: Loved, Liked, Indifferent, Bad and Poor). Figure 9 shows the students' responses.

Figure 8. Questionnaire screens accessed by a mobile device
Source: Authors' elaboration.

Figure 9. Answers to the eleventh question
Source: Authors' elaboration.

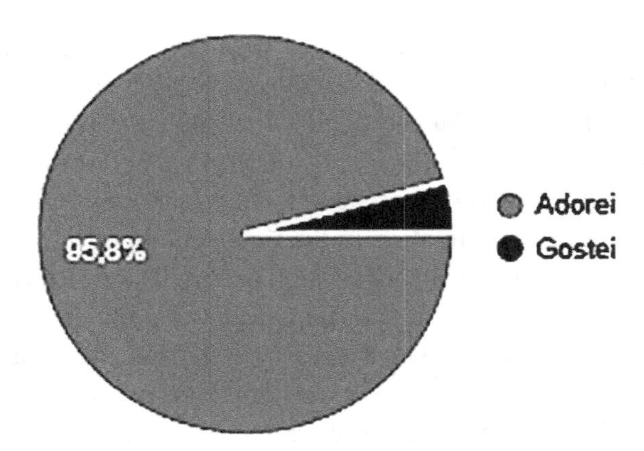

The next question was about student learning, and it was developed to see if the use of tablets facilitated the teaching of content addressed by the teacher responsible for the group of students. The question presented the text "Do you think you have learned the easiest way to use tablets? ". Like the eleventh question, many students said yes, but this time some students (8.3%) stated that they would feel indifferent and would learn the same way through the traditional teaching method; on the other hand, everyone else agreed that learning became easier with the use of tablets (75% loved it and 16.7% liked it). Figure 10 shows the students' answers.

Figure 10. Answers to the twelfth question
Source: Authors' elaboration.

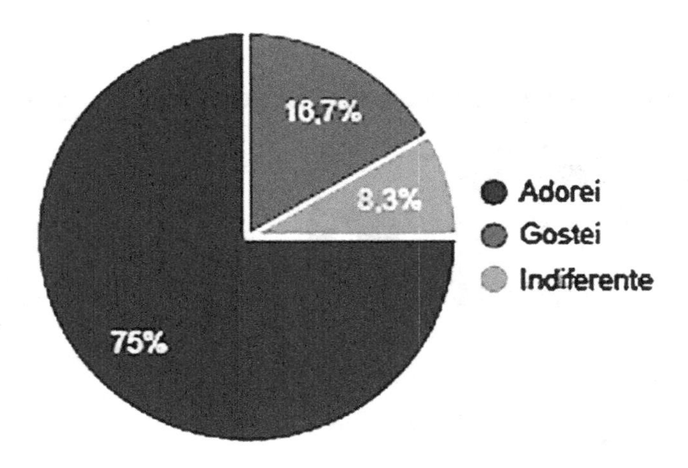

In relation to the thirteenth question, this was related to the fun, aiming to verify if the class was pleasant to the students involved in the study. The text in the question asked the following question: “Did you find the class with the tablets more fun? “. According to the responses reported by the students participating in the case study, the acceptance was very positive overall (95.8% loved and 4.2% liked it). Figure 11 presents the answers.

The purpose of the fourteenth question was to verify whether students preferred the traditional teaching method, ie without integration of digital technologies such as tablets. The question asked: “Do you prefer the lesson with content only on the board, without the tablets? “. Regarding this question, the answers were less homogeneous. In general, students would not like to attend classes based on the traditional method (54.2% think this would be bad and 29.2% think it would be awful), but some felt that it would not make a difference (8.3%) and others said they like of the traditional teaching method (8.3%). Figure 12 presents the answers to the fourteenth question.

The next question is the statement “How would you feel if the teacher sent you to research something or do some Internet activity as a homework” in order to check if students would like to use digital technologies such as the internet, notebooks, mobile devices and the like when doing homework assignments. Again, the question obtained divided answers that generated a homogeneous result, but most would like (70.8% would love and 12.5% would like) to use the Internet when doing homework. 8.3% would feel indifferent and 8.4% would not like it (4.2% would think it was bad and 4.2% would feel terrible). Figure 13 shows the students’ answers.

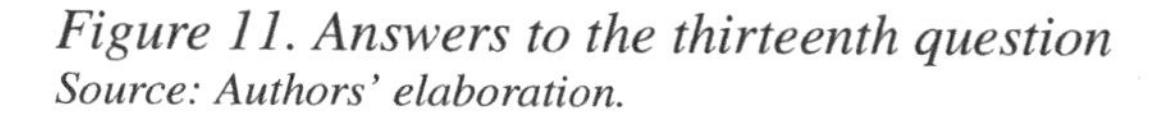

Figure 11. Answers to the thirteenth question
Source: Authors’ elaboration.

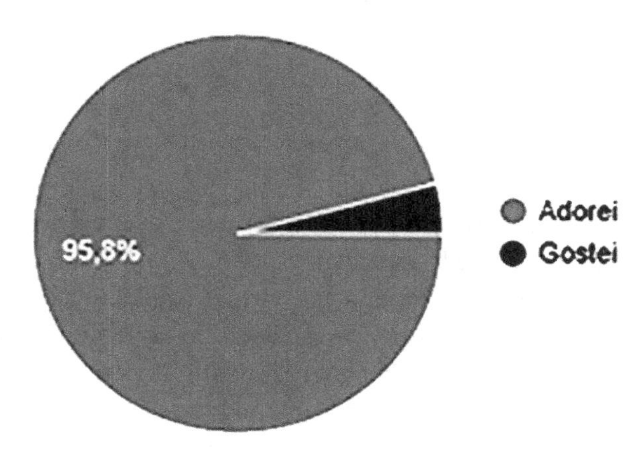

Figure 12. Answers to the fourteenth question
Source: Authors' elaboration.

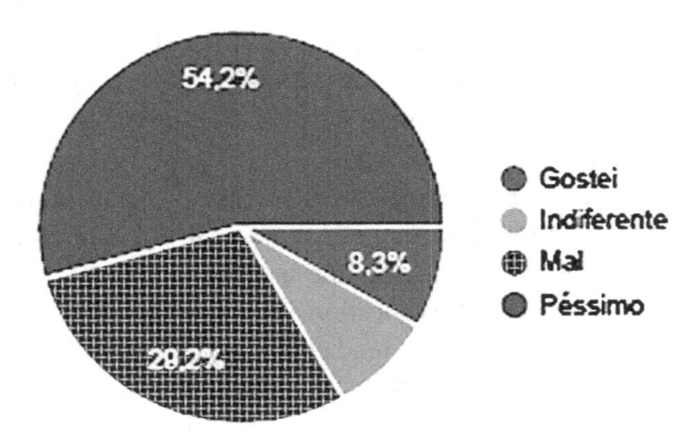

Figure 13. Answers to the fifteenth question
Source: Authors' elaboration.

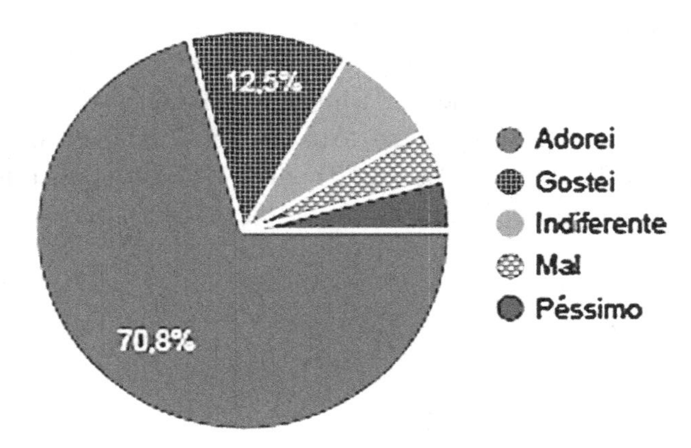

DISCUSSION

This section aims to discuss the results of the research, as the basis of the previous section of this document, called "Results".

From the answers acquired through the virtual questionnaires applied to the students involved in the research, it was possible to observe that these students reacted positively in relation to the integration of digital technologies in the classroom. This acceptance denotes a notable impact of the use of technology in relation to student learning and characterizes an opportunity to continue to apply this type of technology - applications and mobile devices - in science education in basic education, especially in schools similar to the one in that the case study was carried out, located in a region of high social vulnerability, where digital inclusion may be a key element for improving teaching and learning.

When confronted about their satisfaction, the students reacted positively, revealing that they enjoyed using technology in the classroom. In addition, they said that in fact learning was much easier when coupled with tablets and mobile applications and that these made the class more fun. In this way, it is perceived that the use of technology, besides facilitating learning, is also satisfying and pleasurable for students, who feel good when using technological tools in the classroom, something that motivates them to learn more, once who are doing something they like.

In addition, most of the students involved in the case study would like to use digital technologies more frequently in the classroom and beyond, including extra class work. Most likely this was due to the way information is arranged on the network in order to make students learn beyond what is written in textbooks; on the network, information is arranged in a variety of ways - not just texts, videos, audios, images, games, among others, in order to make learning more accessible to all.

In general, in relation to the traditional teaching method, the students stated that this method did not please them, making learning more difficult and less pleasurable, something that discouraged them. Considering that the application was made in the discipline of science, an area that needs to be stimulated due to the high demand of professionals, demotivation should be avoided. According to the responses reported by the students, they would love to attend classes that work in a way that integrates digital technologies into the conventional classroom.

CONCLUSION

The present research aimed to present a case study on the application of applications and mobile devices in the discipline of science in a class of students of the fifth year of elementary education in a public school located in a region of high social vulnerability, thus promoting an inclusion of these students, found in the most rural region of the city, in relation to the New Technologies of Information and Communication. This application was carried out in partnership with RExLab, a research group belonging to the Federal University of Santa Catarina, Brazil; In this partnership, the research group provided tablets and a trainee to assist the teacher responsible for the class during the application. In the course of the study, one can see that there is a great need in relation to the integration of digital technologies in science education in public basic education in Brazil; this necessity is due to difficulties obtained by the students. These difficulties grow over the years, something that discourages them at the end of basic education, when they begin to think about which course to choose when they go to university, and avoid the scientific-technological issues due to lack of interest. After all, it becomes very rare to take an interest in a subject on which you feel difficulties.

Based on data collection through the application of virtual questionnaires to the students participating in the research, it was possible to verify that the use of technological tools during science class had a positive impact on the group involved in the case study, being something well received by the students, who stated that they wish to interact more with this type of practice because of the high participation that digital technologies have in their daily lives.

Through this work, it was possible to perceive that, in these days, technology and school must move together, since the application of digital technologies in the classroom has the power to expand the concept of classroom, providing information in different ways and in a simplified and accessible way, making learning easier and more complete for students. After all, while learning is facilitated with technology integration, it presents tools that complement the content taught in the classroom in order to put students and teachers involved in the world, being able to discover experiences and tactics of other people in the network and even share their own experiences.

In this way, through the research it is possible to plan future work, taking into account the results obtained through the present study. In addition, new applications of digital technologies will be undertaken at the same school, as well as negative responses from students in order to improve the project activities within which the resources applied in this research are owned.

ACKNOWLEDGMENT

The present study was funded by the University Extension Program (PROEX) of the Federal University of Santa Catarina and the Ministry of Education (MEC), the Foundation for Research and Innovation Support of the State of Santa Catarina (FAPESC), the Coordination for Improvement of Higher Education Personnel (CAPES) and the National Research Council (CNPq) Institutional Program for Scientific Initiation (PIBIC).

REFERENCES

Acioli, C. G. (2014). *A educação na sociedade de informação e o dever fundamental estatal de inclusão digital* (Doctoral dissertation). Available from BDTD/IBICT Theses database: http://tede2.pucrs.br/tede2/handle/tede/4283

Ally, M., Balaji, V., Abdelbaki, A., & Cheng, R. (2017). Use of Tablet Computers to Improve Access to Education in a Remote Location. *Journal Of Learning For Development*, *4*(2), 221–228.

Bonilla, M. H. S. (2010). Políticas públicas para inclusão digital nas escolas. *Revista Motrivivência*, 40-60.

Bressler, D. M., & Bodzin, A. M. (2016). Investigating Flow Experience and Scientific Practices During a Mobile Serious Educational Game. *Journal of Science Education and Technology*, *25*(5), 795–805. doi:10.100710956-016-9639-z

Clayton, K., & Murphy, A. (n.d.). Smartphone Apps in Education: Students Create Videos to Teach Smartphone Use as Tool for Learning. *The Journal of Media Literacy Education*, *8*(2), 99–109.

Cofield-Poole, B. (2016). *A touch with technology: Creating a foundation for meaningful digital inclusion through local government* (Order No. 10585321). Available from ProQuest Dissertations & Theses Global. (1880188513). Retrieved from https://search.proquest.com/docview/1880188513?accountid=26642

Cogo, D., Brignol, L.D., & Fragoso, S. (2014). Práticas cotidianas de acceso às TICs: outro modo de compreender la inclusão digital. *Palabra Clave - Revista de Comunicación, 18*(1), 156-183.

Coll, C. (2013). El currículo escolar en el marco de la nueva ecología del aprendizaje. *Aula de innovación educativa, 219*, 31-36.

Fakomogbon, M.A., & Bolaji, H.O. (n.d.). Effects of Collaborative Learning Styles on Performance of Students in a Ubiquitous Collaborative Mobile Learning Environment. *Contemporary Educational Technology, 8*, 268-279.

Fallata, S. M. (2016). *Information and communications technology integrating at tatweer schools: Understanding experiences of Saudi female English as foreign language teachers* (Order No. 10141361). Available from ProQuest Dissertations & Theses Global. (1822226200). Retrieved from https://search.proquest.com/docview/1822226200?accountid=26642

Fundação Lemann e Meritt. (2016). *Censo Escolar*. Retrieved September 14, 2017, from http://qedu.org.br/

Howard, S. K., & Thompson, K. (2015). Seeing the system: Dynamics and complexity of technology integration in secondary schools. *Education and Information Technologies*, *21*(6), 1877–1894. doi:10.100710639-015-9424-2

IBGE. (2016). *Estatísticas*. Retrieved September 12, 2017, from http://downloads.ibge.gov.br/downloads_estatisticas.htm

Ibrahim, N., Ahmad, W. F. W., & Shafe, A. (2016). Practitioners' validation on effectiveness of multimedia Mobile Learning Application for children. *International Conference on Computer And Information Sciences*, *3*, 103-108. 10.1109/ICCOINS.2016.7783197

Kurtz, R.G.M. (2016). *Resistência à atitude e intenção de adoção do m-learning por professores do ensino superior* (tese de doutorado). Disponível na base de dados da Pontífica Universidade Católica do Rio de Janeiro.

Lei n. 12.249, de 11 de junho de 2010. (n.d.). *Criação do Programa um Computador por Aluno*. Retirado em 28 de setembro de 2017, de http://www.planalto.gov.br/ccivil_03/_ato2007-2010/2010/lei/l12249.htm

Lei n. 14.363 de 25 de janeiro de 2008. (n.d.). *Proibição do uso de telefone celular em escolas públicas do estado de Santa Catarina*. Retirado em 28 de setembro de 2017, de http://www.leisestaduais.com.br/sc/lei-ordinaria-n-14363-2008-santa-catarina-dispoe-sobre-a-proibicao-do-uso-de-telefone-celular-nas-escolas-estaduais-do-estado-de-santa-catarina

Lei n. 7.232, de 29 de outubro de 1984. (n.d.). *Criação do Conselho Nacional de Informática*. Retirado em 28 de setembro de 2017, de http://www.planalto.gov.br/ccivil_03/leis/L7232.htm

Li, Y., & Liu, X. (2017). Integration of IPad-Based M-Learning into a Creative Engineering Module in a Secondary School in England. Tojet. *The Turkish Online Journal of Educational Technology*, *16*, 43–57.

Li, Y., & Wang, L. (n.d.). Using iPad-based mobile learning to teach creative engineering within a problem-based learning pedagogy. *Education and Information Technologies*, *22*, 1–14.

Males, S., Bate, F., & Macnish, J. (2017). The impact of mobile learning on student performance as gauged by standardised test (NAPLAN) scores. *Issues in Educational Research*, *1*(27), 99–114.

Ministério da Educação. (2006). *Programa Nacional de Informática na educação*. Retirado em 28 de setembro de 2017, https://www.fnde.gov.br/sigetec/relatorios/indicadores_rel.html#Dois

Minussi, M. M., & Wyse, A. T. S. (2016). Web-Game educacional para ensino e aprendizagem de Ciências. *Novas Tecnologias na Educação*, *14*, 1–10.

Montrieux, H., Courtois, C., Grove, F., Raes, A., Schellens, T., & Marez, L. (2013). Mobile Learning in Secondary Education: Perceptions and Acceptance of Tablets of Teachers and Pupils. *International Conference Mobile Learning*, 204-208.

Neri, M. C. (2009). *O Tempo de Permanência na Escola e as Motivações dos Sem-Escola*. Federal University of Rio de Janeiro.

Pina, F., Kurtz, R., Ferreira, J., Freitas, A., Silva, J. F., & Giovannini, C. J. (2016). Adoção de m-Learning no Ensino Superior: O Ponto de Vista Dos Professores. *Revista Eletrônica de Administração*, *22*, 279–306.

Prensky, M. (2001). Digital Natives Digital Immigrants. On the Horizon, 9.

Presby, B. (2017). *Barriers to reducing the digital-use divide as perceived by middle school principals* (Order No. 10268273). Available from ProQuest Dissertations & Theses Global. (1887122371). Retrieved from https://search.proquest.com/docview/1887122371?accountid=26642

Rodrigues, A. L. (2016). A integração pedagógica das tecnologias digitais na formação ativa de professores. *Atas do IV Congresso Internacional das TIC na Educação: Tecnologias digitais e a Escola do Futuro,* 1320-1333.

Santa Catarina. (2016). *Lei nº 14.363, de 25 de janeiro de 2008. DispÕe Sobre A ProibiÇÃo do Uso de Telefone Celular nas Escolas Estaduais do Estado de Santa Catarina*. Retrieved September 14, 2017, from http://www.leisestaduais.com.br/sc/lei-ordinaria-n-14363-2008-santa-catarina-dispoe-sobre-a-proibicao-do-uso-de-telefone-celular-nas-escolas-estaduais-do-estado-de-santa-catarina

Santos, R. P., & Freitas, S. R. S. (2017). Tecnologias Digitais Na Educação: Experiência do uso de Aplicativos de Celular no Ensino da Biologia. *Cadernos de Educação, 16*(32), 135–150. doi:10.15603/1679-8104/ce.v16n32p135-150

Steinert, M. E. P., Hardoim, E. L., & Pinto, M. P. P. R. C. (2017). De mãos limpas com as tecnologias. *Revista Sustinere*, *4*, 233–252.

Tighe, L. (2016). *Teacher perceptions of usefulness of mobile learning devices in rural secondary science classrooms* (Order No. 10239808). Available from ProQuest Dissertations & Theses Global. (1868429722). Retrieved from https://search.proquest.com/docview/1868429722?accountid=26642

UNESCO. (2014). *O futuro da aprendizagem móvel*. Paris: Unesco.

APPENDIX: TECHNOLOGICAL PROFILE QUESTIONNAIRE

Student Profile Questionnaire and Technological Profile

This questionnaire aims to obtain data on the use of technology by children.

1. How old are you?
    a. 6 years
    b. 7 years
    c. 8 years
    d. 9 years
    e. 10 years
2. You are:
    a. Boy
    b. Girl
3. You own at home:
    a. Laptop
    b. Tablet
    c. Smartphone
    d. Desktop (desktop)
    e. Neither of them
4. Do you have internet at home?
    a. Yes
    b. No
5. Who taught you the most about using computers?
    a. School
    b. Friends
    c. Family
    d. Alone
    e. Lan house
    f. Other
6. Who taught you the most about using the internet?
    a. School
    b. Friends
    c. Family
    d. Alone
    e. Lan house
    f. Other

7. Where do you have access to a computer?
    a. At home
    b. At school
    c. Lan house
    d. Other
8. How long have you been using your computer?
    a. Less than 1 year
    b. 1 to 3 years
    c. 3 to 5 years
    d. More than 5 years
9. How long have you been using Internet?
    a. Less than 1 year
    b. 1 to 3 years
    c. 3 to 5 years
    d. More than 5 years
10. How do you feel about using computers at school?
    a. I loved it
    b. I like it
    c. Indifferent
    d. Bad
    e. Terrible
11. How do you feel about using tablets in class?
    a. I loved it
    b. I like it
    c. Indifferent
    d. Bad
    e. Terrible
12. Do you think you learned the easiest way to use your tablets in class?
    a. I loved it
    b. I like it
    c. Indifferent
    d. Bad
    e. Terrible
13. Did you find the tablet class more fun?
    a. I loved it
    b. I like it
    c. Indifferent
    d. Bad
    e. Terrible

14. Do you prefer content-only content on the board, without the tablets?
    a. I loved it
    b. I like it
    c. Indifferent
    d. Bad
    e. Terrible
15. How do you feel if the teacher sends you something to research, or do some activity, on the internet as a homework assignment?
    a. I loved it
    b. I like it
    c. Indifferent
    d. Bad
    e. Terrible
16. How do you feel if the teacher orders you to research something, or do some internet activity in the classroom (in the school lab or on the tablets)?
    a. I loved it
    b. I like it
    c. Indifferent
    d. Bad
    e. Terrible
17. How would you feel if the teacher told you to do activities on your cell phone?
    a. I loved it
    b. I like it
    c. Indifferent
    d. Bad
    e. Terrible
18. How would you feel if the teacher took the test with the tablets or the school computer?
    a. I loved it
    b. I like it
    c. Indifferent
    d. Bad
    e. Terrible

Chapter 6
Mobile Games for Language Learning

Monther M. Elaish
University of Malaya, Malaysia

Norjihan Abdul Ghani
University of Malaya, Malaysia

Liyana Shuib
University of Malaya, Malaysia

Ahmed Mubarak Al-Haiqi
Universiti Tenaga Nasional, Malaysia

ABSTRACT

Education, including the subset of language learning, has been greatly influenced by information and communication technologies. This influence manifests itself in the form of various paradigms, starting from distance or digital learning (d-learning) to electronic learning (e-learning) then mobile learning (m-learning) and eventually ubiquitous learning (u-learning). The integration of these paradigms with supportive techniques to enhance inclusion, engagement, and to overcome the classic problem of lack of motivation led to a series of innovations, culminated in the notion of educational mobile game applications. This chapter focuses on the roots of this emergent trend, including the elements of mobile technology and the aspect of gaming, and how instrumental are they in empowering and motivating learners. The relationship of mobile games with the concept of gamification is examined, and a few major challenges to building effective mobile game applications for language learners are highlighted for future attention.

DOI: 10.4018/978-1-5225-5270-3.ch006

INTRODUCTION

In a globalized world, the ability to communicate in more than one language is becoming more compelling. For example, in many parts of the globe, English language is the universal business language, and knowing English is a competitive advantage in several aspects. This drives many non-English speaking countries into paying great attention to teaching the English language at an early stage (Kachru, 2006). Apart from the business world, second languages like English can be essential for gaining further and continuous education. For example, most of the academic literature and scientific results are published in English.

In one sense, the ability of teaching and learning another language is becoming an instance of the more general problem of inclusion and empowerment. Those who have access to the necessary resources and enabling forces to learn other languages than their own mother tongues can enjoy more opportunities and prosperous careers. Those who have no access to resources and no exposure to positive forces may be at a disadvantage. Traditional schooling is a case in point. Traditional, face-to-face learning has served well the previous generations, provided that learners have access to local resources such as a school, qualified teachers and quality materials. Even when so, traditional teaching is perceived as boring (Jean & Simard, 2011), and without positive forces that drive their motivation and engagement, more and more learners are still susceptible to suffer from the natural difficulty of learning new languages. This situation leads to the setup where only a small part of the learners – who are gifted with intrinsic motivation and/or raised in a supportive environment – can reap the benefits of traditional settings in teaching and learning.

To remedy this situation, educators did not stop looking for new ways to close the gap and reach out to learners in different places, times, and contexts as well as to raise the learners' level of engagement and motivation. In the context of language learning, people moved from face-to-face learning, to distance and web-based learning, then mobile-assisted language learning (MALL) (Kukulska-Hulme & Shield, 2008) as a derivative of mobile learning, up to the use of games in game-based learning, and finally we are at a point where many of the games are built on mobile platforms and used for teaching languages. This construct of *mobile game applications for language learning* is the subject of this chapter.

Mobile games for language learning (MGLL) is still a new trend. Although it has gained the attention of researchers since the last decade, it is still not ubiquitous in the learning environment. MGLL is an instance of the revolution of mobile apps in almost all other domains, such as mobile banking, mobile governance and mobile health. The authors believe it is similarly going to penetrate society, and may be a valuable addition to the arsenal of teachers and learners if understood correctly and if issues encountered in previous paradigms have been addressed earlier. This chapter aims to shed some light on the concept of MGLL, focusing on the roots of the concept and its potential, as well as issues that need to be addressed, in the hope that this will provide educators and researchers with some insight as well as few clues to issues that need attention.

BACKGROUND

Learning a second language is not easy (Spolsky, 1986). Because it is a tedious and demanding task, governments, educators and researchers spare no effort or tool to assist teaching a second language, especially English. Whenever there is a new technology or an emergent trend that could help in enhancing the performance of learners or overcoming some of the traditional difficulties in learning languages, the communities and individuals are fast to adopt it. This is also true for many other domains, such as health and lifestyle.

Information and communication technology (ICT) is one of the most influential applications of sciences on learning and education in the recent history. ICT extended the traditional modes of learning into new paradigms that did not exist before the twentieth century. For example, the revolution of telecommunications accompanied by the introduction of digital computers enabled learners to be situated remotely from teachers. This, on one side, enabled more inclusion of learners even outside the class boundaries. On the other side, it also excluded parts of the learning community who cannot afford the necessary technology.

Later on, the affordability of computers combined with the ubiquity of the Internet, especially the World Wide Web, enabled many more cross-national learners to gain access into more resources, including high-quality materials. In language as well as other subject learning, the availability of the web and digital storage media like CD ROMs removed the necessity of face-to-face teaching by professional trainers, and it was possible to practice the learning using electronic technology, leading to the paradigm of *e-learning* (Clark & Mayer, 2016).

Afterwards, wireless technologies, combined with the introduction of portable computers and later mobile devices, introduced the concept of mobility and situated learning, so learners can be mobile and learning materials can be mobile (Sharples, Taylor, & Vavoula, 2010). And the notion of mobile learning (*m-learning*) was formed. The power of mobile devices, the pervasiveness of electronic gadgets and the abundance of information made the learning material available everywhere, ubiquitously, forming the notion of ubiquitous learning, u-learning (Liu & Hwang, 2010).

The story, however, does not stop here. The availability of the tools and information everywhere in many shapes does not by itself tempt the learners to use them and overcome the troublesome of getting actually engaged in the learning process. Educators looked around for something else that can assist the learning process beyond the mere availability of technological tools. It is often the case that military and the business sectors are having an edge over the other sectors in taking advantage of new innovations; healthcare and education are usually lagging behind. For example, healthcare industry is always hoping to follow banking systems in the smooth and reliable exchange of information. Similarly, the digital industry created a very lucrative business, video games, that attracted hundreds of millions of people, and drew millions of hours per month from the time of societies all over the world (David Baszucki, n.d.).

Educators started to wonder, why cannot people also engage in learning in the same fashion they engage in games? And they thought that some techniques like gaming might be able to attract learners and alleviate some of the classic problems in education, such as the lack of motivation. Out of these thoughts, new techniques appeared, such as the concept of gamification (Deterding, Dixon, Khaled, & Nacke, 2011; Werbach & Hunter, 2012), which aims to borrow ideas from the game industry and game design into the application of education. In the following section, the roots of mobile game applications are explored as the answer to address two main issues in the context of learning: the problem of inclusion and the lack of motivation.

MOBILE GAMES FOR LEARNING: FROM ISSUES TO SOLUTIONS

Learning in general, and language learning in particular, can be encumbered by many hurdles. Lack of adequate and sustainable motivation is always a major issue (Dörnyei, 1998; Oxford, 1996). Together with problems, many other issues also persist in language learning, such as the lack of qualified teachers, absenteeism, cultural barriers, and even classroom anxiety (Horwitz, Horwitz, & Cope, 1986). The authors adopt the view that all these problems, at some level of abstraction, are but instances of the more general problem of exclusion.

A Problem of Inclusion

Exclusion comes in many forms and targets groups of the society at several levels. One form of exclusion is the inability to access well-qualified teachers. It is very unusual to have native speakers of a second language teaching in other than specialized, and high-cost, language centers. In many parts of the world, the best local teachers would be hired by private schools, while public schools would be left with less-qualified ones. This situation creates a gap between the qualities of teaching that different groups of learners receive. This exclusion problem applies equally to the quality of teaching materials and teaching modalities. Depending on whether the learners are in the right or wrong place (e.g. a good vs. bad school), they will get access to different quality of teaching.

The above problem assumes that learners do attend some sort of schooling, but the issue becomes more obvious when referring to groups of people who attend no school. This can be the case for underserved populations of otherwise potential students. It can also be the case for minorities with cultural barriers to join public schools, or even for actual students who skip learning classes because they suffer from classroom anxiety. Another important group of learners who suffer from a subtle form of exclusion is the group of adult learners who exceeded the schooling age and might be too involved in the workplace that they can afford no time to continue their education in a conventional school or college, or they might be unable to afford the money to continue –or start, for that matter– their learning. This is an exclusion problem in the setting of lifelong learning.

In all those cases, as well as other scenarios, several traditional treatments exist to address the problem of exclusion and attempt to increase the level of inclusion for various groups of learners. Examples of these attempts include the laws of compulsory schooling, programs to qualify language instructors and curriculums, night schools, and the use of technology in the classroom up to the provision of quality materials on digital storage that can be run using computers in home. Nevertheless, these solutions still do not address the basic issue of geographic barrier; for inclusion, learners still have to have access to the place where there is a good teacher or a computer, be it in the home, next village or another country.

To include more learners, therefore, it seems that the next necessary step is to flip the order and bring teachers and materials to the learners, in place. This would allow learners not only to learn wherever they are, but also whenever they have the time and mood, which ought to increase their inclusion in the learning process and increase their chances of gaining its benefits. However, this solution was not possible, and will not be feasible, without an enabling technology. In this case, mobile technology does provide the basic ingredients of the sought learning approach: 1) the means to move the digitized teaching materials, including recorded lectures, over the space and time barriers, and 2) a device that can accompany the learner everywhere and deliver the material in a consumable format. This was the foundation of *mobile learning*.

Mobile Learning for Social Inclusion

Mobile learning (also abbreviated as *m-learning*) can be defined as the use of mobile or wireless devices for the purpose of learning while on the move (Park, 2011). This definition is not universally praised; for example, Winters (2007) criticized the definition as being *Technocentric* (centered around the technology). However, the authors prefer this definition because it links mobile learning directly to mobile technology as a natural progression and a subset of *e-learning*. In any case, the mobility in *m-learning* does refer to both the learning and the learner's mobility, across time and space.

Mobile learning has been studied extensively in the literature, although mostly not in the context of language learning. For a recent and comprehensive survey on mobile language learning, the reader is referred to (Elaish, Shuib, Ghani, Yadegaridehkordi, & Alaa, 2017) where the authors classify a total of 133 works on mobile learning for the English into several categories, draw few statistics based on the type of mobile applications, learning problems addressed by the papers, and the used mobile tools, then provide a set of motivations, challenges and recommendations. Other recent meta-analyses of Mobile-Assisted Language Learning (MALL) can be found in (Bozdougan, 2015; Burston, 2015).

With reference to works cited above as a sample, most of the studies on mobile learning have praised the new technology, focusing on various aspects of its value in boosting the learners' performance or explicitly stating its benefits in increasing the learners' inclusion and empowerment. In a "Policy Brief" report from UNESCO's Institute for Information Technologies in Education, several benefits for using mobile and wireless technologies are identified, some of which are directly related to the benefits of social inclusion are summarized below (Kukulska-Hulme, 2010):

- Improved access to education as a result of using comparatively inexpensive technologies, learning at one's own pace and preferences, and also having a better degree of privacy that may be difficult to enjoy when sharing computer facilities in the classroom. This is particularly important for women and girls.
- Access to more options than those provided in class through podcasts and free online learning materials. This helps the learners explore opportunities beyond what is available in their local environments. For example, learning languages that are not offered in local programs.
- Psychological support for those at risk of dropping out; e.g., through social networks.
- Catering for underserved and disadvantaged social groups of potential students.

- Delivering high teaching quality to more learners regardless of their location or context.
- Supporting continuing education.
- Providing more equitable access to education, for those suffering exclusion for social or economic reasons.
- Building a culture of lifelong and life-wide learning.

One relevant note here is that many of the classic works on mobile learning (e.g. (Kukulska-Hulme, 2007; Kukulska-Hulme & Traxler, 2005; Schwabe & Göth, 2005)) have been published around the mid of the last decade before the introduction of the first smartphones (in 2007 and 2008, for iPhone and Android, respectively). This means that some of the arguments provided in those works are irrelevant in today's technology. For example, some drawbacks of mobile devices mentioned in (Kukulska-Hulme, 2007) include: content and software application limitations; lack of built-in functions; the difficulty of adding applications; and network speed and reliability. Obviously, these types of limitations do not apply to mobile phones today in most of the cases.

A Problem of Motivation

The advantages of mobile learning in social inclusion are obvious, but not conclusive. The availability of mobile devices and the corresponding mobile applications can enhance the inclusion of learners and may encourage them to learn, but it cannot by itself guarantee the sustainable use and engagement, because it does not directly address the issue of motivation. In Sandberg, Maris, and Hoogendoorn (2014), a mobile English learning application was investigated with a group of primary school children, and of the resulting observations was that the time of application use per day diminished gradually. The authored hypothesised that the lack of engagement and flow might be the cause of the decreasing motivation.

In another study of student motivation in the context of mobile learning (Ciampa, 2014), the author reports postitive results on the effect of mobile devices in enhancing student motivations. However, she also provides an insightful discussion in the *Limitations of the study* section. She notes that the "positive outcomes learning from the new medium, having more positive attitudes about learning will tend to decline as the technology becomes more familiar and its novelty wears off." In explaining this notice, a reference is made to the concept of *novelty effect* (Krendl & Clark, 1994), where learners tend to be more motivated to use a new piece of technology for learning because it is new.

Mobile Games for Sustainable Motivation

Indeed, the response to the above observations was soon obvious, because the technology was already available. Many studies have already showed that the use of games can enhance the levels of engaegement and maintain the motivation. Reports from the Europian Commission even attributed to digital games an explicit role in the empowerment and social inclusion of groups at risks of social exclusion (Stewart et al., 2013; Panzavolta & Lotti, 2013). What remains is only to transfer the gaming technology to the realm of mobile learning and closing the inclusion loop. This transfer moved from the use of full-blown games for learning purposes to the use of gaming elements in learning applications, in what is known as *gamification*.

Before presenting the main gaming concepts in brief, a couple of notes are in order. First, the study of games in the education literature can take more than one perspective. This chapter considers the notion of educational mobile games from the standpoint that learners are game players. This is not the only perspective from which mobile games can be used in education. For example, students have been taught computer science subjects through the actual development of mobile games (Kurkovsky, 2009, 2013). Second, it can also be noted here that the two concepts –mobile technology and gamification –are orthogonal to each other, the gamification dimension can be added to other technologies as a factor of engagement, motivation and immersion.

Game-Based Learning

The National Foundation for Educational Research (NFER) published a report on game-based learning in 2013 (Perrotta, Featherstone, Aston, & Houghton, 2013). According to the report, "game-based learning broadly refers to the use of video games to support teaching and learning." This is another mature field in education that have been received positively, many times under the name of Digital Game-Based Learning (DGBL) (Papastergiou, 2009; Prensky, 2003; Van Eck, 2006). This result comes as no surprise given that playing games is by itself an engaging and motivational activity, which can be used to wrap other activities that is of less appeal to learners; i.e. learning. This was the same rationale to establish the more general realm of gamification. Although there is still a distinction in the literature between gamification and DGBL, the authors advocate the view that DGBL is a special case of gamification, which is a more comprehensive and flexible concept. Therefore, the next subsection elaborates further on gamification.

In fact, the NFER research review cited above presents the concept of gamification as "a much newer concept than game-based learning. It is about using 'elements' derived from video-game design, which are then deployed in a variety of contexts, rather than about using individual video games."

Gamification

As mentioned earlier, the idea of playing games for purposes other than the pure joy is not new (Garris, Ahlers, & Driskell, 2002). However, it was only at the turn of the current decade when researchers in academia and industry started to look into contexts totally different than games, especially in business, and to augment those contexts deliberately with elements that are inspired from playing games, for the purpose of increasing engagement and motivation. Most of the resources agree on the basic definition of gamification that differentiates this growing field from the games industry and from the concept of play. According to this definition, gamification is the use of game design elements and game design techniques in non-game contexts (Deterding et al., 2011; Werbach & Hunter, 2012).

A common theme in the gamification field is that of *fun*. From a certain perspective, gamification is a deliberate practice of adding a fun element using techniques from game design to push the customers or learners towards achieving some objective. In this perspective, gamification is the process of manipulating fun to serve real-world objectives (Werbach & Hunter, 2012). The factor of fun in games has intensively been analyzed and appraised, and one of the highly regarded and cited works in game-design literature is titled: "Theory of fun for game design" (Koster, 2013).

For illustrative examples of using fun in gamification as a means to achieve a goal in a non-game context, the reader is referred to the site of www.thefuntheory.com (Volkswagen, n.d.), where Volkswagen initiated a series of experiments in which novel and innovative fun ideas were incorporated in doing tasks as dull as taking the stairs instead of the escalator (they painted the stairs as the keys of a piano and the individual stairs would produce different tones when stepping on them) and throwing the rubbish in the dedicated bins (the bin produces a long sound as if the piece of rubbish was falling along a deep height), in order to encourage people to change their behavior. The point of these examples is to exploit the fun in gaming elements to engage people in doing the tasks and sustain their motivation, because it is entertaining. The same concept can be applied to learning as well.

As an indication of the importance and relevance of gamification, several universities started to offer courses on the subject. One of the first courses ever on gamification was offered at the Wharton School, University of Pennsylvania by Professors Kevin Werbach and Dan Hunter. An online version if this course is still offered online as a MOOC course on the Coursera platform (Coursera, n.d.). Although the general focus of gamification is mostly on the business applications, a systematic mapping of gamification in education can be found in (Dicheva, Dichev, Agre, & Angelova, 2015), and a brief overview can be found in (Lee, & Hammer, 2011).

Mobile Games as a Persuasive Technology

Looking back at the role of gamification to enhance mobile-assisted learning, one can observe the striking similarity to other technologies used for very similar purposes, even though in contexts other than learning. This section discusses the concept of persuasive technology and highlights the similarity to gamification.

Persuasive technology is a more established approach than gamification and refers to any interactive computing system designed to change people's attitudes or behaviors, which can collect information about users' actions to help them achieve their goals (Fogg, 2002). To illustrate the similarities between the concepts of gamification and persuasive technology, Table 1 lists some of the most common game elements from the gamification field, while Table 2 lists some of the persuasive principles from one of the most popular models of persuasive technology (Oinas-Kukkonen & Harjumaa, 2009).

Mental linking of persuasive technology to using mobile games in learning language does not occur immediately due to the different objectives of the two. Persuasive technology aims to change behaviors and attitudes, while technologies for learning languages such as MGLL aim for the acquisition of the language. In the general context of education, changing behavior or attitude is still an important objective, but it is less obvious in language learning. In any case, a question arises from the apparent similarity and overlap in the terminology and meaning of the basic elements in both technologies. The authors suggest that at some point, persuasive technology can complement the techniques of gamification as applied to language learning in mobile settings. This suggestion is yet to be examined in further research, where the authors envisage that it is possible to develop a coherent framework in which all these components play together to initiate the engagement of learners, sustaining their motivation, enhancing their learning experience and convincing them of adopting the right way.

As a simple example of how the basic persuasive principles (Table 1) or game elements (Table 2) can be applied in mobile games for language learning, Figure 1 shows one screen in a smartphone app for teaching the English vocabulary to primary-school students in Malaysia. This app is currently under development by the authors, and is written for the Android platform. The user interface of this screen shows a number of elements that implements few of the persuasive/gaming principles/elements. Each element on the figure is labeled with the corresponding principle. For example, the badges on the middle top of the screen implements the principles of *recognition*, *competition* and *visual status* from the game elements in

Table 1. Persuasive principles

Principles	Description
Reduction	Makes the system simpler.
Tunneling	Guides users.
Tailoring	Design depends on needs, interests, personality, the use of context or any aspect belong to a user group.
Personalization	Offers personalized content or services.
Simulation	Provides simulation (before and after).
Rehearsal	Provides training systems to cope with reality problem.
Self-monitoring	Allow users to track their performance and status.
Praise	Offers praise.
Rewards	Rewards depend on target behavior.
Reminders	Reminders depend on target behavior.
Suggestion	Offers fitting suggestion
Similarity	Offers things reminding the users of themselves
Liking	Offers visual attraction
Social role	Adopts social role.
Trustworthiness	Views as trustworthy.
Expertise	Views as incorporating expertise.
Surface credibility	Has a competent look and feel?
Real world feel	Provides information about the people or organization behind its content and service.
Authority	Leverages roles of authority.
Third party endorsements	Authorization from a respected source.
Verifiability	Easy verifies the accuracy of content via outside sources.
Social learning	Allow users to observe others performance.
Social comparison	Allow comparison
Normative influence	Gather people together who have the same target.
Social facilitation	Shows user others performing the same behavior.
Cooperation	Provides cooperation.
Competition	Provides competition.
Recognition	Offers public recognition.

Source: Adopted from Oinas-Kukkonen & Harjumaa, 2009.

Table 2. Game elements

Game Element	Description/Mechanism
Goals	Specific, clear, moderately difficult, immediate goals
Challenges and quests	Clear, concrete, actionable learning tasks with increased complexity
Customization	personalized experiences, adaptive difficulty; challenges that are perfectly tailored to the player's skill level, increasing the difficulty as the player's skill expands
Progress	Visible progression to mastery Points, Progress bars, Levels, Virtual Goods/Currency
Feedback	Immediate feedback or shorten feedback cycles; immediate rewards instead of vague long-term benefits
Competition and cooperation/social engagement loops	Badges, Leaderboards, Levels, Avatars
Accrual grading	Points
Visible status	Reputation, social credibility and recognition Points, Badges, Leaderboards, Avatars
Access/Unlocking content	
Freedom of choice	Multiple routes to success, allowing students to choose their own sub-goals within the larger task
Freedom to fail	Low risk from submission, multiple attempts
Storytelling	Avatars
New identities and/or roles	Avatars
Onboarding	
Time restriction	Countdown clock

Source: Adopted from Dicheva et al., 2015.

Table 2, and the principles of *competition* and *rewards* from the persuasive principles in Table 1. It should be noted that one feature of the application can implement more than one principle of persuasive technology or gamification, and that the latter wo overlap significantly.

ISSUES FOR FUTURE ATTENTION IN MGLL

Given the power of games and the welcoming reception of gamifying the learning process by the educational community, as well as the great benefits already offered by existing mobile technology, it is an expected progression to move learning games onto mobile platforms. And that is what happened. Referring to the survey in (Elaish et al., 2017), around 10 works described a pure mobile game for learning English,

Figure 1. A screen from a gamified app showing the implementation of persuasive/ gaming principles

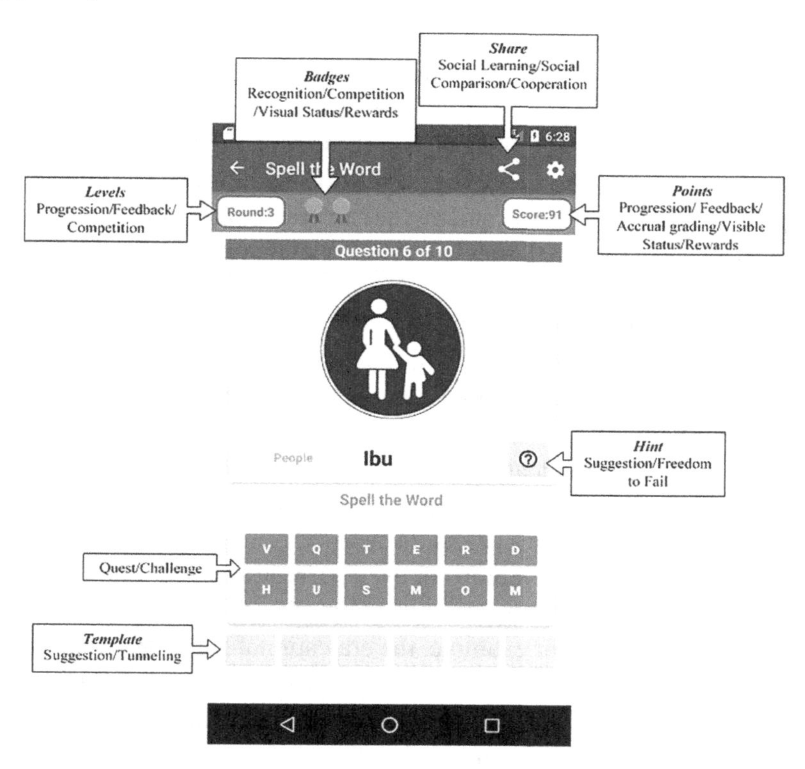

but other works also exist for different learning objectives (e.g. (Holden & Sykes, 2012; Koutromanos & Avraamidou, 2014; Spikol & Milrad, 2008)).

However, the picture is not always bright. There are few caveats ought to be paid attention. First, similar to the case of smartphone medical apps (Hussain et al., 2015), online markets of mobile apps are faster to adopt new trends than academic and regulatory research is able to evaluate, theorize and formulate those trends. A quick search on Android online market (Google Play (Google Inc., n.d.)) using the keywords "language learning games" returns more than 250 apps. Examining the first dozen of these apps, only third of the apps are produced by self-proclaimed educators. The remaining apps are produced by unknown individuals or professional game development companies. The accuracy and pedagogical value of these apps are left for the public learners to find out. This situation calls for a faster move from pilot studies and exploratory case studies into a solid body of research to formalize the theory and practice of producing mobile games for language learning, filtering the existing works based on that basis and potentially leading to the regulation and/ or standardization of the process.

Another problem one often faces in the literature of mobile learning is that most of the focus is inclined towards the mobility of learners and portability of learning devices when defining *m-learning*. Little emphasis is put on the mobile technology itself as a versatile tool. Mobile devices should not be just used as alternative platforms to play games instead of PCs or consoles. Mobile devices, especially smartphones, have novel capabilities, and the full range of these capabilities should be taken into account when designing apps, including games, for teaching and learning languages. These capabilities include sensors, such as the cameras, microphones, accelerometers that can measure the motion of the device, gyroscopes that can measure the inclination of the device, and GPS receivers that can specify the location. Other sensors can also measure other characteristics of the environment like temperature and pressure.

Apart from sensors, smartphones also have the ultimate number of connectivity options, including connection to cellular networks, WiFi networks, WiFi Direct (which does not need access points and allow for peer-to-peer WiFi communication), and recently Near-Field Communication (NFC) connections. Novel use of these features can produce a new landscape of learning that is not only situated and context-aware, but also interactive with the learner and the learner's environment in ways that were not possible using computers, laptops or even current mobile learning apps.

Only few papers pay attention to those aspects of mobile devices, other than their mobility and connectivity. For example, the educational game developed in (Furió, GonzáLez-Gancedo, Juan, Segui, & Costa, 2013) utilized different interaction modalities such as the touchscreen and the accelerometer and combined augmented reality mini-games with non-augmented reality mini-games for better gameplay immersion. As mentioned earlier, the full features of smartphones, including those that provide context awareness and universal connectivity lead to yet another learning paradigm: ubiquitous learning. Although very few, some works have already attempted to combine educational mobile applications with the concept of context awareness, utilizing, for example, the embedded cameras in smartphones (Lu et al., 2011).

On the side of gamification, learning games should not be just a collection of game elements built together to form an app that is then labeled as a language learning tool because it tests vocabulary lists compared, for example, to testing flags of countries. Games for learning languages, similar to the general case of other gamified applications, should be built using a systematic process and should refer to models and frameworks of building apps, which is lacking in the current landscape of MGLL research.

The gamification discipline as well as mobile learning have always been suffering from the lack of formal models and frameworks (Park, 2011). These guidelines can act as a reference with authenticity and evidence-based support to build products that not only engage the learners, but also sustain their motivation and increase their knowledge level. One of the few and good reviews of available models and frameworks for designing mobile learning experiences can be found in (Hsu & Ching, 2015). To gain a sense of what frameworks can offer the process of building or gamifying the learning process, the reader is referred to the gamification hierarchy and the six-step process of gamifying a business given in (Werbach & Hunter, 2012).

Finally, it is worth emphasizing the fact that being mobile, MGLL is not necessarily associated with out-of-class language learning. Mobile games can be and are often used in a class environment (even along lectures, e.g. (Wang, Øfsdahl, & Mørch-Storstein, 2008)).

CONCLUSION

MGLL inherits all the benefits of its constituent elements; i.e. *m-learning* or MALL (including the mobility, connectivity, affordability, context-awareness, situated learning, individuality and the various form factors and input/output modalities) as well as the technique of gamification, which is known to successfully engage learners in contexts that are far from being engaging. MGLL can be seen as an intersection of the affordances of mobile technology and the power of gamification, put together in the context of learning language. In this chapter, these roots of MGLL were presented as a response to the more general problem of learner exclusion due to geographical, social or other reasons. A brief account of mobile learning, game-based learning and gamification was given, as the basis to understand MGLL and understand the challenges that are awaiting to be addressed. The whole of MGLL should be greater than the sum of its parts. This can be achieved if the mobile games are designed and built correctly, and following a systematic and authentic frameworks. In that case, MGLL can produce unique and novel benefits.

REFERENCES

Bozdougan, D. (2015). MALL revisited: Current trends and pedagogical implications. *Procedia: Social and Behavioral Sciences*, *195*, 932–939. doi:10.1016/j.sbspro.2015.06.373

Burston, J. (2015). Twenty years of MALL project implementation: A meta-analysis of learning outcomes. *ReCALL*, *27*(1), 4–20. doi:10.1017/S0958344014000159

Ciampa, K. (2014). Learning in a mobile age: An investigation of student motivation. *Journal of Computer Assisted Learning*, *30*(1), 82–96. doi:10.1111/jcal.12036

Clark, R. C., & Mayer, R. E. (2016). *E-learning and the science of instruction: Proven guidelines for consumers and designers of multimedia learning*. John Wiley & Sons. doi:10.1002/9781119239086

Coursera. (n.d.). *Gamification*. Retrieved December 3, 2017, from https://www.coursera.org/learn/gamification

David Baszucki. (n.d.). *Where are people playing? - Roblox Blog*. Retrieved October 31, 2017, from https://blog.roblox.com/2012/02/where-are-people-playing/

Deterding, S., Dixon, D., Khaled, R., & Nacke, L. (2011). From game design elements to gamefulness: defining gamification. In *Proceedings of the 15th international academic MindTrek conference: Envisioning future media environments* (pp. 9–15). Academic Press. 10.1145/2181037.2181040

Dicheva, D., Dichev, C., Agre, G., & Angelova, G. (2015). Gamification in education: A systematic mapping study. *Journal of Educational Technology & Society*, *18*(3), 75.

Dörnyei, Z. (1998). Motivation in second and foreign language learning. *Language Teaching*, *31*(3), 117–135. doi:10.1017/S026144480001315X

Elaish, M. M., Shuib, L., Ghani, N. A., Yadegaridehkordi, E., & Alaa, M. (2017). Mobile learning for English Language Acquisition: Taxonomy, Challenges, and Recommendations. *IEEE Access: Practical Innovations, Open Solutions*, *5*, 19033–19047. doi:10.1109/ACCESS.2017.2749541

Fogg, B. J. (2002, December). Persuasive technology: Using computers to change what we think and do. *Ubiquity*, *2002*, 5. doi:10.1145/764008.763957

Furió, D., González-Gancedo, S., Juan, M.-C., Seguí, I., & Costa, M. (2013). The effects of the size and weight of a mobile device on an educational game. *Computers & Education*, *64*, 24–41. doi:10.1016/j.compedu.2012.12.015

Garris, R., Ahlers, R., & Driskell, J. E. (2002). Games, motivation, and learning: A research and practice model. *Simulation & Gaming*, *33*(4), 441–467. doi:10.1177/1046878102238607

Google Inc. (n.d.). *Android Apps on Google Play*. Retrieved October 31, 2017, from https://play.google.com/store/apps

Holden, C. L., & Sykes, J. M. (2012). Leveraging mobile games for place-based language learning. *Developments in Current Game-Based Learning Design and Deployment, 27*.

Horwitz, E. K., Horwitz, M. B., & Cope, J. (1986). Foreign language classroom anxiety. *Modern Language Journal*, *70*(2), 125–132. doi:10.1111/j.1540-4781.1986.tb05256.x

Hsu, Y.-C., & Ching, Y.-H. (2015). A review of models and frameworks for designing mobile learning experiences and environments. *Canadian Journal of Learning and Technology*, *41*(3). doi:10.21432/T2V616

Hussain, M., Al-Haiqi, A., Zaidan, A. A., Zaidan, B. B., Kiah, M. L. M., Anuar, N. B., & Abdulnabi, M. (2015). The landscape of research on smartphone medical apps: Coherent taxonomy, motivations, open challenges and recommendations. *Computer Methods and Programs in Biomedicine*, *122*(3), 393–408. doi:10.1016/j.cmpb.2015.08.015 PMID:26412009

Jean, G., & Simard, D. (2011). Grammar teaching and learning in L2: Necessary, but boring? *Foreign Language Annals*, *44*(3), 467–494. doi:10.1111/j.1944-9720.2011.01143.x

Kachru, B. B. (2006). The English language in the outer circle. *World Englishes*, *3*, 241–255.

Koster, R. (2013). *Theory of fun for game design*. O'Reilly Media, Inc.

Koutromanos, G., & Avraamidou, L. (2014). The use of mobile games in formal and informal learning environments: A review of the literature. *Educational Media International*, *51*(1), 49–65. doi:10.1080/09523987.2014.889409

Krendl, K. A., & Clark, G. (1994). The impact of computers on learning: Research on in-school and out-of-school settings. *Journal of Computing in Higher Education*, *5*(2), 85–112. doi:10.1007/BF02948572

Kukulska-Hulme, A. (2007). Mobile usability in educational contexts: What have we learnt? *The International Review of Research in Open and Distributed Learning*, *8*(2). doi:10.19173/irrodl.v8i2.356

Kukulska-Hulme, A. (2010). *Mobile learning for quality education and social inclusion*. Academic Press.

Kukulska-Hulme, A., & Shield, L. (2008). An overview of mobile assisted language learning: From content delivery to supported collaboration and interaction. *ReCALL*, *20*(3), 271–289. doi:10.1017/S0958344008000335

Kukulska-Hulme, A., & Traxler, J. (2005). *Mobile learning: A handbook for educators and trainers*. Routledge.

Kurkovsky, S. (2009). Engaging students through mobile game development. *ACM SIGCSE Bulletin*, *41*(1), 44–48. doi:10.1145/1539024.1508881

Kurkovsky, S. (2013). Mobile game development: Improving student engagement and motivation in introductory computing courses. *Computer Science Education*, *23*(2), 138–157. doi:10.1080/08993408.2013.777236

Lee, J. J., & Hammer, J. (2011). Gamification in education: What, how, why bother? *Academic Exchange Quarterly*, *15*(2), 146.

Liu, G.-Z., & Hwang, G.-J. (2010). A key step to understanding paradigm shifts in e-learning: Towards context-aware ubiquitous learning. *British Journal of Educational Technology*, *41*(2), E1–E9. doi:10.1111/j.1467-8535.2009.00976.x

Lu, C., Chang, M., Kinshuk, D., Huang, E., & Chen, C.-W. (2011). Usability of context-aware mobile educational game. *Knowledge Management & E-Learning: An International Journal*, *3*(3), 448–477.

Oinas-Kukkonen, H., & Harjumaa, M. (2009). Persuasive systems design: Key issues, process model, and system features. *Communications of the Association for Information Systems*, *24*(1), 28.

Oxford, R. L. (Ed.). (1996). *Language learning motivation: Pathways to the new century* (Vol. 11). Natl Foreign Lg Resource Ctr.

Panzavolta, S., & Lotti, P. (2013). *Serious Games and Inclusion. Special Education Needs Network*. European Commission.

Papastergiou, M. (2009). Digital game-based learning in high school computer science education: Impact on educational effectiveness and student motivation. *Computers & Education*, *52*(1), 1–12. doi:10.1016/j.compedu.2008.06.004

Park, Y. (2011). A Pedagogical Framework for Mobile Learning: Categorizing Educational Applications of Mobile Technologies into Four Types. *The International Review of Research in Open and Distributed Learning*, *12*(2), 78–102. doi:10.19173/irrodl.v12i2.791

Perrotta, C., Featherstone, G., Aston, H., & Houghton, E. (2013). *Game-based learning: Latest evidence and future directions. NFER Research Programme: Innovation in Education*. Slough: NFER.

Prensky, M. (2003). Digital game-based learning. *Computers in Entertainment*, *1*(1), 21. doi:10.1145/950566.950596

Sandberg, J., Maris, M., & Hoogendoorn, P. (2014). The added value of a gaming context and intelligent adaptation for a mobile learning application for vocabulary learning. *Computers & Education*, *76*, 119–130. doi:10.1016/j.compedu.2014.03.006

Schwabe, G., & Göth, C. (2005). Mobile learning with a mobile game: Design and motivational effects. *Journal of Computer Assisted Learning*, *21*(3), 204–216. doi:10.1111/j.1365-2729.2005.00128.x

Sharples, M., Taylor, J., & Vavoula, G. (2010). A theory of learning for the mobile age. In *Medienbildung in neuen Kulturräumen* (pp. 87–99). VS Verlag für Sozialwissenschaften. doi:10.1007/978-3-531-92133-4_6

Spikol, D., & Milrad, M. (2008). Combining physical activities and mobile games to promote novel learning practices. In *Wireless, Mobile, and Ubiquitous Technology in Education, 2008. WMUTE 2008. Fifth IEEE International Conference on* (pp. 31–38). IEEE. 10.1109/WMUTE.2008.37

Spolsky, B. (1986). 11 Overcoming language barriers to education in a multilingual world. *Language and Education in Multilingual Settings*, *25*, 182.

Stewart, J., Bleumers, L., Van Looy, J., Mariën, I., All, A., & ... (2013). *The potential of digital games for empowerment and social inclusion of groups at risk of social and economic exclusion: evidence and opportunity for policy*. Joint Research Centre, European Commission.

Van Eck, R. (2006). Digital game-based learning: It's not just the digital natives who are restless. *EDUCAUSE Review*, *41*(2), 16.

Volkswagen. (n.d.). *The Fun Theory*. Retrieved December 3, 2017, from http://www.thefuntheory.com/

Wang, A. I., Øfsdahl, T., & Mørch-Storstein, O. K. (2008). An evaluation of a mobile game concept for lectures. In *Software Engineering Education and Training, 2008. CSEET'08. IEEE 21st Conference on* (pp. 197–204). IEEE. 10.1109/CSEET.2008.15

Werbach, K., & Hunter, D. (2012). *For the win: How game thinking can revolutionize your business*. Wharton Digital Press.

Winters, N. (2007). What is mobile learning. *Big Issues in Mobile Learning*, 7–11.

KEY TERMS AND DEFINITIONS

Game Elements: The basic features of game interface and game experience, such as the points, badges, and quests that implement the overall game mechanisms such as the concepts of rewards, recognition, and challenge.

Gamification: The application of game elements and game design techniques to non-game contexts.

Mobile Learning: The use of mobile and wireless technology, including mobile devices, for the purpose of learning where the learner, material and the learning experience can be on the move.

Mobile-Assisted Language Learning: Based on both mobile learning (m-learning) and computer-assisted language learning (CALL); the use of handheld devices to enhance learning languages.

Persuasive Principles: The general principles of designing interfaces and interactive features to encourage people to behave in certain ways.

Persuasive Technology: Any interactive computing system that is designed to change the behavior or attitude of people. The technology collects information about the users' actions to help them achieve their goals.

Social Inclusion: Involving all groups in the society and providing them with equal chances and opportunities in the various aspects of life, including education.

Chapter 7
Mobile Payment and Its Social Impact

Liguo Yu
Indiana University South Bend, USA

Liping Sun
Harbin University of Science and Technology, China

ABSTRACT

This chapter describes mobile payment, a mobile financial activity born of digital revolution, which is the combination of electronic money and mobile technology. The underlying technologies of mobile payment, its big players, and its status quo and future trend are discussed. In addition, this chapter discusses how mobile payment is related to social equality and social inclusion. Through presenting the historical, technical, economic, and social aspects of mobile payment, this chapter intends to provide readers with a holistic view of one of the fast-evolving financial activities that are transforming business, individuals, and the society.

INTRODUCTION

If you stand in the checkout area of a shopping center and observe how customers make payments for their purchases. You probably could notice that some are using cash, some are using checks, and most are using a plastic debit or credit bankcard, and very few are using their mobile phones. That is the case in the United States and most European countries. However, if you visit some big cities in China, you will find more people are using their smartphones to make payments. Because this kind of payment is made through mobile phones, it is called mobile payment. Although

DOI: 10.4018/978-1-5225-5270-3.ch007

mobile payment is relatively a new payment method, as one of the outcomes of digital revolution, it could have huge impact on our life and society (Haig, 2002).

Currency, as the standard exchange medium between sellers and buyers, was first utilized about 500 BC (Ferguson, 2008). The original forms of currency are coins made of precious metal, such as gold or silver. The usage of banknote (paper currency) began about 700 AD in China. Because paper currency has many advantages over metal currencies, it was gradually adopted nationally in China around 1100 AD (Friedman, 1994). Then, nearly 1000 years later, electronic money (e-currency) was introduced within the modern banking systems (Weatherford, 2009). Examples of e-currency include bank deposits, money transfers, and electronic payments with checks, credit cards, or debit cards. E-currency, money in its digital form, is stored in a computer system that can be accessed through bank tellers, ATM machines, POS (Point of Sale) terminals, landline telephones, or computer network (Gup, 2003). The apparent advantage of e-currency over paper currency is convenience. Therefore by 2000, e-currency became the major payment method in developed countries, such as Europe, North America, and Japan. However, at that time, paper currency was still the major payment method in developing countries, including China.

Mobile payment is the offspring of electronic money and mobile technology. With the increasing usage of smartphones, mobile payment gradually enters into our life. It is a new form of e-currency and is replacing bankcards as the new digital payment method (Deloitte, 2015). It is estimated that worldwide, there are about 4.6 billion mobile phone users by the end of 2016, of which about 2.1 billion are smartphone users. This number is increasing dramatically. It is predicted that by 2020, there will be 6.1 billion smartphone users worldwide. As other business activities are moving online (Cunningham & Fröschl, 2013) and to mobile devices, payment is more of an electronic transaction instead of an exchange of paper currency. Electronic payments, especially mobile payments, represent a new dimension of digital evolution, which might shape our spending habits and social relations.

This chapter describes one area of mobile financial services: mobile payment. The objective of this chapter is to provide users with the latest development in this field, including mobile payment technologies, major institutional players, status quo, potential growth, and economic and social impact. The remaining of this chapter is organized as follows. First, the underlying technologies of mobile payment are described. Second, the major financial institutions and technology giants that support and promote mobile payment are introduced. Finally, the status, social impact, and future trend of mobile payment are discussed.

UNDERLYING TECHNOLOGIES

Currently, there are mainly two techniques supporting mobile payment. They are NFC payment and QR code payment. These two techniques are described below.

Near Field Communication Payment

Near Field Communication (NFC) payment requires users to have their bankcard (credit card or debit card) information stored in their NFC-enabled mobile phones. When a smartphone is placed close enough to a device reader at the checkout terminal, the smartphone and the device reader can communicate. Through this process, the bankcard could be verified and the transactions can be conducted. NFC payment is also called contactless payment, because it differs from the traditional bankcard payment, where a card should be inserted into (or swiped over) a card reader.

NFC is a communication protocol based on the Radio Frequency Identification (RFID), which uses electromagnetic fields to identify and track electronic chips. These chips are used to represent the objects they are attached to (Ahson & Ilyas, 2008; Coskun et al., 2011). RFID has been used in many applications, including name tags, toll collection, logistics, and library book organization, where attached objects could be employee cards, automobiles, shipping packages, books, and so on and so forth. NFC is a specialized high frequency RFID, which is a subset within the family of RFID technology (Thrasher, 2013). In NFC-enabled mobile phones, a RFID tag representing the user's bankcard is embedded as a hardware device inside the phone.

The NFC mobile payments have the same underlying transaction process as payments made with bankcards. In the traditional (contact or contactless) bankcard payment, the device reader reads the card information directly from the card and processes the transaction. In NFC mobile payment, the device reader reads the card information from the RFID tag embedded in the NFC enabled smartphones. Therefore, the RFID tag works like a plastic bankcard.

The transaction process of NFC mobile payment indicates that (1) there is no need for the mobile device to be connected online; (2) bankcard information could be stored in a crypto chip with authenticated access through mobile phone apps, which makes NFC mobile payments relatively secure; and (3) it is easy to use and accordingly considered a user-friendly payment method. In addition, because NFC mobile payment is built on top of the traditional bankcard payment architecture, it is easy for retailers to upgrade their existing payment systems to this new system.

Most smartphones support NFC payment (Alliance, 2011). On nfcworld.com, it lists more than 380 types of NFC-enabled phones. Representatives of NFC-supported mobile payment apps include Google Wallet, Android Pay, and Apple Pay.

As described earlier, NFC payment is based-on RFID technology, which allows two devices to communicate directly. However, RFID is not the only technology available now for contactless communication. Bluetooth, an alternative wireless technology using short-wavelength UHF (Ultra High Frequency) radio waves for exchanging data, is becoming an industry standard and widely used in many applications. For example, Bluetooth allows two mobile devices, such as two smartphones, to communicate; Bluetooth also allows a smartphone to communicate wirelessly with other Bluetooth-enabled devices, such as headphones, speakers, fitness bands, printers, automobiles, and digital cordless phones. All the major mobile operating systems, such as iOS, Android, Windows Phone, and BlackBerry support Bluetooth communication.

Comparing with RFID-based NFC technology, Bluetooth has the following advantages. First, Bluetooth has a longer communication range (over 5 meters) than NFC (about 10 centimeters), which makes the payment transaction a more user-friendly process. Second, Bluetooth payment process is considered faster than NFC, which not only can speed up the transaction process, but also can improve user experiences. Third, Bluetooth is not limited by the one-phone to one-device-reader communication mechanism used in NFC, it can support multiple transactions at the same time from a single device reader, which can further speed up the transaction process for multiple users (Meola, 2016). However, current Bluetooth technology requires a handshaking setup, which could be a tedious process for the customer. In contrast, NFC communication requires no setup time. Accordingly, although NFC is one of the dominant mobile payment technologies, especially in the United States, with the emergence and mature of new technologies, such as Bluetooth, it is not surprising to see that NFC payment will evolve and adapt to the new customer requirement or new business environment in the near future.

QR Code Payment

QR (Quick Response) code is a two-dimensional barcode. It is a machine-readable optical label that contains information about an object (Barrera et al., 2013; Nseir et al., 2013). QR code was initially used in automobile manufacturing industry in Japan to quickly scan and identify automobiles. Because smartphones have built-in cameras, apps can be developed to process the scanned QR code. With the extensive support of smartphone apps, QR code now has a wide range of applications, including package tracking, transportation ticketing, product labeling, information storing, and information identifying.

There are three ways QR code could be used in mobile payment. They are (1) buyer-to-large retailer transactions, (2) buyer-to-small business transactions, and

(3) peer-to-peer transactions (Garg, 2015). These different payment approaches are detailed below.

Buyer-to-large retailer transactions. Suppose a customer is shopping offline at a local store. During the checkout process, a unique QR code for this transaction is generated on the retailer's terminal computer. The customer scans the QR code using his/her mobile payment app and authorizes the payment. Finally, the retailer receives a notification confirming the payment and closing the transaction. Many department stores, supermarkets, and financial institutions have adopted this payment method, such as Walmart, Target, Starbucks, PayPal, Alipay of China, and City Union Bank of India (Garg, 2015). Using this QR code payment method, vendors create different QR codes for different transactions. Therefore, the QR code here is used as the unique transaction ID for the purchase and for the payment of the purchase.

Buyer-to-small business transactions. For a small business or a retail outlet that cannot generate different QR codes for different transactions, they can employ this payment method. To make a payment, using a mobile payment app, the customer first scans a printed QR code provided by the seller. The QR code contains the seller's bank account information. Second, the customer enters the payment amount on his/her smartphone app and authorizes the payment. The money is then transferred to the seller's account. Finally, the seller receives a transaction confirmation on his/her own computer or smartphone indicating the money is received. The major difference between buyer-to-large retailer transactions and buyer-to-small business transactions is the former generates a unique QR code for every transaction while the later has a permanent QR code representing the seller's bank account. This buyer-to-small business transactions payment method is also used worldwide. Representative apps include Chase Pay from Chase bank, Shell Britain, Paytm of India, Zapper, Alipay of China, WeChat Pay of China, and some Bitcoin transactions (Garg, 2015). The QR code used in this type of transaction could be a posted printout or an image displaying on some kind of devices. In any cases, the QR code is a permanent code.

Peer-to-peer transactions. In this type of transaction, a payee could generate a QR code and share it with the payer through an email or a social-networking app. The payer then scans the QR code and completes the transaction. Both the payer and the payee should receive the confirmation. These peer-to-peer transactions are usually conducted using the same mobile payment app. Example apps include PayPal, WeChat Pay of China, Alipay of China, and Paytm of India (Garg, 2015). The QR code used in this peer-to-peer transaction could be permanent, representing the account of the payee, and specific payment amount should be entered by the payer. The QR code could also be generated on the fly, which means it contains both the payee's account information and the payment amount. Different apps may use different mechanisms.

QR code is also considered a secure payment method, because no bank account information is directly transmitted during the transaction (Lee et al., 2011). However, QR code payments require both the payer and the payee to be connected on line so that the transactions can be authorized by the payer and confirmed to the payee.

Comparisons

Both NFC payment and QR code payment are widely used mobile payment technologies (Finžgar & Trebar, 2011). To decide which method to employ in your business, two important factors should be considered: infrastructure and customer habits (Blokdyk, 2017).

To use NFC payment, the merchant should have NFC-enabled device readers. This could be a financial burden for a startup business. In contrast, QR code payments is more affordable and could be the easiest way for a business to go cashless. That is exactly the case in China, even some small outlet vendors have adopted QR code payment.

To use QR code payment, the merchant should make sure the environmental wireless data networks or Wi-Fi networks are reliable and powerful enough to support uneven volume of transactions. Keep in mind, both wireless data network and Wi-Fi network could be vulnerable to the denial-of-service attack, which might paralyze the payment system. In contrast, NFC payment could be more resilient to network attacks (Khalilzadeh et al., 2017).

We are living in a customer-centered business era (Christensen & Raynor, 2013). Customer preference and customer habits are certainly important factors that might influence business decisions (Liébana-Cabanillas et al., 2017). Improving customer shopping experience is the key to achieve our business objectives (Allums, 2014). NFC mobile payment method and QR code mobile payment method might coexist for some time, one might replace another, or they could all be outdated and replaced by some other new technologies. Nobody knows it now. However, one thing is for sure: cash, checks, and plastic cards will be used less and less with the evolution of information technology.

MAJOR PLAYERS

Mobile payments grow rapidly worldwide. The total revenue of global mobile payment of 2015 was $450 billion. It is estimated that this number will be doubled in 2018. Now let us look at some big market players in the United States, China, India, Europe, and beyond. Mobile payment is the application of mobile technology on financial market. Accordingly, the influential players in mobile payment industry are tech giants and big financial institutions.

Google

Google Wallet is Google's first peer-to-peer electronic payment system (Ghag & Hegde, 2012). Each user's Google Wallet is linked to the user's bank account. Fund can be transferred between two Google Wallet accounts through mobile phone apps. Traditionally, Google Wallet allows users to make payment at some physical stores. Making payments through Google Wallet is free of charge for the user. Google's intention about Google Wallet is not making profit, but to build an e-commerce product suite to attract and retain users, because a user's Google Wallet can be linked to his/her Google account, which is a bundle of online Google services, including Gmail, Google+, Google Drive, and Google Doc.

The first Google Wallet app was released in 2011. In 2013, Google introduced a physical wallet card to be used together with Google Wallet. However, this service is phased out in 2016. To expand further into the digital payment business, in 2015, Google launched Android Pay, a built-in feature of Android operating systems. Android Pay is designed for in-store transactions, while the new Google Wallet will handle person-to-person money transfers (Martonik, 2015). It is worth noting that Android Pay utilizes NFC payment method that allows the phone and the payment terminal to communicate when they are placed close enough.

With over 1.5 billion Android devices worldwide, Google wishes that its new payment system, Android Pay, could have a better market position in competing others. Android Pay works with major credit cards and debit cards from most of the top US banks. Currently, near three hundred banks support Android Pay. Android Pay runs in a few countries outside of the United States, including Australia and UK. Given the global market dominance of Android mobile operating systems, it is expected that Android Pay will grow even faster in the next 5-10 years.

PayPal

PayPal was originally launched as a money transfer service in 1998 to facilitate C2C (Consumer to Consumer) e-commerce transactions. When an item is sold online, the payment would be credited to the seller's PayPal account instead of a bank account. This process makes the transaction more convenient and secure. After PayPal was acquired by eBay, the leading C2C e-commerce website in 2002, it became the major payment method of eBay users (Jackson, 2004).

Since partnering with MasterCard in 2007, PayPal could be linked to a MasterCard and used offline at brick-and-mortar stores. Later, peer-to-peer money transfer feature is added to PayPal. In 2016, PayPal's NFC mobile payment is launched to support Visa and MasterCard.

Currently, PayPal has 188 million active accounts. This number doubles what it had in 2010. Figure 1 shows PayPal's annual mobile payment volume. Table 1 compares PayPal's total payment volume with its mobile payment volume in 2014, 2015, and 2016. Although only a small portion of the total payments are conducted with mobile devices, its fast-growing trend should not be underestimated.

PayPal is an international payment platform. Currently, PayPal's service is available in 202 countries with 25 currencies. PayPal's subsidiary services include Braintree (a mobile and web payment system for e-commerce), Xoom (a money

Figure 1. PayPal's annual mobile payment volume
Source: Statista, 2017a.

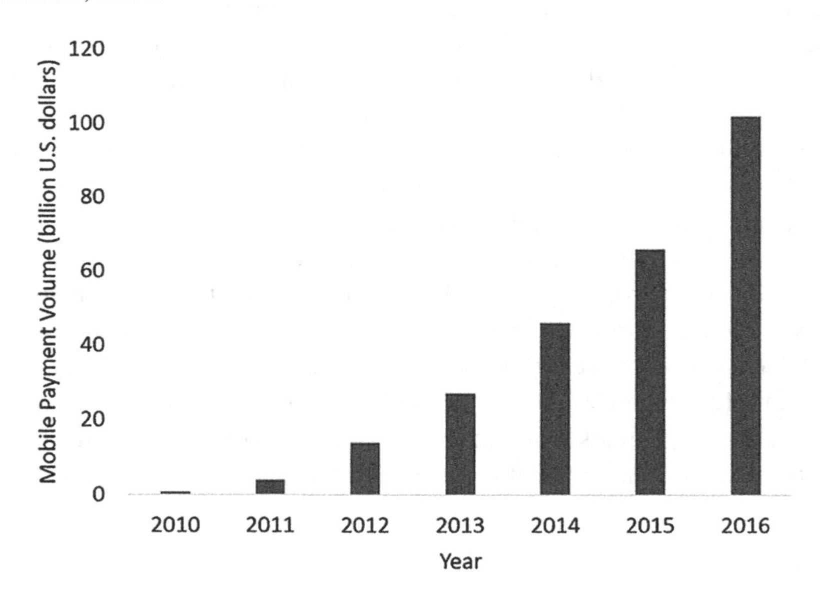

Table 1. Comparison of PayPal's total payment volume and its mobile payment volume

	2014	2015	2016
Total Volume	$234 billion	$291 billion	$354 billion
Mobile Volume	$46 billion	$66 billion	$102 billion
Percentage of Mobile Volume	20%	23%	29%

Source: Statista, 2017a; Statista, 2017b.

transfer system), and Venmo (a mobile peer-to-peer payment system). It is reported that Venmo will support NFC payment and could accordingly be used in offline physical stores like Android Pay. In addition, PayPal also has its own peer-to-peer payment platform, PayPal.Me.

PayPal, as an independent technology-based online payment system, is growing aggressively in several business dimensions through acquiring both startups and potential competitors. Few people have doubt about the future mobile payment market position of PayPal.

Apple

Apple launched its mobile payment system, Apple Pay, in 2014. Apple Pay is designed to provide an electronic alternative to physical bankcard. With stored bankcard, Apple mobile devices, such as iPhone, iPad, and iWatch, could be used to make a payment on NFC-supported POS (Point of Sale) terminals. Like Android Pay, Apple Pay itself does not provide financial services to the customers. Instead, Apple Pay provides mobile payment services for financial organizations. This means Apply Pay will only work on contracted financial institutions. Currently, Apple Pay supports major payment cards (debit or credit) issued in USA, UK, Canada, Australia, China, Singapore, Switzerland, France, and Hong Kong. Table 2. Shows the number of Apple Pay participating banks worldwide.

Although Apple Pay is only about 3 years old. It grows at an unprecedented pace. In 2015, one year after its launch, Apple Pay registered a total of $10.9 billion transactions. In 2016, Apple Pay recorded a 50% growth in the number of monthly credit card transactions over 2015. The latest quarterly result of Apple released in May 2017 indicates that Apple Pay transaction volume is up 450% year over year.

Apple's mobile operating system iOS is second in the mobile operating systems marketplace next to Android. Most importantly, users of Apple products, such as iPhone, iPad, and iWatch are mainly located in developed countries, such as North America, Europe, and Asia-Pacific. It will be relatively easier for those customers to adopt mobile payments. In contrast, besides developed countries, Android is also used in emerging markets, such as Southeast Asia, South Asia, and Africa. In

Table 2. Number of Apple Pay participating banks as of December 2016

	Number of Banks
Asia-Pacific	279
Europe	60
North America	1868

addition, Apple product users are usually loyal Apple fans. They would like to use all apple products (hardware and software) of its ecosystem, including Apple Pay (Bruce, 2016).

Samsung

Samsung is one of the largest smartphone manufacturers worldwide. In 2016, over 306 thousand Samsung smartphones are sold to end users. Samsung entered into the mobile payment market in August 2015, when its first app, Samsung Pay is released. Samsung Pay also utilizes NFC technology. In addition, it can be used in POS terminals that only accept traditional magnetic strip and contactless payments.

As of December 2016, Samsung Pay supports payment cards issued in 19 countries including South Korea, America, China, India, Brazil, Canada, Hong Kong, some European countries, and some Southeast Asian countries. Because Samsung Pay is less than two years old. It has a relatively small market share. Figure 2 shows the estimated monthly active users of Apple Pay, Android Pay, and Samsung Pay as of 2016. It is worth noting that Apple Pay has just been launched for two years, its annual growth rate is about 150%. In the United States, three out of four contactless mobile payments are conducted through Apple Pay. However, Samsung Pay is expected to grow faster in Asia-Pacific regions and make impacts on its local financial systems (Son et al., 2015).

Figure 2. Monthly active users of Apple Pay, Android Pay, and Samsung Pay as of 2016

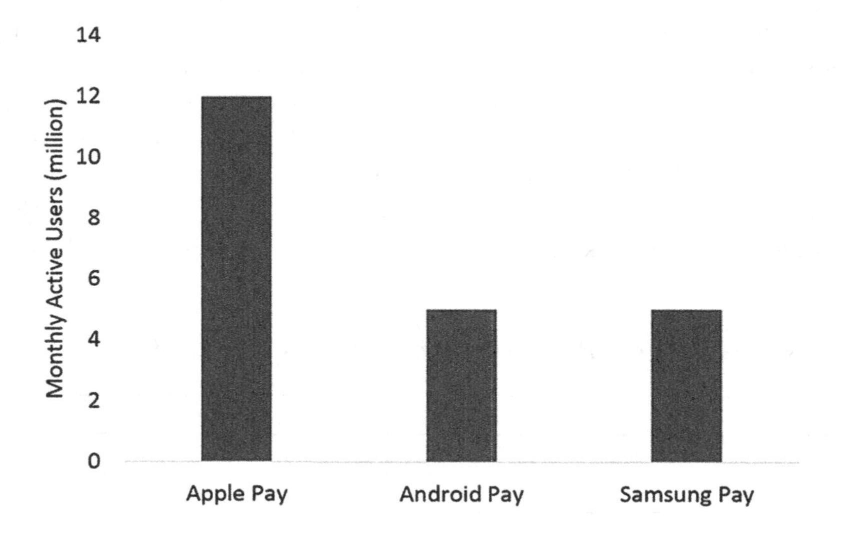

Alipay and WeChat Pay

Alipay is China's e-commerce giant, Alibaba's solution to third party payment. Like PayPal, Alipay was originally launched to support payment in Taobao's C2C e-commerce transactions (Wee, 2012). In C2C business, when an order is placed, the buyer makes the payment and the money is transferred to Alipay. Only after the buyer receives the product and confirms the transaction, the money in Alipay could be transferred to the seller's account. If the order is not delivered or the buyer is not satisfied with the product, the payment will be refunded to the buyer. This transaction mechanism is designed to protect the buyers who make the payment.

In November 2009, Alipay's mobile payment service was launched. Since then, Alipay is no long a pure third-party payment system for e-commerce. After more than ten years growth, Alipay has become the largest mobile payment system in China with about 400 million users. Alipay supports online payment, offline in store payment, and peer-to-peer payment. QR code is mainly used in offline payment and peer-to-peer payment. Alipay's online payment is not limited to e-commerce activities, it is also used to pay utilities bills, medical bills, donations, investment, and tuitions.

Alipay is now one of the many services provided by Ant Financial Group (an affiliate company of Alibaba), which is now valued at $60 billion. Alipay's business target is not limited to China's market. Ant Financial Group is aggressively growing its business internationally. Alipay's Cross-Border E-Payment Service allows travelers to purchase products or services at international partners' physical stores. The travelers make the payment with Chinese currency RMB to Alipay and Alipay will then convert RMB to USD or local currencies and transfer the fund to the product seller or the service provider. The foreign markets open to Alipay include Europe, USA, and many other countries and regions, where Chinese tourists could directly use their Alipay app at Alipay's partnered stores. This reduces the burden of carrying cash and exchanging currencies. Latest information released by Ant Financial Group indicates that as of May 2017, Alipay can be used in over 200 countries and supports 18 currencies.

Alipay is similar to PayPal. Both of them are born out of e-commerce boom. Now, they are all independent technology-based financial institutions. In addition, Alipay is reaching out to offer more financial products, such as insurance, finance management, and social credit scoring.

Backed up by its tremendous users of WeChat, a social networking app, China's Internet Giant, Tencent launched its own money transferring system, WeChat Pay in 2013. As of March 2016, WeChat Pay has registered over 300 million users worldwide.

Because WeChat Pay is built on top of a social-networking platform, it is ideal for small money transfers between friends, relatives, and businesses. In addition, WeChat Pay provides the backbone transaction services for the social e-commerce built on top of WeChat. WeChat Pay also supports cross-border payments. When Chinese shoppers buy goods overseas, they can pay in RMB by WeChat Pay and the sellers could receive the payment in USD or local currencies.

It is worth noting that QR code pay is the major payment method supported by both Alipay and WeChat Pay. There is a good reason why NFC is not widely adopted by Alipay and WeChat Pay: NFC is mainly used to replace bankcard transactions for financial institutions and China has fewer bankcard users; while Alipay and WeChat Pay are registered third party financial institutions, QR code payment is accordingly the perfect alternative to cash transactions. Table 3 compares Alipay and WeChat Pay.

Figure 3 shows the changes of China's mobile payment market share in the past two years. We can see that although WeChat Pay is released much later than Alipay, it caught up fast in the past years. As supported by the most popular social networking app in China, WeChat Pay has become a major player in China's mobile payment market, next to Alipay. Other competitive players in China's mobile payment market include Apple Pay and Samsung Pay.

Paytm

Paytm (Pay through Mobile) is India's largest mobile payment company. It is launched in 2010. Paytm Wallet can be used to pay online purchases, tickets, taxi, toll booth,

Table 3. Comparison of Alipay and WeChat Pay

	Alipay	**WeChat Pay**
Parent company	Ant Financial (an affiliate company of Alibaba)	Tencent
Payment method	QR code	QR code
Release date	2004	2013
Independent App	Yes	No (a function of WeChat app)
E-commerce Partner	Taobao.com, Tmao.com, Aliexpress.com, Alibaba.com	JD.com
Monthly active users	270 million	300 million
Currencies supported	at least 18	at least 9
Global expansion	yes	yes

Figure 3. China's mobile payment market share
Source: Editorial, 2017.

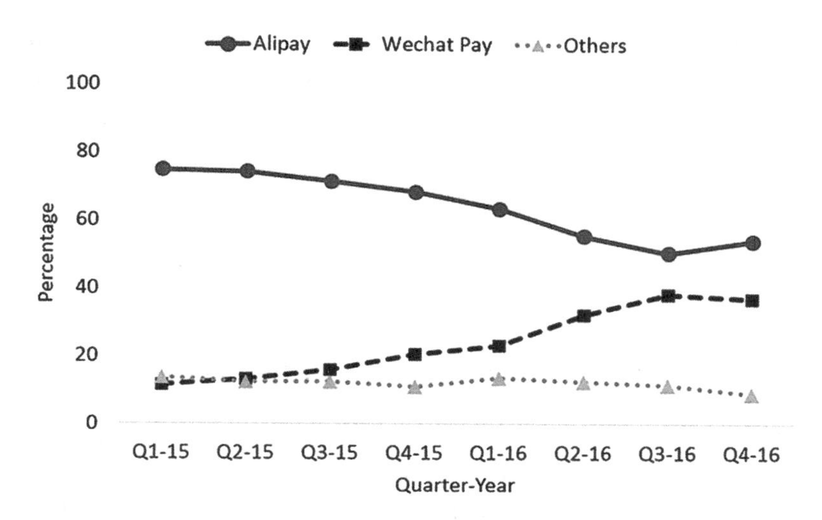

and other services. Like Alipay, Paytm uses the QR code payment feature in its app. Paytm was the first to support cashless and card-less payments for small merchants in India. In addition, Paytm is a licensed bank that offers many other financial services.

It is interesting to know that Alibaba and Ant Financial are major shareholders of Paytm. Betting on the future growth of India's e-commerce and mobile payment, Alibaba and Ant Financial are working with Paytm's parent company, One97 Communications, to develop Paytm into India's Alipay. Table 4 shows the payment data of Paytm. Although Paytm is much smaller than Alipay, given India's market growing potential, it is estimated that Paytm will become a major player in mobile payment industry worldwide (Manikandan & Chandramohan, 2016).

Table 4. Paytm payment data

Total Number of Users	170 million (December 2016)
Monthly Active Users	80 million (2016)
Yearly Transaction	1 billion (2016)
Average Daily Transactions	5 million (2016)
Offline Supporting Merchants	850 thousand (2016)
Valuation	$6 billion (2017)

Others

Besides the aforementioned companies, there are many medium and small size mobile payment service providers in the United States, Europe, and many other countries. These services include Square, Intuit's GoPayment, VeriFone's SAIL, PayAnyWhere, and LevelUp in the United States, Payam and Barclays Pingit in UK, and Klarna, PayU in other European countries. For example, Zapper is a QR code based payment app; when users locate a Zapper QR Code on their bills either online or offline, they can scan it with their Zapper app and make the payment. Zapper could also be used to pay restaurant bills, utility bills, medical bills, taxi, and movie tickets.

Globally, mobile payment grows rapidly in all continents. For example, it is estimated that mobile payment transactions in Western Europe will grow from €52 billion in 2016 to €148 billion by 2021 (Forrester, 2017). Even in less technology-developed African countries, mobile payments are started to be widely adopted. One of such services is M-Pesa, which operates in Kenya, Tanzania, Afghanistan, South Africa, and beyond (Markovich et al., 2017).

Mobile payment is a technology-based payment method. It has a less than 10-year development history. With the evolving of mobile technology, it will be inevitable to see the emergence of new tech companies or financial institutions that may enter this market and influence this market.

SOCIAL IMPACT

Following e-commerce and social-network, mobile payment represents another shockwave of internet-based technology revolution. Its economic impact could be huge (Shaikh et al., 2017; Kremers & Brassett, 2017; Shrier et al., 2016; Omonedo & Bocij, 2017). For example, as China is moving from a cash-dependent society to a cashless society, traditional banks are feeling the pressure of losing customers and revenues. New technology-based financial institutions are transforming all kinds of financial activities online. People are embracing the benefits of mobile payment, such as convenience and security.

At the same time, mobile payment also brings challenges and changes to the society. One of the objectives of prompting mobile payment is to support economic growth. However, we cannot have a sustained growth without social inclusion. The following of this section discusses the social impact of mobile payment.

Status Quo

The United States, China, and Europe are undoubtedly the leading markets in mobile payment, mainly due to their internet booms. Table 5 compares mobile payment usages in the United States, Western Europe, and China. Although the USA is a leading market in many internet-based businesses, such as e-commerce, peer-to-peer lending, peer-to-peer renting, and mobile-enabled transport service, it lost the first place on mobile payment market to China (To & Lai, 2014). There are some good reasons behind this phenomenon.

First, in the past, China was a country that heavily depended on cash transactions. Bankcards were introduced less than twenty years ago and were rarely used. Mobile payment is accordingly becoming a favorite cashless transaction method. In contrast, most Americans are using credit cards, and mobile payment has not shown any significant advantages over credit cards. Therefore, the mobile payment penetration rate of the USA is lower than that of China.

Second, most mobile payment in the USA is just another way of bankcard payment, such as Android Pay and Apple Pay. PayPal is an exception. However, PayPal is mainly used for online transactions. In contrast, Alipay and WeChat Pay of China are licensed financial institutions. It is more convenient to make payment and receive money with Alipay and WeChat Pay, especially for peer-to-peer transactions. In addition, both Alipay and WeChat Pay provide many other financial services, which make them more attractive than cash transactions or bankcard transactions.

Third, both Alipay and WeChat Pay are part of their parent companies' internet-based ecosystems, which make them easier to attract and retain users. Alipay is the official payment method of Alibaba's e-commerce websites and WeChat Pay is based-on one of the most popular social-networking apps. Apparently, Android Pay and Apple Pay of the United States do not have these advantages.

Finally, digital payment, especially mobile payment is part of the government of China's plan to promoting the concept of *Internet+* and *one belt and one road* initiative. Through investing heavily on related projects, the Chinese government is aiming at rewriting the world's business order (Swaine, 2015).

Table 5. Comparison of mobile payment in the USA, Western Europe, and China as of 2016

	USA	Western Europe	China
Yearly transaction volume	$154 billion	$62 billion	$5.5 trillion
Number of users	38.4 million	55 million	469 million
Number of users over population	12%	14%	35%

Another important mobile payment market is India. It is estimated that by 2020, its yearly transaction volume could reach $500 billion. Like China, India was traditionally a cash-based economy. With the increasing usage of smartphones, it is expected that its mobile payment could surpass cash as the major payment method in the next 5-10 years.

Globally, mobile payment is also growing rapidly (Lerner, 2013; Salmony & Jin, 2016; Phonthanukitithaworn et al., 2015). Figure 4 shows the number of mobile payment users of different regions from 2012 to 2016. It can be seen that in most regions, the number of mobile payment users at least doubled in past 4 years.

Despite the tremendous growth, mobile payment does come with issues (Dennehy & Sammon, 2015). The major problem seen now days is fraud. For example, once money is sent out using a peer-to-peer app, it will be hard to get it back. Email scams, phishing messages, malicious QR codes, and many other kinds of virus are causing damages to unwary and inexperienced customers. It is reported that while mobile payments represent 14 percent of transactions among merchants who accept them, they constitute 21 percent of fraud cases, which amounts to $6 billion a year (Kharif, 2015). For example, in China, about 13 percent of mobile payment users have suffered from telemarketing frauds and most of them could not get the lost money back.

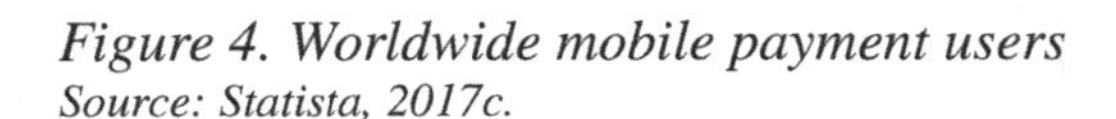

Figure 4. Worldwide mobile payment users
Source: Statista, 2017c.

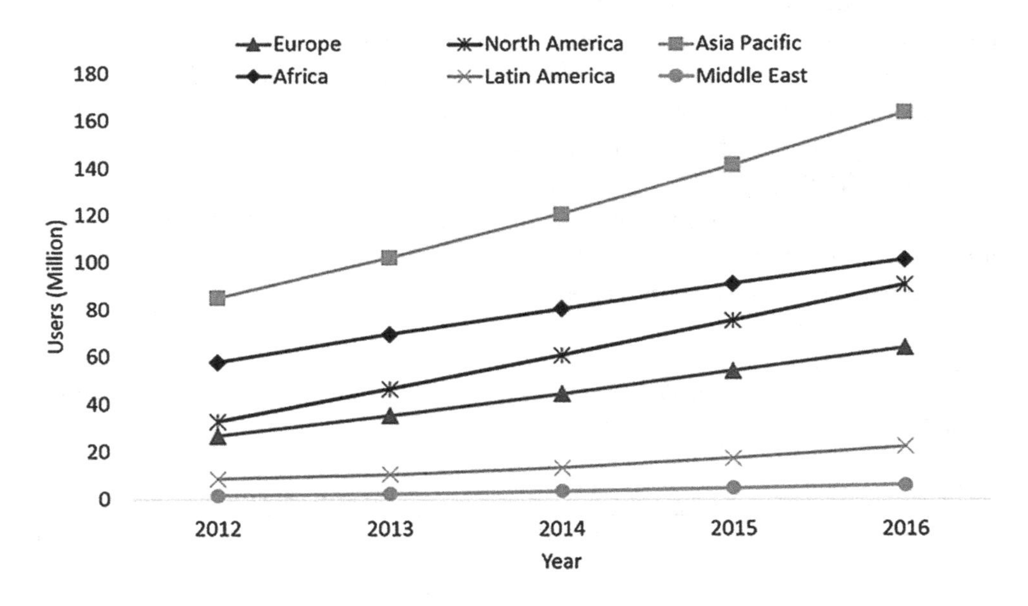

As mobile payment is transforming from the stage of wild growth to the stage of consolidation, some policies are being developed to standardize hardware devices and transaction protocols (Pukkasenung & Chokngamwong, 2016; Duggal, 2013). One of such technical standard is ISO 12812 series. It is expected that a global standard could be developed by 2020. In the United States, although some laws are applicable to cover certain mobile payment issues, there is a lack of comprehensive and consistent laws at the federal level. Apparently, the policy making is lagged behind the technology evolution. To maintain a healthy mobile payment development, the governments should spend more effort in regulating mobile payment industry and mobile payment practices.

Social Inclusion

Technology, in general, is related to social inequality (Arocena & Senker, 2003; Dahlberg et al., 2008). For example, as labor-saving technologies are replacing human workers, the unemployment rate could increase while the producer's profit could increase. On the other side, the cost of the product could be lowered due to technology revolution, which makes products more affordable to low-income consumers. Therefore, technology is a double-edged sword for social inclusion (Warschauer, 2004).

As an emerging technology, mobile payment and its relation to social inclusion is undefined yet. This will largely depend on government policies and international initiatives. To address this issue, technology inequality and social inequality are discussed below.

Technology Inequality

Mobile payment is a smartphone app. The usage of smartphones directly affects the adoption of mobile payments. Worldwide, the smartphone penetration rate varies dramatically from country to country. Table 6 shows the smartphone penetration rates in some developed countries and some developing countries. It can be seen that economic inequality resulted in technology inequality. Smartphones are more accessible and affordable for developed countries. For developing countries, such as those of Africa and South Asia, smartphones are still considered luxury electronics.

Figure 5 shows the percentage of mobile payment users over its population of different regions. Globally speaking, mobile payment usages is still low. Even in North America, less than 16% of its population are mobile payment users. In contrast, 72% of the consumers in the US have at least one credit card. Therefore, technology inequality is still a big hurdle for the wide adoption of mobile payment, even in developed countries.

To reduce technology inequality, we should first wage a campaign to fight economic inequality. Economic inequality is the major barrier to improving social inclusion. This campaign should be the effort of the society as a whole. Second, low-cost smartphones targeting low-income consumers should be promoted by IT industries, even that may lead to less profit. The corporates' social responsibility plays an important role in mitigating technology inequality. Meanwhile, governments should provide legislation support and tax benefits for those engaged corporations.

Social Inequality

As with any other technological innovations, mobile payment might affect social inequality both positively and negatively depending on the regulations. It is worth noting that many of the much-needed regulations have not been created yet.

Mobile payments are creating opportunities for small business, especially those in remoted areas of developing countries. Traditionally, large transactions, especially cross-border transactions are complex, expensive, and slow. Now days, with the support of mobile payments, small businesses are able to access funds and international markets easily. Especially, peer-to-peer payment is considered a convenient and low-cost cross-border transaction method. Therefore, we anticipate that more and more small business will switch to mobile payment. The landscape of international commerce, especially cross-border e-commerce, will be redefined. In this regard, mobile payment could help fight geographic inequality.

Table 6. Countries with the highest smartphone penetration rate and countries with the lowest smartphone penetration rate as of 2016

Rank	Country	Percentage of Population Owning a Smartphone	Region
High	South Korea	88	Asia Pacific
High	Australia	77	Asia Pacific
High	Israel	74	Middle East
High	United States	72	North America
High	Spain	71	Europe
Low	Burkina Faso	14	Africa
Low	Pakistan	11	South Asia
Low	Tanzania	11	Africa
Low	Ethiopia	4	Africa
Low	Uganda	4	Africa

Figure 5. Percentage of mobile payment users over its population of different regions
Source: Statista, 2017c.

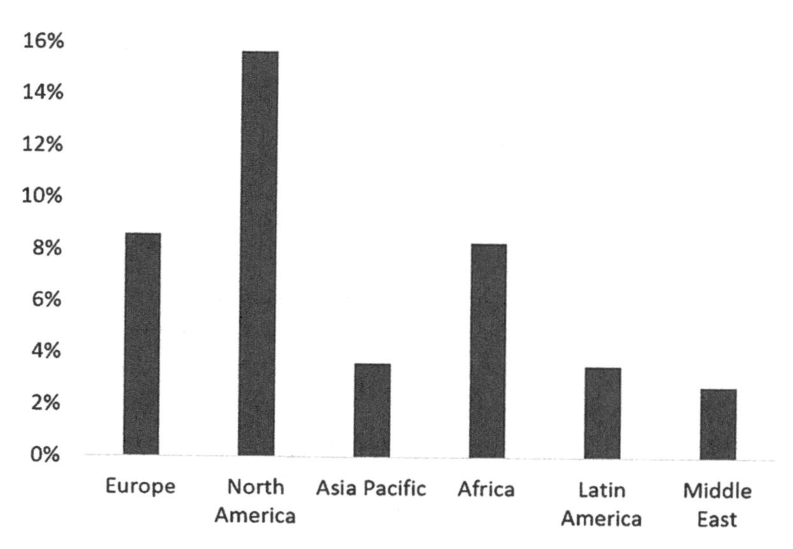

In addition, mobile payment could be extended to pay utility bills, hospital bills, tuitions, and miscellaneous fees. With its fast transaction, model payment could help low-income consumers avoid paying late fees. Mobile payment could also make it more convenient for elderly and disabled consumers to manage their bills. In this respect, mobile payment is considered a social-friendly technology.

Despite the potential positive impacts, there are also potential negative impacts. The winners of mobile payment revolution are no doubt these service providers, such as Apple, Google, PayPal, Samsung, Ant Financial, and Tencent. The losers could be banks, big and small. To contract these mobile payment providers to support their bankcards, the traditional financial institutions need to pay a portion of the transaction fees to the service providers. In this competing process, the technology is expected to win and the traditional business is expected to lose. We will not be surprised to see business closed down and employees laid off. To avoid the worsening of economic inequality, this transition should be carried out smoothly to support workforce's retraining and reemployment.

In addition, governments should regulate mobile payment to avoid or limit its negative impacts. First, only zero or near zero additional fees are allowed to be collected from the mobile payment users. This is especially important for offline in store payment: just like payments made by cash or bankcard, no service fee should be charged to the consumer for mobile payment. For online transactions, mobile payment should incur no more fees than its underlying transactions, such as bankcard payment or e-check payment.

Second, traditional payment methods should also be supported wherever mobile payment is made available. Consumers should always be allowed to use cash and major credit cards. If a vender only accepts mobile payment, it will be detrimental to technology-disadvantaged groups, which can deteriorate social inclusion. Although no federal laws in the US mandating that a private business must accept paper currency for goods or services, some state laws consider refusing to accept cash illegal. For example, in Massachusetts, a state law affirms that no retailer shall discriminate against cash buyers. Similar laws should be passed to provide equal payment options for cash users, credit card users, and mobile users.

To summarize, a healthy business environment should address diverse interests of diverse social groups, especially those low-income groups and knowledge-disadvantaged or technology-disadvantaged groups. Without addressing technology inequality and social inequality, mobile payment revolution could lead to deteriorated social inclusion. On the contrary, when the needs of these less-represented communities are carefully addressed, we not only can improve social relations, but also can create opportunities to grow our customer base and markets.

Future Trend

Looking forward, let us examine the three areas that mobile payment could play its role and make a significant impact: in-store payment, online payment, and peer-to-peer payment.

With the common usage of smartphones worldwide, it is expected that the long used physical currency, paper money, will be gradually replaced by electronic currency. Bankcards and mobile payments could coexist for a long period of time. It is estimated that in-store mobile payments could reach $500 billion in the United States by 2020. The potential winners could be Apple Pay and Android Pay, because both of them support contactless NFC transactions. In China, the future could belong to Alipay and WeChat Pay. Nevertheless, it is safe to say that nobody is one hundred percent sure about the future.

For online shopping, mobile payments provide some, but not significant advantages over bankcards. However, to make the transactions more convenient, a QR code for an order could be generated on the seller's website during the checkout process and the buyers could scan the code and make the payment with their smartphones. This will save buyers' time from entering bankcard data. Due to the dominance of PayPal in e-commerce transactions, PayPal could still be the major online mobile payment method in the USA. In China and beyond, Alipay's influence could grow bigger.

Peer-to-peer payments could also be switched to mobile, which has seen in both China and the USA. In China, both Alipay and WeChat Pay support peer-to-peer payments seamlessly. In the USA, PayPal and Google Wallet are the most popular ones. However, this area of payment will be closely regulated and monitored. In the United States, greater than certain amount of peer-to-peer payment is required to be reported in the tax return. Internationally, precautions will be taken to prevent money laundry with peer-to-peer payment.

The success of a mobile payment service will largely depend on two factors: user experience and security. On one side, making payment a seamless part of the consumer experience is the driving demand for business to switch to mobile payment (Zhou, 2014; Ahmad et al., 2016; Liébana-Cabanillas et al., 2014). Convenience is the characteristic catalyst for currency revolution, which is seen in the transition from precious metals to bank notes, from bank notes to bankcards, and from bankcards to mobile devices. As mobile devices are getting smaller, mobile payment could be conducted with wearable devices. Internet of Things (IoT) is expanding its applications to many domains, including banking industry. It is expected that mobile payment could be extended to remote gadgets, such as smart devices at homes, internet-connected automobiles, trains, and airplanes.

On the other side, security is always the primary concern when adopting a new payment method. For service providers, there are three security levels: data storage security, transaction protocol security, and user device security. Industrywide, these levels of security will be strengthened. For end users, the major security concerns are identity protection and fraud prevention. Therefore, educating customers and setting up fraud alert mechanisms are important steps in reducing fraudulent transactions, increasing customer satisfactions, and improving customer confidence.

CONCLUSION

This chapter described mobile payment, its underlying technologies, big players, status, and social impact. Technologies are changing the business models and the business landscape and redefining social relations. As a new offspring of the business and technology revolution, mobile payment is still in its early development age. Many things are going to change. It is important for business to take this great opportunity to revolutionize their payment systems, improve their customer relations, and redefine their market positions. It is equally important for governments and international organizations to actively support and regulate mobile payment to provide a healthy environment for its growth and to improve social inclusion.

Because mobile payment is relatively a new applied technology, its impact, especially social impact, is unknown yet. We are actively observing its effect and collecting related data now. The main purpose of this chapter is to discuss the potential impact of mobile payment on social inclusion and provide guidance and suggestions for stakeholders, such as governments. Meanwhile, this chapter intends to call the attentions of researchers on the far-reaching impacts of mobile payments.

When metal currencies were used thousands of years ago, few people could have pictured paper currencies, electronic currencies, or mobile payments. Same thing holds true now. The payment methods we are using today will certainly not be able to last another one thousand years or one hundred years. No one can predict the future; however, the future is shaped and influenced by our decisions today.

REFERENCES

Ahmad, A., Farid, M. S., Ismail, Y., & Kadir, B. (2016). Review of customer adoption on mobile payment. *Journal Postgraduate, 1*(2).

Ahson, S. A., & Ilyas, M. (2008). *RFID handbook: applications, technology, security, and privacy*. CRC Press. doi:10.1201/9781420055009

Alliance, S. C. (2011). The mobile payments and NFC landscape: A US perspective. *Smart Card Alliance*, 1–53.

Allums, S. (2014). *Designing mobile payment experiences: Principles and best practices for mobile commerce*. O'Reilly Media, Inc.

Arocena, R., & Senker, P. (2003). Technology, inequality, and underdevelopment: The case of Latin America. *Science, Technology & Human Values*, *28*(1), 15–33. doi:10.1177/0162243902238493

Barrera, J. F., Mira, A., & Torroba, R. (2013). Optical encryption and QR codes: Secure and noise-free information retrieval. *Optics Express*, *21*(5), 5373–5378. doi:10.1364/OE.21.005373 PMID:23482108

Blokdyk, G. (2017). *Mobile payment complete self-assessment Guide*. CreateSpace Independent Publishing Platform.

Bruce, E. (2016). *Apple Pay essentials*. Packt Publishing.

Christensen, C., & Raynor, M. (2013). *The innovator's solution: Creating and sustaining successful growth*. Harvard Business Review Press.

Coskun, V., Ok, K., & Ozdenizci, B. (2011). *Near field communication (NFC): From theory to practice*. John Wiley & Sons.

Cunningham, P., & Fröschl, F. (2013). *Electronic business revolution: opportunities and challenges in the 21st century*. Springer Science & Business Media.

Dahlberg, T., Mallat, N., Ondrus, J., & Zmijewska, A. (2008). Past, present and future of mobile payments research: A literature review. *Electronic Commerce Research and Applications*, *7*(2), 165–181. doi:10.1016/j.elerap.2007.02.001

Deloitte. (2015). *Contactless mobile payments (finally) gain momentum*. Available at https://www2.deloitte.com/content/dam/Deloitte/fpc/Documents/secteurs/services-financiers/deloitte_tmt-predictions-2015-contactless-mobile-payments_en.pdf

Dennehy, D., & Sammon, D. (2015). Trends in mobile payments research: A literature review. *Journal of Innovation Management*, *3*(1), 49–61.

Duggal, P. (2013). *Mobile payments & mobile law*. Saakshar Law Publications.

Editorial. (2017). Alipay vs WeChat Pay – who is winning the battle? *ASEAN Today*. Available at https://www.aseantoday.com/2017/02/alipay-vs-wechat-pay-who-is-winning-the-battle/

Ferguson, N. (2008). *The ascent of money: A financial history of the world*. Penguin.

Finžgar, L., & Trebar, M. (2011). Use of NFC and QR code identification in an electronic ticket system for public transport. In *The 19th International Conference on Software, Telecommunications and Computer Networks* (pp. 1–6). IEEE.

Forrester. (2017). *European mobile payments will almost triple by 2021*. Available at https://www.forrester.com/report/European+Mobile+Payments+Will+Almost+Triple+By+2021/-/E-RES137528

Friedman, M. (1994). *Money mischief: Episodes in monetary history*. Houghton Mifflin Harcourt.

Garg, G. (2015). *QR Code Payments: Everything you need to know*. Scanova. Available at https://scanova.io/blog/blog/2015/04/08/qr-code-payment/

Ghag, O., & Hegde, S. (2012). A comprehensive study of google wallet as an NFC application. *International Journal of Computers and Applications*, *58*(16).

Gup, B. E. (2003). *The future of banking*. Greenwood Publishing Group.

Haig, M. (2002). *Mobile marketing: The message revolution*. Kogan Page Publishers.

Jackson, E. M. (2004). *The PayPal wars: Battles with eBay, the media, the mafia, and the rest of planet Earth*. World Ahead Publishing.

Khalilzadeh, J., Ozturk, A. B., & Bilgihan, A. (2017). Security-related factors in extended UTAUT model for NFC based mobile payment in the restaurant industry. *Computers in Human Behavior*, *70*(C), 460–474. doi:10.1016/j.chb.2017.01.001

Kharif, O. (2015). Fraudulent smartphone payments are becoming a pricey problem. *Bloomberg Businessweek*. Available at https://www.bloomberg.com/news/articles/2015-02-13/mobile-payment-fraud-is-becoming-a-pricey-problem

Kremers, R., & Brassett, J. (2017). Mobile payments, social money: Everyday politics of the consumer subject. *New Political Economy*, 1–16.

Lee, J., Cho, C. H., & Jun, M. S. (2011). Secure quick response-payment (QR-Pay) system using mobile device. In *The 13th International Conference on Advanced Communication Technology* (pp. 1424–1427). IEEE.

Lerner, T. (2013). *Mobile payment*. Springer.

Liébana-Cabanillas, F., Muñoz-Leiva, F., & Sánchez-Fernández, J. (2017). A global approach to the analysis of user behavior in mobile payment systems in the new electronic environment. *Service Business*, 1–40.

Liébana-Cabanillas, F., Sánchez-Fernández, J., & Muñoz-Leiva, F. (2014). The moderating effect of experience in the adoption of mobile payment tools in Virtual Social Networks: The m-Payment Acceptance Model in Virtual Social Networks (MPAM-VSN). *International Journal of Information Management*, *34*(2), 151–166. doi:10.1016/j.ijinfomgt.2013.12.006

Manikandan, M., & Chandramohan, S. (2016). Mobile wallet: A virtual physical wallet to the customers. *PARIPEX-Indian Journal of Research, 4*(9).

Markovich, S., Markovich, S., Snyder, C., & Snyder, C. (2017). M-Pesa and Mobile Money in Kenya: Pricing for Success. Kellogg School of Management Cases, 1–17.

Martonik, A. (2015). *What's the difference between Android Pay and the new Google Wallet?* Available at http://www.androidcentral.com/whats-difference-between-android-pay-and-new-google-wallet

Meola, A. (2016). Mobile payments technology and contactless payments explained. *Business Insider*. Available at http://www.businessinsider.com/mobile-payment-technology-contactless-payments-explained-2016-11

Nseir, S., Hirzallah, N., & Aqel, M. (2013). A secure mobile payment system using QR code. In *The 5th International Conference on Computer Science and Information Technology* (pp. 111–114). IEEE. 10.1109/CSIT.2013.6588767

Omonedo, P., & Bocij, P. (2017). Potential impact of perceived security, trust, cost and social influence on m-commerce adoption in a developing economy. *WORLD (Oakland, Calif.)*, *7*(1), 147–160.

Phonthanukitititihaworn, C., Sellitto, C., & Fong, M. (2015). User intentions to adopt mobile payment services: A study of early adopters in Thailand. *Journal of Internet Banking and Commerce*.

Pukkasenung, P., & Chokngamwong, R. (2016). Review and comparison of mobile payment protocol. In Advances in Parallel and Distributed Computing and Ubiquitous Services (pp. 11–20). Springer Singapore. doi:10.1007/978-981-10-0068-3_2

Salmony, D. M., & Jin, B. (2016). Pan-European mobile peer-to-peer payment–and beyond. *Journal of Digital Banking*, *1*(1), 85–96.

Shaikh, A. A., Hanafizadeh, P., & Karjaluoto, H. (2017). Mobile banking and payment system: A conceptual standpoint. *International Journal of E-Business Research*, *13*(2), 14–27. doi:10.4018/IJEBR.2017040102

Shrier, D., Canale, G., & Pentland, A. (2016). *Mobile money & payments: Technology trends*. Academic Press.

Son, I., Lee, H., Kim, G., & Kim, J. (2015). The effect of Samsung Pay on Korea equity market: Using the Samsung's domestic supply chain. *Advanced Science and Technology Letters*, *114*, 51–55. doi:10.14257/astl.2015.114.10

Statista. (2017a). *Statista data*. Available at https://www.statista.com/statistics/277819/paypals-annual-mobile-payment-volume/

Statista. (2017b). *Statista data*. Available at https://www.statista.com/statistics/277841/paypals-total-payment-volume/

Statista. (2017c). *Statista data*. Available at https://www.statista.com/statistics/279957/number-of-mobile-payment-users-by-region/

Swaine, M. D. (2015). Chinese views and commentary on the 'One Belt, One Road' initiative. *China Leadership Monitor*, *47*, 1–24.

Thrasher, J. (2013). RFID vs. NFC: What's the difference? *RFID Insider*. Available at http://blog.atlasrfidstore.com/rfid-vs-nfc

To, W. M., & Lai, L. S. (2014). Mobile banking and payment in China. *IT Professional*, *16*(3), 22–27. doi:10.1109/MITP.2014.35

Warschauer, M. (2004). *Technology and social inclusion: Rethinking the digital divide*. MIT press.

Weatherford, J. (2009). *The history of money*. Crown Business.

Wee, W. (2012). China's Alipay has 700 million registered accounts, beats PayPal? *Tech in Asia*. Available at https://www.techinasia.com/chinas-alipay-700-million-registered-accounts-beatspaypal

Zhou, T. (2014). An empirical examination of initial trust in mobile payment. *Wireless Personal Communications*, *77*(2), 1519–1531. doi:10.100711277-013-1596-8

Chapter 8
Factors for Resistance to the Use of Mobile Banking:
A Study on the Resistance of Brazilian Internet Users Aged 45 Years or Older

Juliana Yamaguchi Neves da Rocha
Mackenzie Presbyterian University, Brazil

Valéria Farinazzo Martins
Mackenzie Presbyterian University, Brazil

ABSTRACT

In spite of the great potential for the development of mobile banking in Brazil, since the banking index reaches more than half of the population and the number of internet users is even higher, this potential is not evenly distributed among the age groups in the country. Taking into account the tendency to resist new technologies as one ages, this chapter aimed to identify the factors that lead the Brazilian population aged 45 years or over to use the internet and, within this spectrum, identify the barriers to the adoption of mobile banking technology. A questionnaire was applied and 113 responses were analyzed and categorized between functional and psychological aspects in these barriers. This chapter presents the results of this research.

INTRODUCTION

In the last decade, most consumers have adopted new mobile devices at a very fast pace. The Internet made it possible to access personal and financial information, gave access to communication interactively through social networks and made

DOI: 10.4018/978-1-5225-5270-3.ch008

life more efficient overall. The technology of mobile devices (cell phones, PDAs, smartphones) not only provides voice or text messages, but also multimedia files, navigation services (GPS) and even financial transactions such as e-commerce and banking (CRUZ *et al.*, 2010).

Mobile phones are rapidly replacing computers. Sales of smartphones – handsets with an operating system, such as computers – surpassed for the first time the number of "common" mobile phones sold in the first quarter of 2013 in Brazil, accounting for 54% of the total in the country (TELECO, 2016a). The transformation of mobile phones into computers is just one part of the technology convergence phenomenon, which also involves video, audio, among others. The challenge now is to deliver services in line with the perception of value and consumers' confidence.

Mobile banking, also known as m-banking, is the evolution of Internet banking, passing on financial transactions through a computer to through a mobile device. Although ATMs (Automatic Teller Machines), call centers and Internet banking can offer effective delivery established by retail and microfinance banks in the world, according to Safeena *et al.* (2012), m-banking brings significant effects on the market. This is related to the fact that m-banking provides a relatively large number of banking operations, with the advantage of immediate access with the independence of the consumer on the wireless networks or Internet providers, since they are used through the telephony channels.

According to Shaikh (2013) and Shaikh and Karjaluoto (2015), the expanded uses of smartphones in many developed and developing countries have instigated the emergence of new m-banking services, demanding that many more financial institutions offer this type of service together with their sets of products to extend their client reach (including unbanked populations), to retain customers and to increase market share.

On the other hand, despite such benefits mentioned above, the use of mobile phones to conduct banking transactions, such as payment or access to financial information is not as widespread as might be expected (Kleijnen; De Ruyter; Wetzels, 2007; Dineshwar; Steven, 2013, Shaikh; Karjaluoto, 2015). It seems, therefore, the mobile use for banking purposes has not yet reached a maturity level (CRUZ *et al*, 2010).

The strategy to boost the adoption of m-banking was to enable the banking of the low-income segment of the population, initiated in North America and, more recently, followed in Brazil, rather than merely complementing the channels already available. In addition, the Brazilian government has also acted in this direction, achieving in 2013 the approval by the Chamber of Deputies and the Senate the Provisional Measure 615/2013 which deals with the implementation of mobile payments.

According to the Brazilian Federation of Banks (Febrabran, 2017), the Brazilian banking system reached 60% of the population in 2016, meaning a great potential for the development of m-banking in Brazil (CRUZ *et al*, 2010). Considering that

Brazilian Internet users totaled 83 million people in 2013 – that is, 46.5% of the urban population over 10 years old (Telecon, 2016b) – it is possible to imagine that the process of banking can be accelerated by m-banking. However, at the current pace, it is estimated that Brazil reaches a level of banking of more than 90% only in 2023 (Febrabran, 2017). Besides that, Internet usage varies by age group: people aged 45 years or older – population range that was born before the invention of Personal Computer (PC), and therefore have to adapt to technology – represent 31.45% of the urban population over 10 years of age. However, only 21.8% of the mentioned group declared themselves to be Internet users (data based on IBGE (2010) and Teleco (2016c)). IBGE (2010) estimates show that the Brazilian male population from 45 years old adds more than 27.97 million people. In the same estimate, the Brazilian female population within the same age group surpasses 31 million people.

Considering the complexity of mobile technology and the variety of services being offered, this chapter seeks to contribute to the m-banking literature by conducting a research, in a Brazilian context, about determining factors for resistance to the m-banking use by Brazilian Internet users aged 45 years or older.

This chapter is organized as follows. In the Background section, the definition of m-banking is presented, as well as the adoption and use of m-banking and also the factors which contribute to the non-use of m-banking. In the following section, the characteristics of the research done are presented through the Materials and Methods. Continuing, the Results and Discussions section is raised. Finally, this chapter draws the conclusions of the research.

BACKGROUND

Definition of Mobile Banking

Also known as mobile payment, m-banking provides an interaction channel between client and bank, bringing together a set of financial operations provided to the former through a mobile phone with WAP (Wireless Application Protocol), which serves as a payment tool (Laukkanen, 2007; Cruz, 2008).

M-banking services allow customers to conduct banking transactions without temporal and spatial constraints and connect banking services conveniently, easily and quickly using mobile devices. The device is the means to interact with banking applications and the communications network is the way to send/receive information and transactions to/from the bank (Laukkanen, 2007), (Baptista, 2015; Oliveira, 2015). Low fees, time savings, ease of use of the service, speed of service delivery, convenience and compatibility with lifestyle and freedom from time and place are the best advantages of this technology (Laukkanen, 2007), (Lu *et al.*, 2015).

It is also possible to say that m-banking improves customer service, enhances the company's modernity and further elevates the brand image, offering convenience and cost savings for both parties (Laukkanen, 2007).

Initially, m-banking was made available through the exchange of short message services (SMS) between client and bank with limited possibility to execute banking transactions (Zhou, 2012). After that, with the arrival of the WAP, it was possible to use a browser to perform a wider set of operations through the Internet. More recently, software applications (apps) have emerged with smartphones and provide the great majority of banking services (Lu *et al.*, 2015). In this way, banks from several countries did not take long in developing their own financial apps to inform or provide their services to users (Domingos, 2012).

With the development of information technologies, human presence has been less and less necessary in the context of financial transactions. Among these technologies, we can highlight electronic banking, also known as e-banking, home banking, Internet banking, web banking and online banking (Ferreira, 2010). E-banking brings together several banking services that previously required physical presence in banks and are now offered through electronic or digital access.

Originally, the electronic websites of banks on the Internet were exclusively informational, being used only for the dissemination of their services and products (Ferreira, 2010; Rahi, Ghani, & Alnaser, 2017). With the advancement of technology, organizations have gradually adopted this way to transform it into "virtual agencies", allowing their clients to carry out various banking operations electronically, such as consultation of transactions, balance account, payment of services, investments and transfers.

M-banking is an evolution from e-banking and the two of them are different in several aspects (Cruz *et al.*, 2010). Comparatively, financial transactions are computer-centered on e-banking, while, on m-banking, they are user-centered, due to the mobility that their electronic device provides (Lu *et al.*, 2015; Cruz *et al.*, 2010). This device can be either a smartphone or a tablet or PDA (Personal Digital Assistant).

Adoption and Use of Mobile Banking

In the last decades, mobile services have developed with great speed (Shaikh & Karjaluoto, 2015). In addition to the customers' lack of time to the contact and performance of service offered, it is possible to observe that m-banking is a promising and growing area (Lu *et al*, 2015), even though the utilization rates are below expected (Kim, Ferrin, & Rao, 2008). For this reason, it is fundamentally necessary for these systems to mature so that the m-banking market can progress (Cruz *et al.*, 2010).

With the spread of electronic devices around the world, not excluding the poorest areas, it is possible to use them not only as telecommunications networks, but also in an economically beneficial way: to access financial services. Considering this fact, and for example the African scenario, in which only 15 to 20% of households have a formal bank account (Comninos *et al.*, 2008), it is evident the inclusion of the most disadvantaged populations in the formal economy that the mobile services can provide (Cruz *et al.*, 2010). When checking countries from continents with better economic conditions, such as Brazil, there are, on average, four people out of ten without a bank account (Cruz *et al.*, 2010). Also taking into account the fact that there are considerably more mobile phones connected to the Internet than computers, it is possible to glimpse the enormous potential of the development of m-banking in the country (IBGE, 2010).

However, the low growth of m-banking adoption due to its low investment is not unique to the Brazilian market – the same is true, contradictorily, in countries known for their technologically advanced markets, such as Finland (Cruz *et al.*, 2010).

In order to have the m-banking service successfully implemented, all stakeholders must have their interests met. Thus, not only end users but also merchants, financial institutions, mobile operators, credit card administrators and manufacturers of the necessary equipment should be considered. Nevertheless, the complexity of the theme is even greater if different aspects, such as social, technological and political, are taking into account (Akhlaq & Ahmed, 2013).

M-banking operations offer benefits to the most diverse parties, without focusing solely on the consumer. For mobile operators, the service can increase traffic and generate extra revenue from billing services and sales commissions. For credit card administrators, adopting technology can mean survival rather than innovation. For banks, it may be a matter of advantage ahead of their competitors. As for mobile device manufacturers, the opportunity arises to produce equipment with favorable and supportive features for the execution of m-banking operations. Thus, it is clear the possible attractions for the different involved in the service adoption (Akhlaq & Ahmed, 2013).

From the customer's point of view, there are several pieces of research that try to explain the factors which lead them to adopt new information technologies. Some point out that the most propitious individuals to adopt new technologies are mostly based on the young male profile with advanced education and high income (Christensen, 2013). Others prefer to disregard socioeconomic variables, concluding that factors such as relative advantage, compatibility and perceived complexity are the main determinants of m-banking use (Domingos, 2012).

Barriers to the Adoption and Use of Mobile Banking

If, on one hand, the use of mobile devices as a means of performing banking operations presents a huge perspective, on the other hand, this can only occur on a large scale if certain conditions are met. The simple fact of offering services with advanced technology is not enough to guarantee their success since a common point with the users' interests is needed (Laukkanen, 2007) (Lu *et al.*, 2015).

The literature identifies several factors that hinder the adoption of mobile technologies to perform banking transactions. For Rao and Troshani (2007), the main resistance is the perceived financial cost. According to Sohail and Al-Jabri (2014), users' lack of understanding of the benefits of using technology is one of the main barriers. Kuisma, Laukkanen and Hiltunen (2007) argue that there are functional and psychological barriers to the adoption of new technologies: functional barriers include barriers of use, value and risk, while those of a psychological nature are linked to tradition and image. For Laukkanen (2007), the users are very sensitive to the potential invasion of their privacy, being necessary to predict and to divulge what type of safeguards are adopted for its preservation.

Cruz *et al.* (2010) argue that resistance to innovation is related to the socio-demographic characteristics of individuals, such as gender, age, monthly income and educational level. Many others also consider that the limitations of mobile devices compared to computers are one of the main barriers to using m-banking (Cristino, 2012). These limitations can be found in screen size, keyboard keys size and even processor capacity (Sripalawat *et al.*, 2011).

Another critical factor is information security. The lack of experience of the population, in general, in using mobile devices – especially smartphones – for financial transactions, has contributed to the increase of frauds (Shaikh & Karjaluoto, 2015). There is still no culture of caution regarding the use of access passwords, causing many people to adopt passwords that can easily be deduced by a fraudster. Due to negligence or naivety, there are also inadvertent installations of malicious programs (viruses) specially developed for these devices. Mobile phone theft – which has increased more than 60% in Brazil between 2015 and 2016 (TELECO, 2016a), following the pace of its adoption as an important form of communication – or even the simple loan of the device to third parties can also be sources of fraud.

Related to the lack of user experience is their lack of information. Both the ignorance of the existence of the m-banking tool as well as its operation and its benefits present themselves as barriers to its adoption. For these reasons, it would be interesting if banks took a more active position in regards to the dissemination of information about their services, including any kind of doubts their customers may have and comparisons with other access channels (Cristino, 2012).

However, even if the offer of such tool presents several attractions for users, they will show resistance if they have not yet built trust on the mobile device. Because of virtuality, anonymity and physical and temporal separation, Internet banking transactions have both perceived and uncertain risk. Added to this, there is also the risk that the operations by mobile device present. Like any kind of trust, it also needs to be built and maintained (Zhou, 2012). Therefore, Xin, Techatassanasoontorn and Tan (2015) divided it into initial trust and continuous trust.

Nonetheless, the other blocking factors affect user confidence both in their construction and maintenance and are closely correlated. Aesthetics has an influence on the reliability of the cellular system in the questions of ease of use, usefulness and customization, as pointed out by Yuan et al. (2016). On the other hand, Zhou (2012) states that trust interferes significantly in user satisfaction and loyalty, while Domingos (2012) concludes that individual characteristics such as willingness to trust and self-efficacy also affect the trust factor positively.

Laukkanen et al. (2007) and Kuisma, Laukkanen and Hiltunen (2007) also determined another type of risk: perceived consumer risk. Such risk refers to the individual perception of each consumer regarding possible damages or losses and, thus, negatively affects the intention to purchase (Kim, Ferrin, & Rao, 2008). For Kuisma, Laukkanen and Hiltunen (2007), risk can be divided into physical, economic, functional and social risks: physical risk is related to the users or to their assets, while the economic risk is represented by the hasty decision to adopt an innovation rather than waiting for a better and more economical solution. Functional risk refers to the inability to use innovation properly and social risk, to the fear of the adopter being viewed negatively by those close to him, linked to the use of innovation.

MATERIALS AND METHODS

Based on the research question "what are the determining factors for the resistance to the use of m-banking by Internet users aged 45 years or older?", it was considered more appropriate to use the qualitative research methodology, in order to allow the understanding of the social and human phenomena studied (Flick, 2014) from the point of view of Internet users aged 45 years or older, since the best way to reach reality is the one that enables the researcher to 'put themselves in the role of the other', seeing the world through the eyes of those surveyed (Taylor, Bogdan, & Devault, 2015; Silverman; 2016). Qualitative research allows the context where the Internet user is to be a direct source of data, having the researcher as the main instrument (Taylor, Bogdan, & Devault, 2015; Silverman, 2016). This chapter will have an exploratory character, as defined by Sekaran and Bougie (2016), to give the researcher initial knowledge about the phenomenon, opening up opportunities for the latter to be studied in depth and in a more structured way in the future.

Data were collected from primary and secondary sources. Primary sources are the interviewees who fill out structured questionnaires designed specifically for this purpose. Secondary sources are sources other than those from the actors of the phenomenon in question, such as IBGE, 2010 and TELECO, 2013a, 2013b, 2013c. The questionnaire was designed based on the definitions of m-banking adoption factors (functional and psychological) by Laukkanen, et al. (2007) and Kuisma, Laukkanen and Hiltunen (2007). It was developed electronically, via Google Forms (Appendix), and an Internet access address was made available to the participants for access and completion. A pilot was made with few participants, which provided opportunity for improvement of the clarity of questionnaire questions. The questionnaire consists of closed-ended multiple-choice questions and open-ended questions. All questions are mandatory, except for the last one, which requests additional, optional comments from participants. Potential participants were invited to participate in the survey through contact on the social networks Facebook and LinkedIn, chosen for providing rapid diffusion at a very low cost. In addition, being users of social networks, the potential participants would be in line with the profile of Internet users – a basic requirement to participate in the research.

The analysis and interpretation of data were made using categorization techniques for content analysis, as proposed by Neuendorf (2016), Denzin and Lincoln (2008), Riff, Lacy, and Fico, (2014). According to the barrier criteria for adoption and m-banking use from the previous section, Tables 1 and 2 were prepared. Based on definitions by Laukkanen, et al. (2007) and Kuisma, Laukkanen and Hiltunen (2007), two categories were created to analyze the determinants of the resistance of this population segment to the use of m-banking, as well as the factors that facilitate adopting this technology: the first one is functional factors, which are the aspects related to the user experience with the services available via smartphone. It is divided into the subcategories use (ease of use, features, etc.), value (cost) and

Table 1. Relationship among categories of factors which facilitate adoption

Functional			Psychological	
Use	**Value**	**Risk**	**Tradition**	**Image**
Ease of use Breadth of options Speed (time consumption) Customizability Usefulness	Price (of transaction)	High amount of passwords Security	Banking company (with long market history)	Good aesthetics of the app State-of-the-art technology

Source: Prepared by the authors.

risk (vulnerability). The second one is the category of psychological factors, related to the emotional aspects of the user linked to the use of the services. This one, in turn, is divided into the subcategories tradition (trust generated by the perception of the institutional solidity of the provider) and image (aesthetics or technological perception). Table 1 presents the first category, while Table 2 presents the second one.

The complete questionnaire used for the research can be found under Appendix.

RESULTS AND DISCUSSION

The survey was provided for a universe of participants in several locations in Brazil, being available for 30 days and obtaining the return of 113 valid answers, which constitute a non-probabilistic sample by rational choice. Table 3 lists the demographic information of the sample.

Although Brazil has more than 200 million people and half of them have access to the Internet, the more the age increases, the smaller the number of Internet users. From the age of 50, the percentage of those who do not use the Internet dominates the Brazilian scene. Between 51 and 60 years, approximately 10% of this population uses the Internet; from 60 years or more, this percentage drops to approximately 4% according to PNAD (Pesquisa Nacional por Amostra de Domicílios, 2014). Another important point in this research, in regards to the size of the sample, refers to the resistance of this population to answer questionnaires, especially when not having gratifications.

Table 2. Relationship among categories of factors which hinder adoption

Functional			Psychological	
Use	**Value**	**Risk**	**Tradition**	**Image**
I do not know how to conduct transactions through mobile phone I did not know I could conduct transactions through mobile phone Lack of options (if compared to other means)	Mobile phone price (device) Mobile Internet price (service)	Mobile phones are not reliable If I lose my mobile phone, someone will be able to access my account My computer is more secure than my mobile phone My mobile phone is shared	I feel more secure in a bank branch to conduct my transactions	Banking company (is not perceived as having reliable technology)

Source: Prepared by the authors.

From the age range specified for the research, from 45 to 80 years or older, it was verified that the great majority of participants are in the first half of the spectrum, from 45 to 59 years of age (86.6%). In other words, it is the range which involves the youngest in the sample. The demographic information also shows that most of the participants (92.9%) have a high level of education (incomplete or complete high education) and high income (73.5% earn from R$5,100 per month).

The participants declared themselves to be regular Internet users, with 95.6% accessing the network at least once a day and the remaining 4.4%, at least once a week.

Most of the participants reported having smartphones (86.4%), which confirms the trend revealed by TELECO (2013a), but not all of them use their smartphones to perform their activities on the Internet. Figure 1 shows the main activities conducted by the participants over the Internet (regardless the device used).

Adoption and Use of Mobile Banking

Most of the participants (91 individuals, 80.5%) stated that they perform banking or financial activities over the Internet. Of these, 67% (61 individuals) use the smartphone for this purpose, being 54 of them (59.3% of the total) in the age group from 45 to 59 years old, with the predominance of men (30 men and 24 women). There is a certain variation in the frequency of m-banking usage, as shown in Figure 2.

The use of m-banking is very intense by the participants of the survey, in which 80.3% of the participants use the services daily or weekly, although they declare that they also use other means to perform their banking or financial activities, such as bank branch (32 people, 52.5%), ATMs outside bank branches (40 people, 65.6%), desktop (51 people, 83.6%) and tablet (11 people, 18%). On the other hand, only 3.3% of participants (2 people) perform these activities exclusively via smartphone.

There are several reasons why the respondents use m-banking. Figure 3 summarizes these reasons. The reasons for adopting m-banking were divided into the categories created to analyze the factors for the resistance of the population aged 45 or older to the use of m-banking. Table 4 shows the relationship between the categories linked to the adoption of m-banking and the survey questions.

The analysis of the reasons stated by the participants from the perspective of the categories by Laukkanen, et al. (2007) and Kuisma, Laukkanen and Hiltunen (2007) shows that m-banking users are strongly motivated by the functional aspects (83.8% of the answers), with the most important factor being related to the features and eases of use (68.2% of the answers), followed by the aspects linked to risk (14.7% of the answers). On the other hand, the psychological factors have a relatively small weight (16.1% of the answers), in which the impact of the image generated by the perception of state-of-the-art technology used for m-banking services and the impact of the aesthetics of the app are the most relevant (12.3% of the answers). The provider's

Table 3. Demographic information of participants

	Option	Participants	%
Gender	Male	59	52,2
	Female	54	47,8
	Not Informed	0	0
Age	45-49 years old	37	32,7
	50-54 years old	24	21,2
	55-59 years old	37	32,7
	60-64 years old	8	7,1
	65-69 years old	4	3,5
	70-74 years old	1	0,9
	75-79 years old	2	1,8
	80 years or older	0	0
Education	Incomplete elementary school	1	0,9
	Complete elementary school	0	0
	Incomplete high school	2	1,8
	Complete high school	5	4,4
	Incomplete higher education	11	9,7
	Complete higher education	94	83,2
Monthly Income	From US$0 to US$325	4	3,5
	From US$325 to US$750	6	5,3
	From US$750 to US$1075	20	17,7
	From US$1075 to US$ 3250	29	25,7
	Above US$ 3251	54	47,8

Source: Prepared by the authors

tradition – the bank – had a smaller weight, with 3.8% of the participants. Figure 4 summarizes this analysis.

Barriers to the Adoption and to the Use of Mobile Banking Among Internet Users Who Do Not Perform Banking or Financial Activities Over the Internet

Only a small minority (22 people; 19,5%) stated that they do not perform banking or financial activities over the Internet. In this small group, 19 of them said they have own smartphones (86.4%). Despite having the device to use m-banking, they

Figure 1. Activities of Internet users in the network
Source: Prepared by the authors.

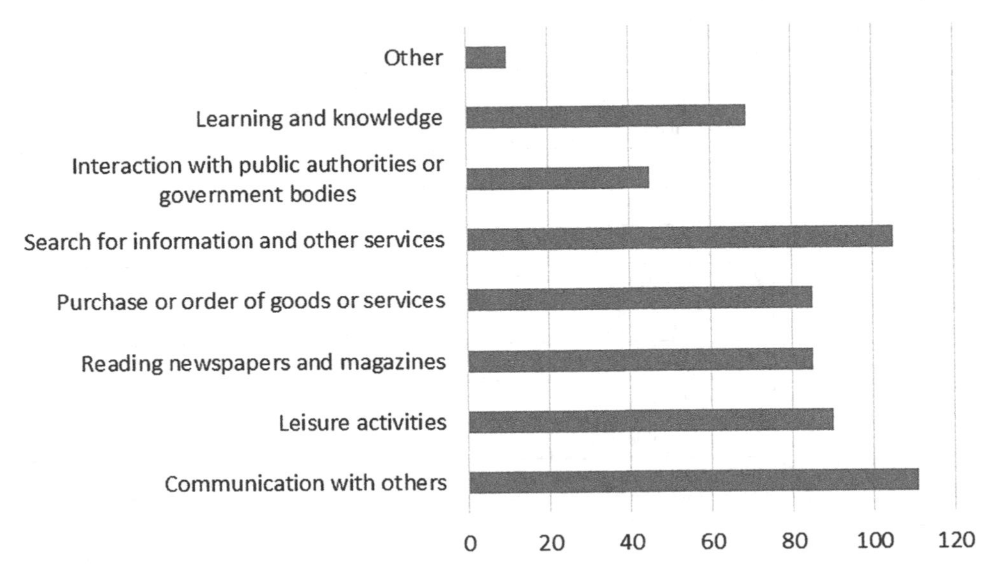

Figure 2. Frequency of performing mobile banking activities
Source: Prepared by the authors.

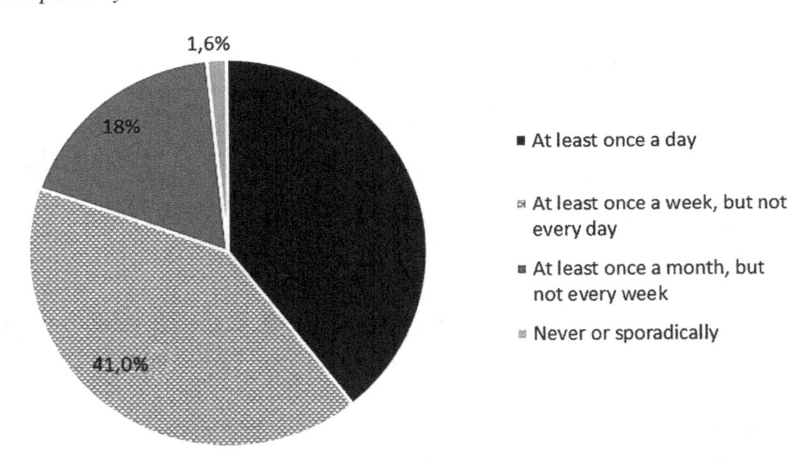

pointed out several reasons why they avoid these services. Figure 5 summarizes these reasons.

When categorizing the answers of this group of participants, it is observable that the latter, like those who adopt m-banking, are also strongly motivated by the functional aspects (71 answers, 96.1%). However, the lack of security perceived

Figure 3. Reasons for adopting mobile banking
Source: Prepared by the authors.

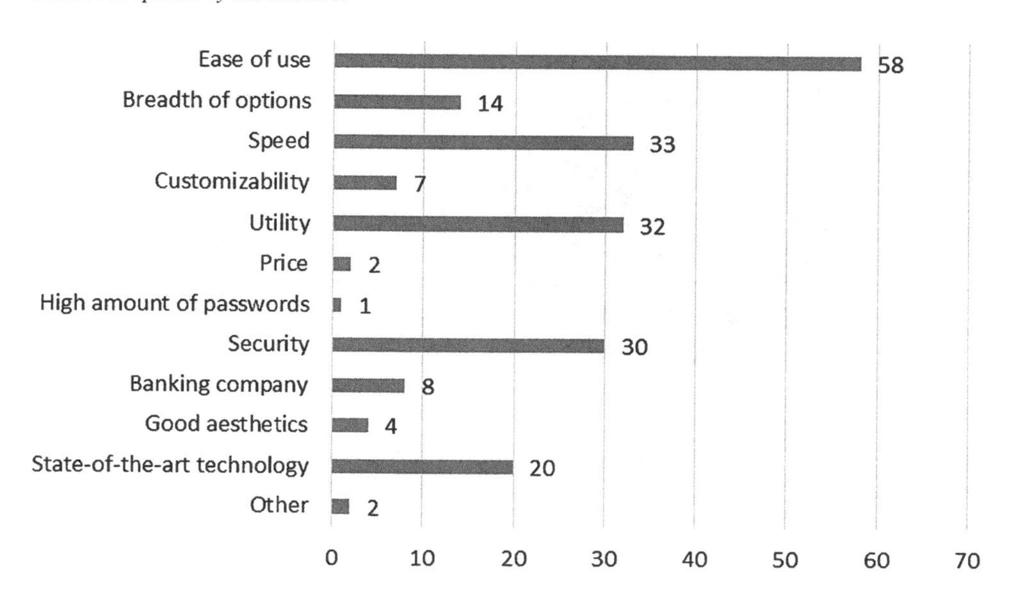

Table 4. Relation among categories of factors which facilitate adoption

Functional			Psychological	
Use	**Value**	**Risk**	**Tradition**	**Image**
Ease of use	Price (of transaction)	High amount of passwords	Banking company (with long market history)	Good aesthetics of the app
Breadth of options		Security		State-of-the-art technology
Speed (time consumption)				
Customizability				
Usefulness				

Source: Prepared by the authors.

in the use of the smartphone, that is, the risk in using it is the strongest reason to avoid its use (49 answers, 55.7%). Part of these participants avoids using banking or financial services via smartphone due to the value (high cost) of both the device and the data plan of their mobile carriers (21 answers, 13.6%). Finally, there is the question related to the use (10 answers, 11.4%), in which a small part of this group claimed to be unaware of the existence of m-banking apps or how to use them.

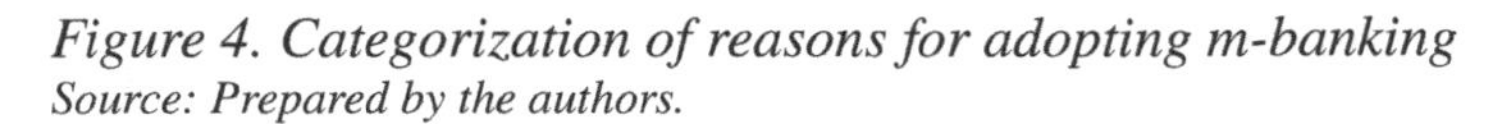

Figure 4. Categorization of reasons for adopting m-banking
Source: Prepared by the authors.

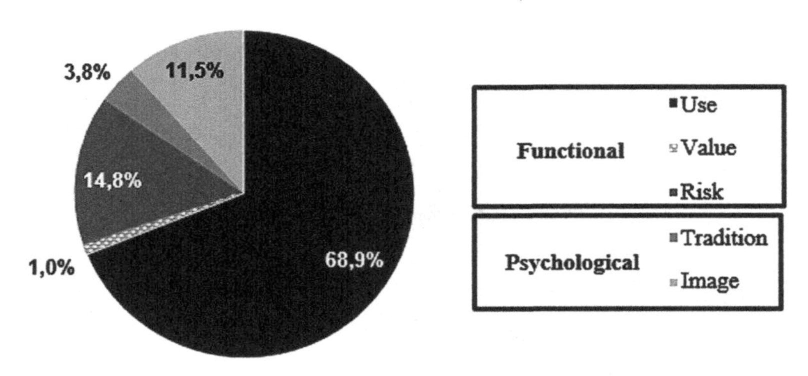

Figure 5. Reasons to avoid the adoption of mobile banking among Internet users who do not perform banking or financial activities over the Internet
Source: Prepared by the authors.

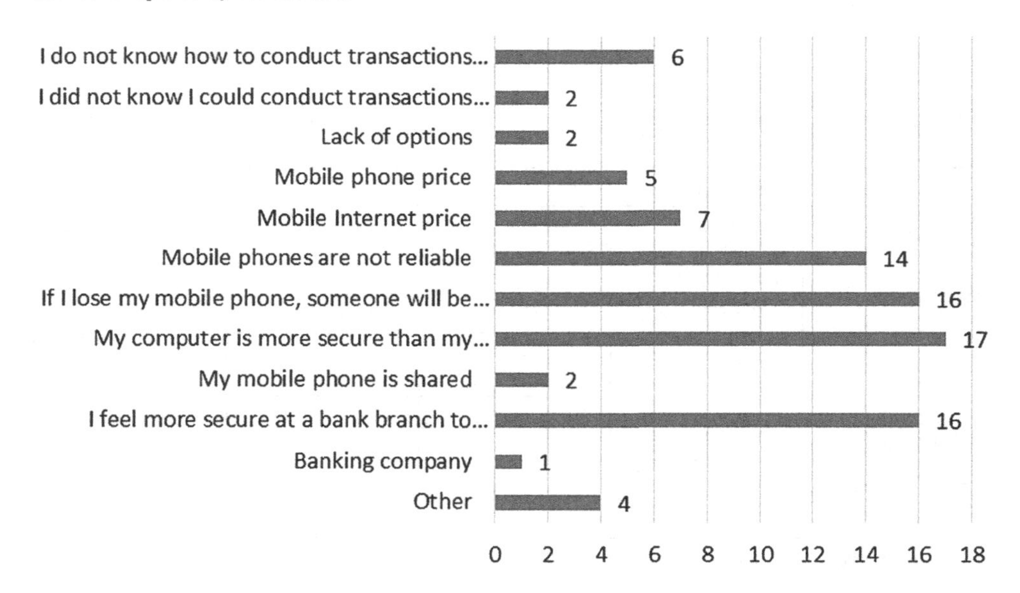

Regarding the psychological factors, only one of the participants in this group says to be influenced by the image generated by the perception of the state-of-the-art technology used for m-banking services and by the aesthetics of the app. However, the transition is an important burden as a barrier to the adoption of m-banking (16 answers, 18.2%). In this respect, the interpretation of the answers was made not only in relation to the bank reputation (market history) and the relationship with its customers, but also in terms of the participants' habit to go to bank branches to conduct financial transactions instead of doing it electronically. Table 5 shows the relationship among the categories related to the barriers to the m-banking adoption and the survey questions. Figure 6 summarizes this analysis.

Barriers to the Adoption and to the Use of Mobile Banking Among Internet Users Who Perform Banking or Financial Activities Over the Internet

Out of the participants who declared performing banking or financial activities over the Internet, almost a third (29 people, 32.2%) reported they do not use their smartphone for this purpose. This group, when on the Internet, conducts their transactions via computer (25 people, 96.2%) or tablet (5 people, 17.2%). Outside the Internet, this group attends bank branches (17 people, 58.6%) and ATMs (15 people, 51.7%). The participants pointed out their reasons to avoid the use of m-banking. Figure 7 summarizes these reasons.

Table 5. Relationship among categories of factors which hinder adoption

Functional			Psychological	
Use	**Value**	**Risk**	**Tradition**	**Image**
I do not know how to conduct transactions through mobile phone	Mobile phone price (device)	Mobile phones are not reliable	I feel more secure at a bank branch to conduct my transactions	Banking company (is not perceived as having reliable technology)
I did not know I could conduct transactions through mobile phone	Mobile Internet price (service)	If I lose my mobile phone, someone will be able to access my account		
Lack of options (if compared to other means)		My computer is more secure than my mobile phone		
		My mobile phone is shared		

Source: Prepared by the authors.

Figure 6. Categorization of reasons for avoiding the adoption of mobile banking among Internet users who do not perform banking or financial activities over the Internet
Source: Prepared by the authors.

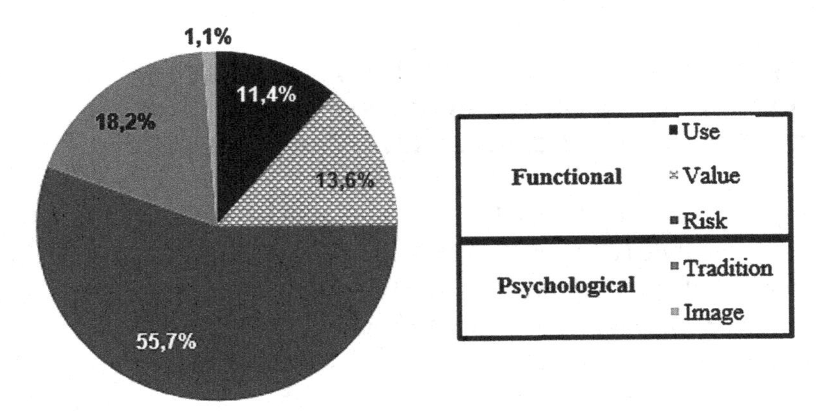

Figure 7. Reasons to avoid the adoption of mobile banking among Internet users who perform banking or financial activities over the Internet
Source: Prepared by the authors.

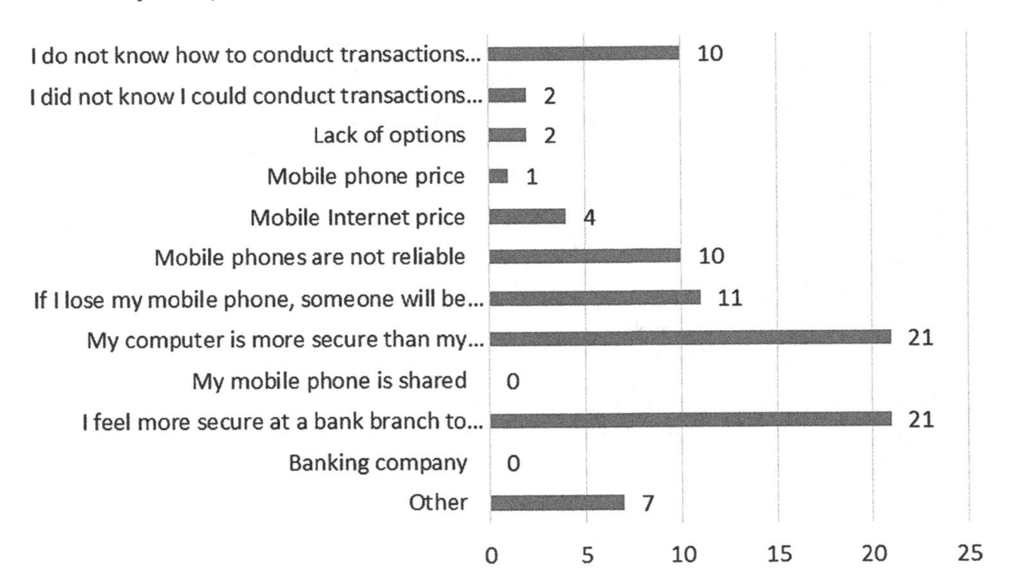

Based on the categorization of the reasons which lead the participants not to use m-banking (see Table 5), it is possible to notice a certain similarity among this group and the Internet users who do not perform banking or financial activities over the Internet at all. The functional aspects influence both groups. In the former group, 61 answers (74.4% out of the total) had the risk of using the smartphone as the main reason to avoid accessing m-banking (42 answers, 51.2% out of the total). But unlike the group that does not perform banking or financial activities over the Internet, this same group points out that the factors related to the subcategory use, namely the lack of knowledge about the existence of m-banking apps or how to use them, obtained 14 answers (17.1%). The factors related to value (cost) had a relatively small weight (5 answers, 6.1%).

In psychological factors, there is again an alignment between the two groups with respect to tradition. The Internet users who conduct electronic transactions indicated a representative sense of security in making bank branches transactions instead of doing it through the smartphone (21 answers, 25.6%). However, in this group, none of the participants said to be influenced by the image generated by the perception of the state-of-the-art technology used for m-banking services and the aesthetics of the app. Figure 8 summarizes this analysis.

Figure 8. Categorization of the reasons to avoid the adoption of mobile banking among Internet users who perform banking or financial activities over the Internet
Source: Prepared by the author.

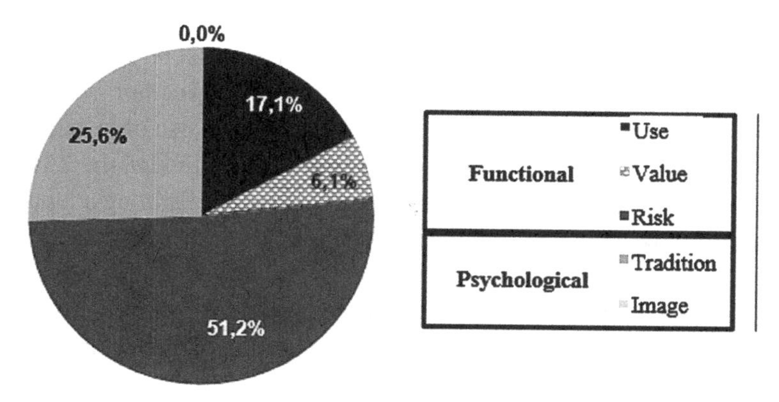

CONCLUSION

This chapter has provided a study on the determinants of resistance to the use of mobile banking by Brazilian Internet users who are, at least, 45 years old. Based on the fact that smartphone sales exceeded the number of mobile phones sold still in the first four months of 2013 in Brazil (TELECO, 2013a) and on the fact that financial institutions have been increasingly promoting the use of m-banking for convenience and cost reduction for both companies and customers (Cristino, 2012; Domingos, 2012), it is not difficult to infer that the alternative of using smartphones to conduct banking or financial transactions would be easy to adopt in general. However, this adoption faces greater resistance by Internet users as the age increases. This research aimed to identify the determinant factor for the resistance to the use of m-banking by the Internet-using population in the spectrum between 45 and 80 years or older in Brazil.

The analysis of the data collected through the consultation of 113 people through a structured questionnaire indicates that there is basically a balance among those who adopt and those who resist the use of m-banking. There is also a gender balance: women represent 52.5% and men, 47.8% out of the total. In other demographic questions, there is a relative standardization. It was found that 92.9% of the participants have a high education level (incomplete or complete higher education) and have a high income (73.5% earn from R$5,100 per month – about US$1,630). There is also a concentration in a certain age group within the spectrum: it was verified that the great majority of participants are the youngest, from 45 to 59 years of age (86.6%).

Thus, the study could not find support in the literature that socio-demographic characteristics are determinant for the adoption of new technologies or for their rejection. The relative homogeneity of age, education and income would suggest a consistent homogeneity in the adoption or rejection of the use of m-banking, but it could not be proven in this study. From the gender point of view, there is a balance with a small advantage for the female, and there is a balance among adepts and non-adherents of m-banking (with a small advantage for those who adopt it), which could suggest that women are more prone to the adoption of m-banking. However, in the age group from 45 to 59 years, the study showed that there is a male predominance in adoption (30 men and 24 women), thus contradicting this hypothesis.

The study was able to clearly point out the reasons for adoption or rejecting the use of m-banking. Based on the definitions by Kuisma, Laukkanen and Hiltunen (2007), the reasons were classified into two categories: 1) functional factors, which refers to the aspects related to the user experience with m-banking, divided into the subcategories use (ease of use, features, etc.), value (cost) and risk (vulnerability) and 2) psychological factors, with respect to the user's emotional aspects related to the use of the services, also subdivided in tradition (trust in the provider) and image (perception about aesthetics or technology).

The adepts of m-banking represent 54% of the respondents. The m-banking adepts do it primarily because of the good user experience: ease of use, less time consumed with banking transactions (speed) and the perception as a handful resource (usefulness). In regards to risk, the adepts feel secure when conducting transactions via smartphone. The perception of the benefits with the use of m-banking (and not due to the demographic characteristics) is in line with what is established by Domingos (2012) on the determinants of the use of m-banking.

Curiously, the remaining 46% of Internet users who do not use m-banking to conduct banking or financial transactions also refer to the functional aspects. In this group, however, the perception that there is a high risk for transactions via smartphone is preponderant, considering the perception that traditional computers are more secure than smartphones. Data violation via smartphone when it is lost is a big concern along with lack of trust in the smartphone as a secure device. This perception of risk found in the study is in line with the barriers to the adoption of m-banking cited in the literature. The preference for computers over smartphones comes from the perception that the computer provides greater security over information and that it is less limited than the smartphone (Cristino, 2012). The possibility of third party violation of smartphone data is also related to information security and to the possibility of damages and losses, called "physical risk" by Laukkanen, et al. (2007) and Kuisma, Laukkanen and Hiltunen (2007). The lack of trust in smartphones is especially addressed by Zhou (2012).

Side by side with fear of use is the ignorance about its use. Part of the respondents lacked knowledge about how to use the m-banking app, much in line with what is called "functional risk" by Laukkanen *et al.* (2007) and Kuisma, Laukkanen and Hiltunen (2007), meaning "the inability to use innovation properly" (p.8).

On the psychological factors side, Internet users who do not use m-banking rely on the habit of conducting banking or financial transactions at a bank branch, due to the fact that they feel more secure in doing it this (traditional) way.

Based on this study, it is possible to say that a greater adoption of m-banking involves the work of reinforcing the security of information in smartphones, perhaps with the availability of specific antivirus and mechanisms which prevent the violation of data by third parties, besides the wide dissemination to the users so that trust on this type of device can be developed. Another point to emphasize is that m-banking app providers, notably financial institutions, need to give more support to their customers in order to make them able to use m-banking properly and thereby extract the benefits that this medium can provide.

This work had an exploratory character to provide an initial knowledge about the phenomenon. The research agenda can continue in the sense of deepening the knowledge of this phenomenon from the statistical and probabilistic points of view.

As future work, this chapter proposes that other criteria for the adoption or the resistance to the adoption of m-banking by the target audience are analyzed, such as usability evaluation of the main apps of Brazilian m-banking, how much the usability of these applications interfere with its adoption, street safety issues, the decrease in the number of attendants at the bank branches, among other criteria.

ACKNOWLEDGMENT

This research was supported by Mackenzie Presbyterian University.

REFERENCES

Akhlaq, A., & Ahmed, E. (2013). The effect of motivation on trust in the acceptance of internet banking in a low income country. *International Journal of Bank Marketing*, *31*(2), 115–125. doi:10.1108/02652321311298690

Baptista, G., & Oliveira, T. (2015). Understanding mobile banking: The unified theory of acceptance and use of technology combined with cultural moderators. *Computers in Human Behavior*, *50*, 418–430. doi:10.1016/j.chb.2015.04.024

Christensen, C. M. (2013). *The innovator's dilemma: when new technologies cause great firms to fail*. Harvard Business Review Press.

Comninos, A., Esselaar, S., Ndiwalana, A., & Stork, C. (2008). *M-banking the Unbanked*. Academic Press.

Cristino, P. C. N. V. (2012). *Mobile Banking: fatores determinantes na adesão* (Doctoral dissertation). Escola Superior de Comunicação Social.

Cruz, P., Barretto Filgueiras Neto, L., Munoz-Gallego, P., & Laukkanen, T. (2010). Mobile banking rollout in emerging markets: Evidence from Brazil. *International Journal of Bank Marketing*, *28*(5), 342–371. doi:10.1108/02652321011064881

Denzin, N. K., & Lincoln, Y. S. (2008). *Collecting and interpreting qualitative materials*. Los Angeles, CA: Sage.

Dineshwar, R., & Steven, M. (2013, February). An investigation on mobile banking adoption and usage: A case study of Mauritius. In *Proceedings of 3rd Asia-Pacific Business Research Conference* (pp. 25-26). Academic Press.

Domingos, C. R. R. (2012). *Mobile Banking: factores influenciadores da utilização de APPS bancárias* (Doctoral dissertation). Instituto Superior de Economia e Gestão.

Febraban. (n.d.). Retrieved from https://portal.febraban.org.br/

Ferreira, C. E. C. (2010). *Os bancos brasileiros na internet: um estudo de caso sobre os motivadores para a sua forma de atuação neste ambiente* (Doctoral dissertation).

Flick, U. (2014). An introduction to qualitative research. *Sage (Atlanta, Ga.).*

IBGE. (n.d.). *Resultados do Universo do Censo Demográfico 2010*. Retrieved from http://www.ibge.gov.br/home/

Kim, D. J., Ferrin, D. L., & Rao, H. R. (2008). A trust-based consumer decision-making model in electronic commerce: The role of trust, perceived risk, and their antecedents. *Decision Support Systems*, *44*(2), 544–564. doi:10.1016/j.dss.2007.07.001

Kleijnen, M., De Ruyter, K., & Wetzels, M. (2007). An assessment of value creation in mobile service delivery and the moderating role of time consciousness. *Journal of Retailing*, *83*(1), 33–46. doi:10.1016/j.jretai.2006.10.004

Kuisma, T., Laukkanen, T., & Hiltunen, M. (2007). Mapping the reasons for resistance to Internet banking: A means-end approach. *International Journal of Information Management*, *27*(2), 75–85. doi:10.1016/j.ijinfomgt.2006.08.006

Laukkanen, T. (2007). Internet vs mobile banking: Comparing customer value perceptions. *Business Process Management Journal*, *13*(6), 788–797. doi:10.1108/14637150710834550

Laukkanen, T., & Cruz, P. (2008, July). Comparing consumer resistance to mobile banking in Finland and Portugal. In *International Conference on E-Business and Telecommunications* (pp. 89-98). Springer.

Laukkanen, T., Sinkkonen, S., Kivijärvi, M., & Laukkanen, P. (2007). Innovation resistance among mature consumers. *Journal of Consumer Marketing*, *24*(7), 419–427. doi:10.1108/07363760710834834

Lu, M. T., Tzeng, G. H., Cheng, H., & Hsu, C. C. (2015). Exploring mobile banking services for user behavior in intention adoption: Using new hybrid MADM model. *Service Business*, *9*(3), 541–565. doi:10.100711628-014-0239-9

Neuendorf, K. A. (2016). The content analysis guidebook. *Sage (Atlanta, Ga.)*.

Pesquisa Nacional por Amostra de Domicílios. (2014). Available at: https://ww2.ibge.gov.br/home/estatistica/pesquisas/pesquisa_resultados.php?id_pesquisa=40

Rahi, S., Ghani, M. A., & Alnaser, F. M. (2017). The Influence of E-Customer Services and Perceived Value on Brand Loyalty of Banks and Internet Banking Adoption: A Structural Equation Model (SEM). *Journal of Internet Banking and Commerce*, *22*(1), 1–18.

Rao, S., & Troshani, I. (2007). A conceptual framework and propositions for the acceptance of mobile services. *Journal of Theoretical and Applied Electronic Commerce Research*, *2*(2).

Riff, D., Lacy, S., & Fico, F. (2014). *Analyzing media messages: Using quantitative content analysis in research*. Routledge.

Safeena, R., Date, H., & Kammani, A. (2011). Internet Banking Adoption in an Emerging Economy: Indian Consumer's Perspective. *Int. Arab J. e-Technol., 2*(1), 56-64.

Sekaran, U., & Bougie, R. (2016). *Research methods for business: A skill building approach*. John Wiley & Sons.

Shaikh, A. A. (2013). Mobile banking adoption issues in Pakistan and challenges ahead. *J. Inst. Bank. Pak*, *80*(3), 12–15.

Shaikh, A. A., & Karjaluoto, H. (2015). Mobile banking adoption: A literature review. *Telematics and Informatics*, *32*(1), 129–142. doi:10.1016/j.tele.2014.05.003

Silverman, D. (Ed.). (2016). *Qualitative research*. Sage.

Sohail, M. S., & Al-Jabri, I. M. (2014). Attitudes towards mobile banking: Are there any differences between users and non-users? *Behaviour & Information Technology*, *33*(4), 335–344. doi:10.1080/0144929X.2013.763861

Sripalawat, J., Thongmak, M., & Ngramyarn, A. (2011). M-banking in metropolitan Bangkok and a comparison with other countries. *Journal of Computer Information Systems*, *51*(3), 67–76.

Taylor, S. J., Bogdan, R., & DeVault, M. (2015). *Introduction to qualitative research methods: A guidebook and resource*. John Wiley & Sons.

Teleco. (2016a). *Smartphones no Brasil*. Retrieved from www.teleco.com.br/smartphone.asp

Teleco. (2016b). *Perfil dos Usuários de Internet no Brasil*. Retrieved from www.teleco.com.br/internet_usu.asp

Teleco. (2016c). *Internet no Brasil*. Retrieved from www.teleco.com.br/internet.asp

Xin, H., Techatassanasoontorn, A. A., & Tan, F. B. (2015). Antecedents of consumer trust in mobile payment adoption. *Journal of Computer Information Systems*, *55*(4), 1–10. doi:10.1080/08874417.2015.11645781

Yuan, S., Liu, Y., Yao, R., & Liu, J. (2016). An investigation of users' continuance intention towards mobile banking in China. *Information Development*, *32*(1), 20–34. doi:10.1177/0266666914522140

Zhou, T. (2012). Examining mobile banking user adoption from the perspectives of trust and flow experience. *Information Technology Management*, *13*(1), 27–37. doi:10.100710799-011-0111-8

KEY TERMS AND DEFINITIONS

Banking: The business of a bank or a banker.

Barriers to the M-Banking Use: Factors that prevent the use of mobile banking by the population.

Brazilian Banking System: Banking system practiced in Brazil.

Elderly: In Brazil, anyone aged 60 or older is considered elderly.

Functional Aspects: Aspects linked to use, value, and risk of the adoption of new technologies.

Psychological Aspects: Aspects linked to tradition and image of the adoption of new technologies.

Social Impacts: Impacts on society caused by banking or m-banking.

Survey: To query (someone) in order to collect data for the analysis of some aspect of a group or area.

APPENDIX

Consolidation of valid survey answers generated by Google Forms.

Table 6. What is your gender identity?

Female	59	52.2%
Male	54	47.8%
I prefer not to inform it	0	0%

Table 7. How old are you?

45-49 years old	37	32.7%
50-54 years old	24	21.2%
55-59 years old	37	32.7%
60-64 years old	8	7.1%
65-69 years old	4	3.5%
70-74 years old	1	0.9%
75-79 years old	2	1.8%
80 years or older	0	0%

Table 8. What is your level of education?

Incomplete elementary school	1	0.9%
Complete elementary school	0	0%
Incomplete high school	2	1.8%
Complete high school	5	4.4%
Incomplete higher education	11	9.7%
Complete higher education	94	83.2%

Table 9. What is your monthly income?

R$0 to R$1,020	4	3.5%
More than R$1,020 to R$2,040	6	5.3%
More than R$2,040 to R$5,100	20	17.7%
More than R$5,100 to R$10,200	29	25.7%
More than R$10,200	54	47.8%

Table 10. How often do you use the Internet?

At least once a day	108	95.6%
At least once a week, but not every day	5	4.4%
At least once a month, but not every week	0	0%
Never or sporadically	0	0%

Table 11. Which of the following activities do you perform on the Internet?

Communication with others	111	98.2%
Leisure activities	90	79.6%
Reading newspapers and magazines	85	75.2%
Purchase or order of goods or services	85	75.2%
Search for information and other services	105	92.9%
Interaction with public authorities or government bodies	45	39.8%
Learning and knowledge	69	61.1%
Other	10	8.8%

Table 12. Do you perform banking or financial activities over the Internet?

Yes	91	80.5%
No	22	19.5%

Table 13. How often do you perform banking or financial activities over the Internet?

At least once a day	34	37.4%
At least once a week, but not every day	39	42.9%
At least once a month, but not every week	16	17.6%
Never or sporadically	2	2.2%

Table 14. Do you own a smartphone?

Yes	90	98.9%
No	1	1.1%

Table 15. By which means do you perform your banking or financial activities?

Bank branches	15	68.2%
ATMs outside bank branches	16	72.7%

Table 16. Do you own a smartphone?

Yes	19	86.4%
No	3	13.6%

Table 17. Since when do you own a smartphone?

2013
2012
2010
2014
2000
2008
2009
2007
2011
3 years ago
4 years ago
2015
2013+/-
Since the first iPhone
since the release of iphone 3gs
release iphone
Since the release of samsung galaxy note 1
Since 2011
since the iphone was released
since 2010
10 years
+ 3 months
1999
Since 2012/2013
8 or 10 years
Can't recall
Since 2015
Since it was released
1year
three years
About 5 years
2014 May
3 years
Since about 5 years
When the S4 was released
Since the Pt 550
2005
2002
+ or - 8 years
07/2012
2016
2006
Since the iPhone was released
Always
Two years
About 4 years
Since BlackBerry
5 years
6 years
03/2015
2years

Table 18. Which of the following devices do you use more often?

Desktop	21	23.3%
Laptop/notebook	42	46.7%
Tablet (without 3G)	14	15.6%
Mobile phone (smartphone)	74	82.2%

Table 19. Which of the following activities do you perform via smartphone?

Communication with others	89	98.9%
Leisure activities	68	75.6%
Reading newspapers and magazines	51	56.7%
Purchase or order of goods or services	40	44.4%
Search for information or other services	67	74.4%
Interaction with public authorities or government bodies	18	20%
Learning and knowledge	31	34.4%
Other	5	5.6%

Table 20. Do you perform banking or financial activities over the smartphone?

Yes	61	67.8%
No	29	32.2%

Table 21. Since when do you own a smartphone?

2010 2011 2013 2012 Five years 6 years february 2014 Almost 2 years 2015 approximately 10 years ago 5years 1 year march 2010 2014 December 2008

Table 22. Which of the following devices do you use more often?

Desktop	4	21.1%
Laptop/notebook	6	31.6%
Tablet (without 3G)	7	36.8%
Mobile phone (smartphone)	18	94.7%

Table 23. Which of the following activities do you perform via smartphone?

Communication with others	19	100%
Leisure activities	14	73.7%
Reading newspapers and magazines	13	68.4%
Purchase or order of goods or services	3	15.8%
Search for information or other services	15	78.9%
Interaction with public authorities or government bodies	3	15.8%
Learning and knowledge	9	47.4%
Other	0	0%

Table 24. Which of the following devices do you use more often?

Desktop	0	0%
Laptop/notebook	3	75%
Tablet (without 3G)	1	25%

Table 25. If you owned a smartphone, would you use it to conduct banking transactions?

Yes	2	50%
No	2	50%

Table 26. How often do you perform banking or financial activities over the smartphone?

At least once a day	24	39.3%
At least once a week, but not every day	25	41%
At least once a month, but not every week	11	18%
Never or sporadically	1	1.6%

Table 27. Since when do you conduct banking transactions via smartphone?

2014
2015
2010
2013
2012
2016
1 year
End 2015
5 years
Since the first iPhone
since my 1st smartphone
2 years ago
Last year, when my daughter introduced our bank app to me
When I bought the smartphone
+ 3 months
14-05-2015
Since I have a smartphone
2010
2000
2 years
2008
1year
the beginning of the month
About 3 years ago
Since the PT 550
2011
3 years
07/2014
For over 10 years
Since it has emerged
3 years
2years

Table 28. By which other means do you perform banking or financial activities?

Bank branches	32	52.5%
ATMs outside bank branches	40	65.6%
Computer	51	83.6%
Tablet (without 3G)	11	18%
No other besides the smartphone	2	3.3%

Table 29. For which of the following reasons do you use banking mobile apps?

Good aesthetics of the app	4	6.6%
Ease of use	58	95.1%
Breadth of options	14	23%
High amount of passwords	1	1.6%
Speed	33	54.1%
Banking company	8	13.1%
Customizability	7	11.5%
Price	2	3.3%
Usefulness	32	52.5%
Security	30	49.2%
State-of-the-art technology	20	32.8%
Other	2	3.3%

Table 30. Which of the following aspects do you dislike in a banking mobile app?

Mobile phones are not reliable	5	8.2%
I feel more secure at a bank branch to conduct my transactions	0	0%
If I lose my mobile phone, someone will be able to access my account	11	18%
I do not know how to conduct transactions through mobile phone	0	0%
My computer is more secure than my mobile phone	9	14.8%
I did not know I could conduct transactions through mobile phone	1	1.6%
Banking company	1	1.6%
Lack of options	14	23%
Mobile phone price	9	14.8%
Mobile Internet price	13	21.3%
My mobile phone is shared	0	0%
Other	13	21.3%

Table 31. By which means do you perform banking or financial activities?

Bank branch	17	58.6%
ATMs outside bank branches	15	51.7%
Computer	25	86.2%
Tablet (without 3G)	5	17.2%

Table 32. For which of the following reasons you do NOT use banking mobile apps?

Mobile phones are not reliable	10	20.8%
I feel more secure at a bank branch to conduct my transactions	21	43.8%
If I lose my mobile phone, someone will be able to access my account	11	22.9%
I do not know how to conduct transactions through mobile phone	10	20.8%
My computer is more secure than my mobile phone	21	43.8%
I did not know I could conduct transactions through mobile phone	2	4.2%
Banking company	0	0%
Lack of options	2	4.2%
Mobile phone price	1	2.1%
Mobile Internet price	4	8.3%
My mobile phone is shared	0	0%
Other	7	14.6%

Table 33. Which of the following items would make you use banking mobile apps?

Good aesthetics	0	0%
Ease of use	25	52.1%
Breadth of options	7	14.6%
High amount of passwords	2	4.2%
Speed	14	29.2%
Banking company	3	6.3%
Customizability	5	10.4%
Price	6	12.5%
Usefulness	10	20.8%
Security	36	75%
State-of-the-art technology	8	16.7%
Other	3	6.3%

Table 34. For which of the following reasons would you use a banking mobile app?

Good aesthetics	0	0%
Ease of use	23	46%
Breadth of options	5	10%
High amount of passwords	1	2%
Speed	20	40%
Banking company	4	8%
Customizability	4	8%
Price	4	8%
Usefulness	14	28%
Security	35	70%
State-of-the-art technology	11	22%
Other	4	8%

Table 35. For which of the following aspects you would NOT use a banking mobile app?

Mobile phones are not reliable	14	26.9%
I feel more secure at a bank branch to conduct my transactions	16	30.8%
If I lose my mobile phone, someone will be able to access my account	16	30.8%
I do not know how to conduct transactions through mobile phone	6	11.5%
My computer is more secure than my mobile phone	17	32.7%
I did not know I could conduct transactions through mobile phone	2	3.8%
Banking company	1	1.9%
Lack of options	2	3.8%
Mobile phone price	5	9.6%
Mobile Internet price	7	13.5%
My mobile phone is shared	2	3.8%
Other	4	7.7%

Chapter 9
Monitoring the Physical Activity of Patients Suffering From Peripheral Arterial Disease

Dennis Paulino
INESC TEC, Portugal & University of Trás-os-Montes e Alto Douro, Portugal

Arsénio Reis
INESC TEC, Portugal & University of Trás-os-Montes e Alto Douro, Portugal

Joao Barroso
INESC TEC, Portugal & University of Trás-os-Montes e Alto Douro, Portugal

Hugo Paredes
INESC TEC, Portugal & University of Trás-os-Montes e Alto Douro, Portugal

ABSTRACT

The peripheral arterial disease (PAD) is characterized by leg pain during walking, and a recommended treatment for this disease is to perform supervised physical activity. In this chapter, a system that monitories the physical activity containing one application for smartwatch, one application for smartphone, and a back-end webservice is presented. The applications collect heart rate, GPS locations, step count, and altitude data. The methodology used for the development of the system was based on the agile method with the production of prototypes. In this chapter, four development cycles, which cover the users' and researchers' needs, are presented. In this work, the main objective is to evaluate the current mobile technologies on the physical activity data collection and the development of a system that assists the users to maintain an active life.

DOI: 10.4018/978-1-5225-5270-3.ch009

INTRODUCTION

The main symptom of peripheral arterial disease (PAD) is intermittent claudication, that is characterized by a leg pain throughout walking, which compels the people suffering PAD to stop walking temporarily. These individuals progressively show a diminution on physical aptitude. In Portugal, it is estimated that 3% at 10% of people suffer from PAD (Menezes et al., 2009). The individuals with PAD shows a change in the blood flow at the arterials of the inferior members, which is caused by blockages in the arteries. In extreme situations, this disease can cause wounds and the necessity of amputations. A recommended treatment for PAD is to perform supervised physical activity (Gornik & Beckman, 2005). The research conducted by Garg et al. (2006) concluded that people with PAD which perform physical activity regularly had lower mortality rate comparing to people with PAD whom were less active. The control of the beginning of pain, the magnitude, the number of times that the patient needs to stop and how much time the pain takes to disappear, are very important factors for the control of the progression on the intensity of recommended exercise. The exercise program for patients with PAD is applied in a Portuguese hospital, located in Vila Real, Portugal. Some of these patients must get transport from longer distances three times a week.

With the appearance of mobile devices, it is possible through the sensors that come built-in, collect a panoply of data from users like their health status and physical parameters during the performing of exercise. This information is useful for the patients, because they can have a better perception of their fitness status. If this information is also available to the health professionals, it can be used to be part of a support tool to keep tracking of the evolution of their patients. This approach in the application of health information systems, has introduced benefits and rationalities in these processes (Reis et al., 2016b).

This chapter is divided in the following sections: "Related works", presents some studies related to PAD treatment and ambulatory monitorization systems; "Background", supplies a contextual frame of the work and it is presented a comparison on the smartwatches and fitness applications currently available; "Design of proposed system", describes the developed system; "Methodology", describes the method used in the elaboration of user tests and collection of results; "Results", describes the results obtained; "Analysis", analysis made upon the results obtained; "Conclusion", it is summarized the work elaborated; "Future work", it is presented perspectives about the future.

For this work, the objectives are: the construction of a system for the monitorization of physical activity and health data to users that suffer PAD and support the health professionals in keep tracking of the current user´s health and physical status.

RELATED WORKS

In PAD, the recommended treatment is the performing of supervised physical activity, which normally is done in hospital environment. The study conducted by Regenstein et al. (1997), concluded that patients which entered in a hospital exercise program to perform physical activity had better results comparing to patients who perform physical activity unsupervised at home. To avoid the patient going to the hospital to perform supervised exercise, it is necessary to allow the health professionals to supervise the patient´s physical activity at home. There are some studies that approaches this problem, presenting systems that collects health data or physical activity data.

The study conducted by Fokkenrood et al. (2014) was based on a validation of an activity monitor that stores offline the step count and the type of activity performed from the patient (e.g. sitting, resting, locomotion) to a maximum time of seven days. This activity monitor could be useful to help health professionals evaluate the activity performed by the PAD professionals, but has the disadvantage that the patient has the responsibility to send the collected data to the health professional. In this study the target population is patients with PAD.

The study conducted by Anliker et al. (2004), described an advanced care and alert portable telemedical monitor (AMON), which monitories the patient collecting health data like heart rate and blood pressure. In this study the target population is high-risk cardiac/respiratory patients. It also gathered the information from acceleration sensors on the patient current activity (walking, running or resting). This research focused more on the patient´s health information but lacked the physical activity information like counting the number of steps.

The study made by Jovanov et al. (2005), showed a prototype composed of physiological sensors integrated with a wearable Wireless Body Area Network (WBAN), with the objective of collecting physical activity data (e.g. step count) and health data (e.g. electrocardiogram). It is mentioned that the data retrieved could be sent to a medical server for the health professionals keep tracking of current patients´ health status, but in this article was not described the mechanism to send the retrieved data to the medical servers. In this study the target population was ambulatory patients.

The system developed by Butussi and Chittaro (2008), is composed by wearable devices that collects the user´s physical activity and heart rate. It uses context-aware and user-adaptive techniques to motivate and supervise the user, offering an exercise program suitable to the user´s goals. In this study the target population is persons who wants to upkeep a healthy life. This system doesn't have the management feature for the health professionals to keep tracking the user´s current physical activity and health status.

In Table 1, it is made an overview of the many studies mentioned before in this section and that is related on monitories people collecting health or physical activity data.

The study conducted by Anliker et al., (2004) approaches the problem of keeping track of patients´ health and physical activity status in an autonomous way, but didn´t focused on people with PAD. In patients with PAD, there are some parameters important to measure like the distance walked, the elevation on walk (to calculate with more effectiveness the user´s effort in the physical activity) or the user´s feedback about pain level. The studies mentioned before in this section didn´t collected in their developed system the elevation on walk or the user´s feedback about pain.

BACKGROUND

This paper results of previous work made in the scope of project NanoSTIMA (Paulino et al., 2016). The project NanoSTIMA has two main objectives: explore and develop micro and nano technologies for monitorization of health parameters and the promoting of welfare of population.

Currently, there are many technologies and applications that collects several types of health and physical activity data. In this work, it was chosen a smartwatch and smartphone as main devices for the system that will monitor the user in a

Table 1. Overview of relative works

Author	Target Population	Remote Monitorization	Measure Distance	Observations
Fokkenrood et al., 2014	Patients with PAD	Yes, but the patient has to send the data.	Yes	Although this study is focused on people with PAD, the patient is responsible of delivery the collected data.
Anliker et al., 2004	High-risk cardiac/ respiratory patients	Yes	No	This study is focused on remote monitorization of health data, but has the disadvantage of not measuring the distance walked
Jovanov et al., 2005	Ambulatory patients	No	Yes	In the prototype of this article, it can collect health and physical data but doesn't have an autonomous way to send it to the medical servers.
Butussi et al., 2008	Persons who wants to upkeep a healthy life	No	Yes	This study is focused on retrieving physical activity data, but it lacks the feature of keeping updated the health professional about the user status.

continuous way. These devices can accompany the user in their daily lives and has the necessary sensors and interfaces.

The research team of this proposed system has researchers from the area of Informatic (responsible of the proposed system development) and researchers from the area of Sports & Health (coordinates an exercise program in which is made the supervision of physical activity and health parameters in patients with PAD).

Evaluation of Smartwatches

In this work, the following smartwatches were evaluated: Samsung Gear S2 (Samsung, 2017a); Apple Watch (Apple, 2017); Moto 360 Sport (Motorola, 2017); and Garmin VivoActive HR (Garmin, 2017). These smartwatches have the following basic features: Wifi, Bluetooth, heart rate monitor, notifications in real time, and pedometer. The evaluation was made upon the information retrieved in the official website of each device. These devices were analyzed gathering each one their advantages and disadvantages:

- The Samsung Gear S2 it is a smartwatch with an innovative interface with a new rotating bezel that helps controlling the device. One disadvantage is the lack of applications.
- The Apple Watch is a smartwatch with many built-in fitness applications. If the user buys the Dexcom G5 sensor (costs 1308 euros), it can monitor in a continuous way the level of glycoses. It has the disadvantage of not having GPS, barometer and have low autonomy.
- The Moto 360 Sport is a smartwatch with design and specific applications for sports athletes. It has a panoply of sensors, including barometer and GPS. The disadvantage is related with the low autonomy.
- The Garmin VivoActive HR is a smartwatch with many built-in features and suitable for many different sports.

In Table 2, it is shown a summary of the analysis on the smartwatchs, which had excluded the features previously considered as basic feature (Wifi, Bluetooth, heart rate monitor, notifications in real time, and pedometer). The feature "Autonomy" presented in each smartwatch on Table 2 consists on the smartwatch battery life.

From the comparison made on Table 2, it is possible to check that the Garmin VivoActive HR smartwatch has better autonomy comparing to the other devices in the table (7 days of battery life) and is the smartwatch with most features. The Samsung Gear S2 has some interesting features but what is highlighted is the design, which is more intuitive.

Table 2. Overview of the smartwatch analysis (the price shown is in euros and it is rounded to the units)

Smartwatch	Features	Price	Key Factors
Samsung Gear S2	Barometer Autonomy: 2 days	317 €	+Features, Design - Apps
Apple Watch	Autonomy: 1 day	529 €	+ Apps - Autonomy, GPS, Barometer
Moto 360 Sport	GPS, Barometer Autonomy: 1 day	317 €	+Features, Design - Autonomy
Garmin VivoActive HR	GPS, Barometer, Quality of sleep Autonomy: 7 days	264 €	+ Features + Autonomy ± Design

Evaluation of Fitness Applications

The fitness applications chosen for this evaluation were: Nike + Run Club (Nike, 2017); Adidas Train (Adidas, 2017); Google Fit (Google, 2017); and S Health (Samsung, 2017b). These applications were analyzed gathering from each one their features:

- The Nike + Run Club application measures the distance walked, the elevation and the calories burnt. It can be used with Android Wear.
- The Adidas Train application measures the distance walked, the elevation and the calories burnt. It can be used with Android Wear.
- The Google Fit application measures the distance walked, elevation, calories burnt and the sleep quality. It can be used with Android Wear, Nike+, Runkeeper, Strava, MyFitnessPal, Lifesum, Adidas train & run, Basis, Withings and Xiaomi Mi SmartBands. It can register the quality of sleep, nutrition and weight.
- The SHealth application measures the distance walked, elevation, calories burnt and the sleep quality. This application can register several sports and give audio feedback (important with people that has vision limitations).

All these fitness applications allow the integrations with other fitness applications and to share records on the socials networks. In Table 3, it is shown an overview of the features available in the fitness applications previously analyzed.

Table 3. Overview of the features in the fitness applications

App	Compatibility	Distance	Several Sports	Audio Feedback	Sleep Quality	Calories	Maps
Nike +Run	✓	✓	✗	✗	✗	✓	✓
Adidas Train	✓	✓	✓	✓	✗	✓	✓
Google Fit	✓	✓	✓	✗	✓	✓	✓
S Health	✓	✓	✓	✓	✓	✓	✓

Monitorization System

The objective of this work, is to build a monitorization system with two applications: one for smartwatch and other for smartphone. To begin it was necessary to choose the smartwatch and to choose a fitness application to help collect the physical activity data. The smartwatch chosen was the Samsung Gear S2, for his simple design and intuitive interface. The fitness application chosen was the SHealth, that allows the integration with other fitness application, and allows the collection of several types of data related to the physical and health of the user. This system was developed for people with PAD, allowing them to register in the application the level of pain and the monitorization of their physical activity. This system can be used by people without PAD that want to maintain an active life. When building a system that will have some interaction with users, it is important to assess the usability (Abreu et al., 2017; Gonçalves et al., 2017; Reis et al., 2016a). There was a special care to guarantee the accessibility and usability in the system (Reis et al., 2013).

DESIGN OF PROPOSED SYSTEM

To build the proposed system in this chapter it was developed: a smartwatch application (it was selected the operative system Tizen because of the chosen smartwatch Samsung Gear S2), a smartphone application (operative system Android) and a back-end webservice in PHP.

The application developed for smartwatch had two objectives: (1) Collect heart rate, barometer, and step count from user; (2) Send the collected data to the smartphone application.

The application developed for smartphone had eight objectives: (1) Collect the data from smartwatch; (2) Collect the GPS data; (3) Integrate the developed application with the previously chosen fitness application; (4) Show the calendarization of tasks created at Google Calendar, with the aim to schedule objectives for user to do physical activity; (5) Save the data collected in the smartphone and when the network connectivity is suitable, synchronized the saved data with the webservice; (6) Implement a mechanism for the users notify a predefined contact in case of an emergency, sending their last known location; (7) Collect the feedback from users with PAD about his current pain level; (8) Monitories walks, with the implementation of chronometers.

The back-end webservice had two objectives: (1) make available a database that allows the saving of the collected data by the smartphone application; (2) make available features, using REST architecture, to make the webservice more interoperable, becoming easier the CRUD (Create, Read, Update, Delete) operations over the database.

In Figure 1, it is shown the activity diagram, using Unified Modeling Language (UML), which illustrate the functioning of the principal module from the main application for smartphone. When the application starts, first it will check if the GPS is active for requesting GPS update (receiving the current location). At the same time, the application is ready to receive the data from the developed smartwatch application, containing mainly the heart rate from the user. If the user starts walking with the smartphone application launched, it will start receiving data send from the S Health application, containing the information related with the step count. In each 5 minutes, the application will synchronize with the webservice if it has new data collected and if the smartphone has a network connection. The user can do several actions in the application like start the chronometer for a walk, register the pain level or sending an emergency notification if necessary.

In Figure 1, the walk module shows the request GPS activity, the retrieval of data from SHealth application, the register of pain level, the mechanism of notification in case of emergency and other activities.

In Figure 2, it is shown an activity diagram UML which illustrates the functioning from history module in the smartphone application, which allows the user to check their records and publish them in the social networks.

METHODOLOGY AND DEVELOPMENT

The system was developed following the agile method, through the development of prototypes as a first approximation of the final system, that should be tested and reformulated, in successive iterations, until it arrives to a complete system (Pressman

Figure 1. Activity diagram of walk module

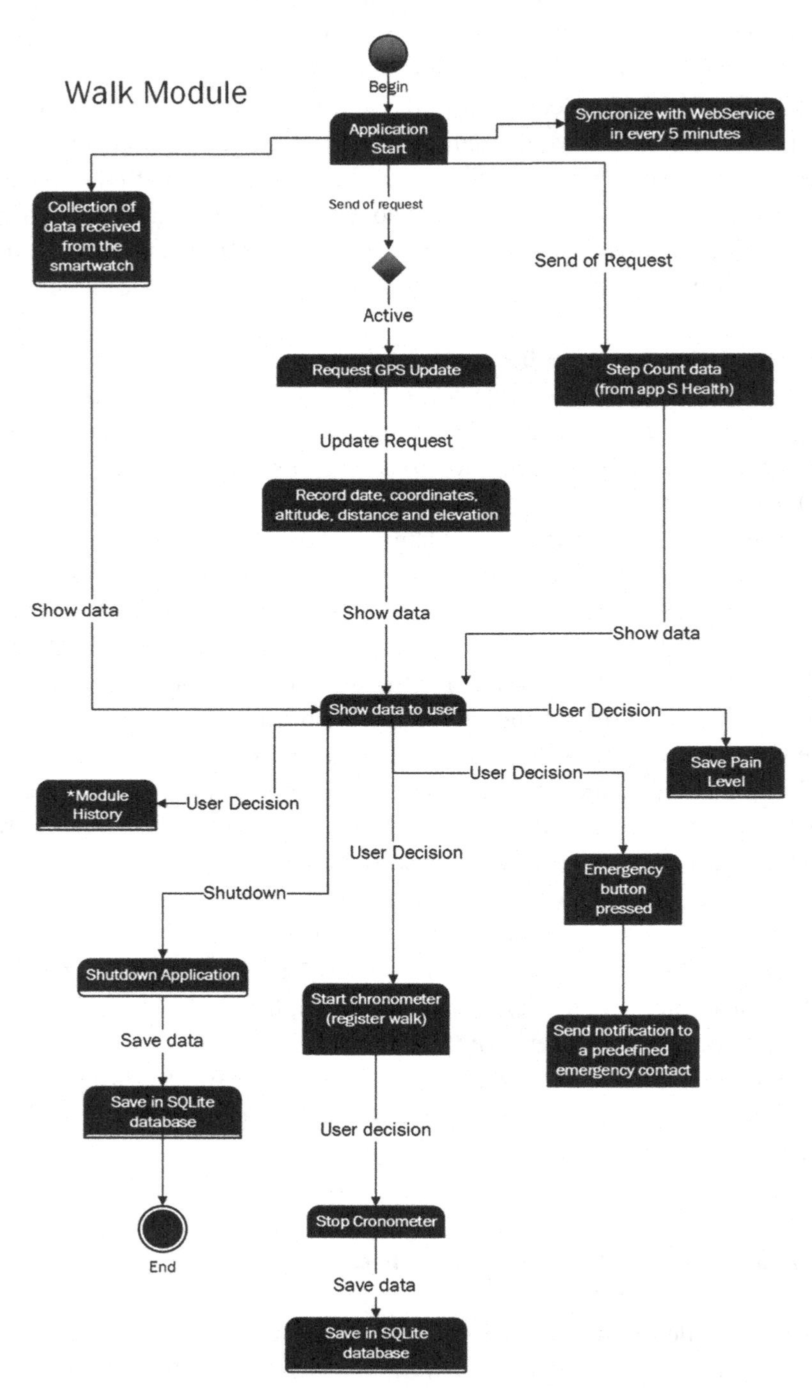

Figure 2. Activity diagram of history module

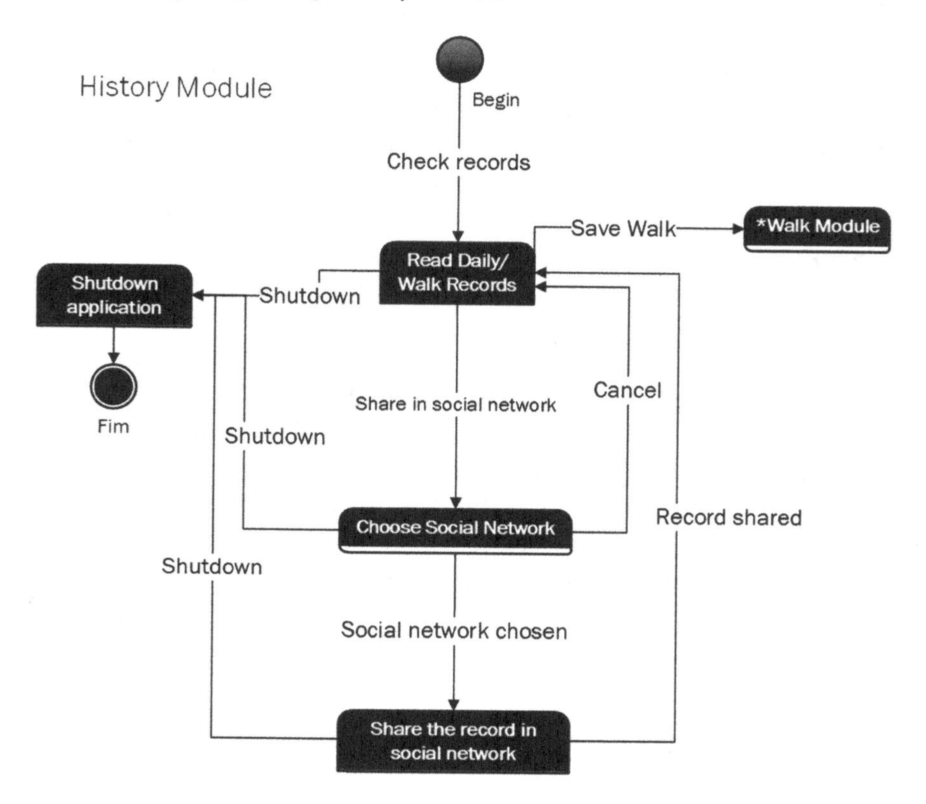

et al., 2005). In each iteration, it is made an analysis from the researchers in specifying the necessary requirements to have a system suitable to the user´s needs. Then, it is build a prototype based on the specified requirements. After prototype´s conclusion, it is made an analysis by the researchers on the prototype to assess the fulfilment of the requirements. If the researchers doesn´t agree that the prototype accomplishes the requirements proposed, it is necessary to continue developing the prototype. But if the prototype accomplishes the requirements proposed then it is evaluated with real users identifying possible errors and collecting the users´ feedback. After this evaluation, the researchers make an analysis on the obtained results and if necessary starts other iteration in the system development cycle.

When the system development cycle has a iteration that has produced a functional prototype, it is tested on patients with PAD using usability tests.

In this approach, it was developed the three main components of the system: the smartwatch application, the smartphone application, and the back-end webservice.

In this chapter, it will be presented four cycles of the smartphone application´s development.

Development of the Smartphone Application

The first phase for the construction of the system was the development of the smartphone application, since it is the core of this presented system because it must support many features and present to the user the main interface. The application was developed through the cycle of four iterations, which in each cycle was developed one prototype. The analysis in each prototype by the investigation team and authors of this work, would serve as a starting point for the next iteration.

The first prototype presented in Figure 3, had the chronometer and collected the user location through GPS.

The prototype presented in Figure 3, was shown to the investigators team of this project, which suggested some changes: (1) the design should be based on symbols and not in text; (2) it should be introduced a new feature to monitories walk activity indoor through the measuring of step count.

In Figure 4, it is shown the second prototype developed, which had the new following features: (1) pain level buttons; (2) reception of the data collected by the smartwatch; (3) integration with the fitness application SHealth to collect the step count.

Figure 3. First prototype of smartphone application

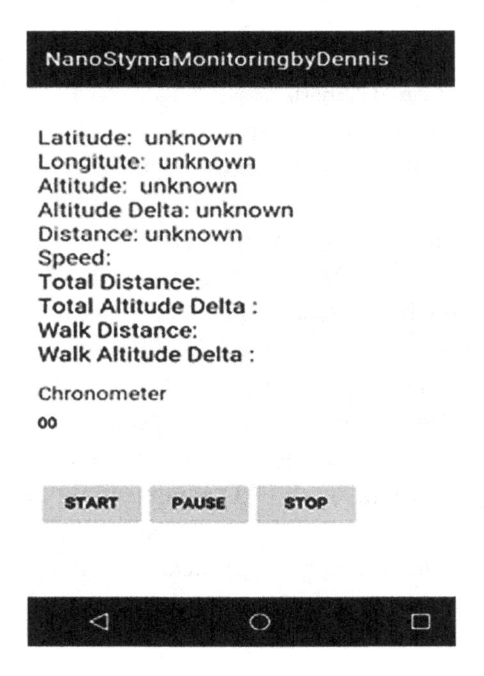

Figure 4. Second prototype of smartphone application

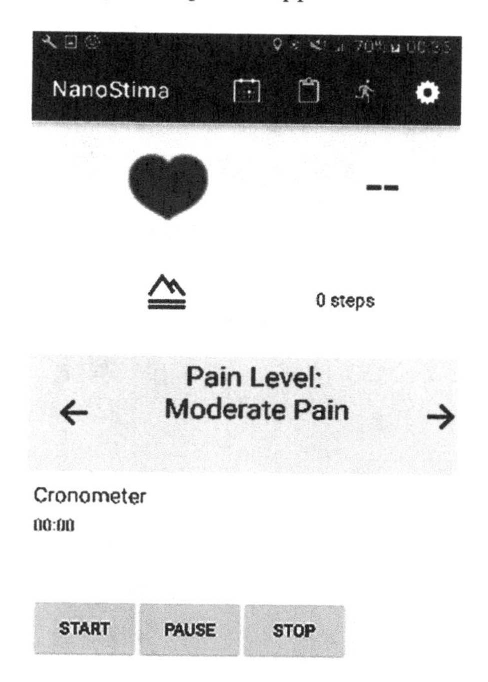

The second prototype shown in Figure 4 was analyzed with the researchers team of this project with the following recommendations: (1) the interface of pain level selection should be change to a more accessible interface for the patients with PAD; (2) the application should include a chronometer with real time (time of walk including pauses); (3) addition of an emergency button which allows the send of a notification to a predefined contact.

In Figure 5, it is shown the third prototype, with two chronometers (first one is for walk time without pauses and the second one is for walk time with pauses) and a new interface with tabs for pain level selection.

The third prototype shown in Figure 5, was tested in an outdoor hospital environment with 2 users with PAD and the results analyzed with the investigators team of this project with the following recommendations: (1) Change the location of the emergency button to the top bar, because in the current location the users could by mistake press the emergency button; (2) Change the colors of the pain level interface to colors with bigger contrast, because during the tests the sunlight interfered with the correct reading of the pain level tabs.

Figure 5. Third prototype of smartphone application. Translation "Portuguese"–English: "passos"–steps; "nível de dor"–level of pain; "Início"–Start; "Pausa"-Pause, "Fim"- End

In Figure 6, it is shown the fourth and last prototype, including the images of the "Walk/Daily Record" and "History" modules respectively. In the module "Walk/Daily Record", it was changed the colors of the pain level for colors with more contrast and it was changed the location of the emergency button to the top bar.

In Figure 7, it is shown the images of the "Calendar" and "Configurations" modules respectively of the fourth prototype. The module Calendar is integrated with the Google Calendar API. This API allows the scheduling of events, which in the developed application shows the scheduled events. The module "Settings" allows the user to insert the emergency contact.

Development of the Smartwatch Application

The developed application for smartwatch had the main purpose of gathering, in a continuous, the heart rate from user, and send them to the developed smartphone application.

Figure 6. Fourth prototype of the smartphone application, "Walk/Daily Record" (left), "History" (right) modules respectively. Translation "Portuguese"–English: "passos"–steps; "nível de dor"–level of pain; "Início"–Start; "Pausa"-Pause, "Fim"- End; "Registos diários"- daily records

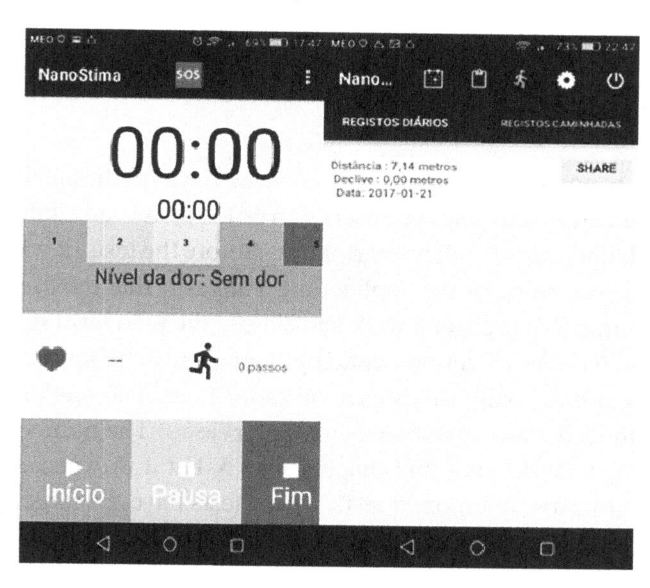

Figure 7. Fourth prototype of the smartphone application, "Calendar" (left), "Settings" (right) modules respectively

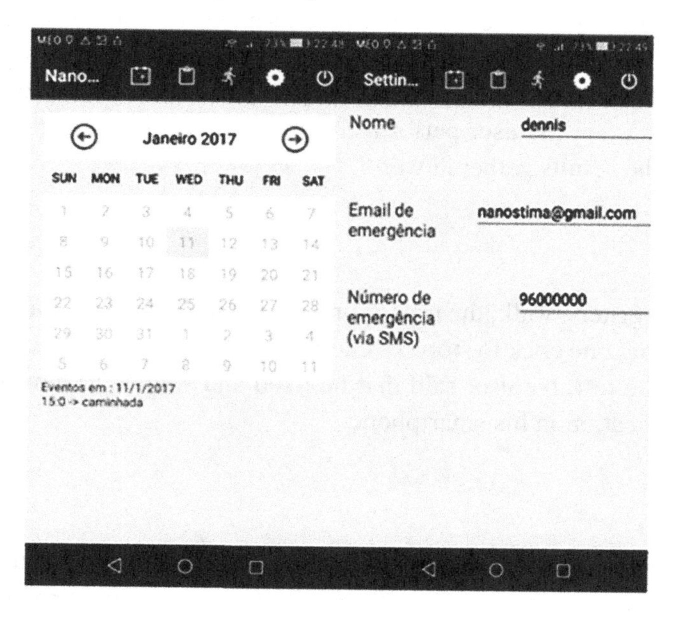

Development of Back-End Webservice

The back-end webservice was developed to store the data send by the application in smartphone, and allowing the interoperability with other systems using REST architecture.

Usability Tests

The usability tests on users with PAD will evaluate the hypothesis of the developed smartphone application is usable. The users will do the tests in an outdoor circuit with the application launched in a given smartphone. Before the tests, it will be explained to the users the functioning of the application. The tasks that the user need to do is using the chronometer to register a walk and choose the pain level if necessary. The variables are the number of actions done by the users in the application, the time in each action and how many times users asks for help. The methods for the data collection are: think-a-loud, questioners and interviews. The number of tests done at the moment of the written of this chapter were 5, but it is expected until the end of the year 2017 to carry out more 5 tests. These tests have the target population of the patients with PAD in a Portuguese hospital.

RESULTS

The users (from this point they will be called user n°1, user n°2, user n°3, user n°4, and user n°5), obtained a 100% score in the effectiveness on the proposed tasks. In the next description of results, it is mentioned the number of clicks, referring to the number of clicks that the user performs on the smartphone screen to accomplish some action. The results gathered were:

User n°1

In the task of register a walk, the user n°1 needed two clicks on the interface to start the walk and only one click to stop. To choose the pain level only needed one click. In the end of the test, the user said that he liked and wanted to have installed the developed application in his smartphone.

User n°2

In the task of register a walk, the user n°2 needed four clicks on the interface to start the walk and only one click to stop. To choose the pain level he need help and he needed two clicks. In the end of the test, the user said that he liked but didn´t want to have install the application because he didn´t had any smartphone.

User n°3

In the task of register a walk, the user n°3 needed two clicks on the interface to start the walk and two clicks to stop. To choose the pain level he needed between 4 and 5 clicks. In the end of the test, the user said that he liked but didn´t want to have install the application because he didn´t had any knowledge on informatics.

User n°4

In the task of register a walk, the user n°4 needed one click on the interface to start the walk and another one to stop. He didn´t choose any pain level. In the end of the test, the user said that the application was useful and he liked to try it at home.

User n°5

In the task of register a walk, the user n°5 needed one click on the interface to start the walk and another one to stop. He chooses for several times different pain level, each one with one click. In the end of the test, the user said that the application was easy and he liked to try it at home.

ANALYSIS

The results obtained are scarce to get any effective conclusion on the analysis of the developed smartphone application. From the results obtained, the users fulfill with 100% effectiveness the proposed tasks, although three users needed more than one click to accomplish some actions like starting the walk or pause it. The interaction with the pain level buttons were positive but the users had some difficulty to see the text within the buttons.

Overall the feedback from the users was positive, with three users wanting to try the proposed application at home. The results obtained indicates that developed application is usable and fulfill the initial proposed objectives.

CONCLUSION

In this work, it was presented a system that retrieve several data from user like the heart rate or the step count. This work had the objective of constructing a system that monitories the users with PAD and provides the means necessary to upkeep the health professionals updated about the user´s health and physical activity status. For this system, it was developed: one application for smartphones in Android, one application for smartwatch in Tizen and one back-end webservice in PHP. After the system was developed, the smartphone application was tested using usability tests. This system fulfills the initial proposed objectives, gathering information that can help the users with PAD getting a more active life. It can also assist the professionals that observing the data retrieved from the users, specify the necessary help based on the health parameters and evolution of each user.

FUTURE RESEARCH DIRECTIONS

For future work, it is necessary the elaboration of more user tests, to evaluate with more accuracy the system usability and effectiveness. It is also important to build an online platform for the health professionals monitories the health and physical data retrieved from users.

ACKNOWLEDGMENT

This work was supported by Project "NanoSTIMA: Macro-to-Nano Human Sensing: Towards Integrated Multimodal Health Monitoring and Analytics/NORTE-01-0145-FEDER-000016" financed by the North Portugal Regional Operational Programme (NORTE 2020), under the PORTUGAL 2020 Partnership Agreement, and through the European Regional Development Fund (ERDF).

REFERENCES

Abreu, J., Rebelo, S., Paredes, H., Barroso, J., Martins, P., Reis, A., ... Filipe, V. (2017). *Assessment of Microsoft Kinect in the Monitoring and Rehabilitation of Stroke Patients. In Recent Advances in Information Systems and Technologies* (pp. 167–174). Cham: Springer; doi:10.1007/978-3-319-56538-5_18

Anliker, U., Ward, J., Lukowicz, P., Trster, G., Dolveck, F., Baer, M., ... Vuskovic, M. (2004). AMON: A wearable multiparameter medical monitoring and alert system. *IEEE Transactions on Information Technology in Biomedicine*, *8*(4), 415–427. doi:10.1109/TITB.2004.837888 PMID:15615032

Apple Inc. (2017). *Apple Watch Series 1 - Apple*. Retrieved March 21, 2017, from http://www.apple.com/apple-watch-series-1/

Buttussi, F., & Chittaro, L. (2008). MOPET: A context-aware and user-adaptive wearable system for fitness training. *Artificial Intelligence in Medicine*, *42*(2), 153–163. doi:10.1016/j.artmed.2007.11.004 PMID:18234481

Fokkenrood, H., Verhofstad, N., van den Houten, M., Lauret, G., Wittens, C., Scheltinga, M., & Teijink, J. (2014). Physical activity monitoring in patients with peripheral arterial disease: Validation of an activity monitor. *European Journal of Vascular and Endovascular Surgery*, *48*(2), 194–200. doi:10.1016/j.ejvs.2014.04.003 PMID:24880631

Garg, P., Tian, L., Criqui, M., Liu, K., Ferrucci, L., Guralnik, J., ... McDermott, M. (2006). Physical activity during daily life and mortality in patients with peripheral arterial disease. *Circulation*, *114*(3), 242–248. doi:10.1161/CIRCULATIONAHA.105.605246 PMID:16818814

Garmin. (2017). *vívoactive® HR | Garmin*. Retrieved March 21, 2017, from https://buy.garmin.com/en-US/US/p/538374

Gonçalves, C., Rocha, T., Reis, A., & Barroso, J. (2017). *AppVox: An Application to Assist People with Speech Impairments in Their Speech Therapy Sessions*. Recent Advances in Information Systems and Technologies. doi:10.1007/978-3-319-56538-5_59

Google. (2017). *Fit – Google*. Retrieved March 21, 2017, from https://www.google.com/fit/

Gornik, H., & Beckman, J. (2005). Peripheral Arterial Disease. *Circulation*, *111*(13), 169–172. doi:10.1161/01.CIR.0000160581.58633.8B PMID:15811861

Jovanov, E., Milenkovic, A., Otto, C., & de Groen, P. C. (2005). A wireless body area network of intelligent motion sensors for computer assisted physical rehabilitation. *Journal of Neuroengineering and Rehabilitation*, *2*(1), 16–23. doi:10.1186/1743-0003-2-6 PMID:15740621

Menezes, J., Fernando, J., Carvalho, C., Barbosa, J., & Mansilha, B. (2009). Estudo da Prevalência da Doença Arterial Periférica em Portugal. *Angiologia e Cirurgia Vascular*, *5*(2), 59–68.

Motorola Mobility, L. L. C. (2017). *Moto 360 Sport - Sports Smartwatch powered by Android Wear - Motorola*. Retrieved March 21, 2017, from https://www.motorola.com/us/products/moto-360-sport

Nike, Inc. (2017). *Nike Running. Nike.com*. Retrieved March 21, 2017, from http://www.nike.com/us/en_us/c/running

Paulino, D., Reis, A., Paredes, H., & Barroso, J. (2016). Usage of mobile devices for monitoring and encouraging active life. *1st International Conference on Technology and Innovation is Sports, Health and Wellbeing.*

Pressman, S., Fachinger, J., den Exter, M., Grambow, B., Holgerson, S., Landesmann, C., & Titov, M. (2005). *Software engineering: a practitioner's approach*. Palgrave Macmillan.

Regensteiner, J., Meyer, T., Krupski, W., Cranford, L., Hiatt, W., & Regensteiner, J. G. (1997). Hospital vs. home based exercise rehabilitation for patients with peripheral arterial occlusive disease. *Angiology*, *48*(4), 291–300. doi:10.1177/000331979704800402 PMID:9112877

Reis, A., Barroso, J., & Gonçalves, R. (2013). Supporting Accessibility in Higher Education Information Systems. *Proceedings of the 7th international conference on Universal Access in Human-Computer Interaction: applications and services for quality of life*. 10.1007/978-3-642-39194-1_29

Reis, A., Lains, J., Paredes, H., Filipe, V., Abrantes, C., Ferreira, F., ... Barroso, J. (2016a). *Developing a System for Post-Stroke Rehabilitation: An Exergames Approach. In Universal Access in Human-Computer Interaction. Users and Context Diversity* (pp. 403–413). Springer International Publishing. doi:10.1007/978-3-319-40238-3_39

Reis, A., Morgado, L., Tavares, F., Guedes, M., Reis, C., Borges, J., Gonçalves, R., & Cruz, J. (2016b). Gestão de listas de espera para cirurgia na rede hospitalar pública portuguesa - O sistema de informação dos programas de recuperação de listas de espera. *CISTI, 11.ª Conferência Ibérica de Sistemas e Tecnologias de Informação*. DOI:10.1109/CISTI.2016.7521612

Samsung. (2017a). *Samsung Gear S2 - The Official Samsung Galaxy Site*. Retrieved March 21, 2017, from http://www.samsung.com/global/galaxy/gear-s2/

Samsung, (2017b). *S health | Start a Health Challenge*. Retrieved March 21, 2017, from https://shealth.samsung.com/

Chapter 10
Continuous Stress Assessment:
Mobile App for Chronic Stress Prevention

Luís Daniel Simões
Polytechnic Institute of Cávado and Ave, Portugal

Joaquim Sílva
Polytechnic Institute of Cávado and Ave, Portugal

Joaquim Gonçalves
Polytechnic Institute of Cávado and Ave, Portugal

ABSTRACT

Chronic stress is a spreading disease that affects millions of individuals with an enormous economic and social impact. Its prevention is, increasingly, a fundamental aspect for the improvement of the quality of life of individuals and the overall society. This chapter aims to understand how stress can be continuously monitored with the goal of predicting and alerting the occurrence of chronic or pathological stress and burnout situations. For this purpose, a non-invasive individual measurement instrument was developed to measure biometric signals through a wearable device that is connected to a mobile device. The prototype consists of a mobile application that gets the signals from a smartband and sends the data to an information system, tracking the individual physical condition to calculate the risk of entering the state of chronic stress. Continuous assessment of signs of stress is a key aspect for early detection of distress and effective intervention.

DOI: 10.4018/978-1-5225-5270-3.ch010

INTRODUCTION

In the Horizon 2020 Research and Innovation Program, the European Commission (EC) has proposed to invest more than €2 billion for providing better health for all (European Commission, 2017). This work addresses one of the main goals of this challenge: to improve the ability to monitor health and to prevent, detect, treat and manage disease.

To prevent and detect disease, we need a continuous monitoring approach, which is not compatible with invasive and disturbing measuring instruments. It requires a measuring instrument that becomes "transparent" to the user. Wearable devices are used increasingly, especially for fitness purposes, and their accuracy and reliability is improving. Enhanced quality and a higher number of biometric sensors are turning wearable devices suitable for health purposes.

This work aimed to develop a mobile application, to be used in smartphones or tablets for collecting data from a wearable device and integrate it in a Quality of Life Information System (QoLIS). The data will be processed for producing a set of reports to be made available to health professionals and users, i.e., the individuals under stress assessment.

Healthcare professionals will get access to the data generated by the wearable device by requesting their permission to the user under assessment. The data will be provided in an easily understandable format for supporting the decision-making of health professionals. The QoLIS platform will allow the continuous monitoring of each individual enabling an effective treatment in case early detection of pathological stress.

It is intended to provide to the user relevant data, in a friendly and perceptible format, about his/her health condition, including the current stress level. The mobile application will also inform the user about the appropriated procedures to keep the stress values at a recommended level interval. A good stress level can have a positive impact on the health condition, the quality of life, the professional performance and in the family and social relationships.

Of course, data collection is only important if the data are analysed with consequent production of results for the individual and health professionals. Since the perception of chronic stress is not immediate, continuous assessment of signs of stress is a key aspect for early detection of distress and efficient intervention when the individual is at risk of chronic stress. The mobile application will present information about the stress level using appropriated dashboards, considering the health professionals and users feedback about the data formats and user interface. Users and professional will have access to distinct dashboards. The aggregated results of this projects will be shared with health professionals to ensure the quality of the process used and data collected.

In the next section is described the state of the art of the information systems for stress monitoring. The third section, presents the technologies, the concepts, and the requirements considered for the development of the mobile application, including the its overall ecosystem. The penultimate section presents the results and future development perspectives. The last section consists of the chapter conclusion.

BACKGROUND

Chronic Stress

Stress is one of the most worrying concerns of occupational safety and health, area where there is an emerging need to find mechanisms to prevent harm to the individual (Juliet Hassard & Tom Cox, 2015). Stress arises, usually, based on the person's lifestyle or from the constant professional life challenges. Stress can manifest itself in very distinct forms (Murcho, Jesus, & Pacheco, 2009): it can be a source of well-being or, paradoxically as a source of malaise. An adequate level of stress at work is a positive and desirable factor (Wendy Taormina-Weiss, 2012) and can positively influence the way an individual feels, thinks and decides (Stansfeld, Fuhrer, Shipley, & Marmot, 1999). However, a level of excessive stress interferes with productivity, and has an negative impact on the individual's physical and emotional health (Hicks & Caroline, 2007; Stansfeld et al., 1999; Wendy Taormina-Weiss, 2012). Therefore, learning new and better ways to deal with today's job pressures is a key aspect in public health.

Most of these changes are easily reversible, representing a mild dysfunction and some associated discomfort. Nevertheless, a particularly intense and prolonged stress experience can have consequences on professional performance and lead to physical and psychological health problems and social dysfunctions (Devereux, Buckle, & Vlachonikolis, 1999).

In high reliability jobs, such as air traffic controllers, military, and health professionals, workers are often exposed to high-pressure situations, which often lead to depressions in a professional context, also known as burnout (Institute for Quality and Efficiency in Health Care, 2017), or stress-related mental disorders. Furthermore, the environment and workplaces can influence stress due inappropriate lighting, temperature or noise conditions. In these cases, stress should be controlled, since the decision making of the individual will affect third parties.

Stress Assessment

Continuous exposure to stressors at work can lead to emotional exhaustion, and in the latter case, chronic stress increases the risk of burnout, resulting in personal, professional, family, social and economic losses (Maslach & Leiter, 2008). These can occur both in the professional domain and in the health sector and may result in a variety of reactions, such as substance abuse, high absenteeism, accompanied by an increase in staff turnover and a decrease in productivity levels (Carvalho, 2011). This diagnosis is based on the fact that stress reactions manifest themselves in a multimodal way through psychological, physiological, behavioural and / or performance symptoms. The cardiovascular, respiratory, endocrine, gastrointestinal and immune systems are usually affected by excess stress (De Beeck, Hermans, & European Agency for Safety and Health at Work., 2000).

The European Agency for Safety and Health at Work (EU-OSHA), in a study conducted in 2014, warns that 25% of workers claim to suffer from work stress (European Agency for Safety and Health at Work., 2015). The research work that has been developed, especially in the last years, indicates some strategies for the prevention of excessive levels of stress, empathizing the continuous monitoring of the signs of stress as the key for early diagnosis, making strategies defined for stress management more effective and anxiety. In our research, we did not found any system of continuous monitoring of stress and anxiety based on non-intrusive instruments that predicts stress levels according to the individual's profile.

The transactional stress model of Lazarus and Folkman defines stress as a psychological state that results from the process of interaction between the person and his work environment. The model consists of a sequence of relations between the work environment and workers' perceptions, between these perceptions and the experience of stress and between the psychological state that characterizes this experience and the changes that occur at the behavioural, emotional, physiological and physical levels (Lazarus & Folkman, 1984). This process makes its measurement an extremely difficult and complex task because no single measure is sufficiently valid to measure stress (European Agency for Safety and Health at Work, 2000). An effective and scientifically acceptable stress measure requires the characterization of the work environment, a survey to evaluate the worker's perception and the identification of some of his reactions at work, namely behavioural, physiological or physical symptoms. The actions to be triggered for the measurement require the worker involvement, are intrusive to the worker routine and may increase the stress and anxiety levels.

From the point of view of the organism, the human being interprets stress as a threatening stimulus and creates defense mechanisms of biological and behavioural nature. The main biological agents are the release of catecholamines, such as epinephrine (adrenaline), and corticosteroids (cortisol). The positive effects of acute stress are attributed, above all, to the effect of catecholamines, while exposure to chronic stress results in a persistent activation of the hypothalamic-pituitary-adrenal axis, leading to cortisol secretion and an elevation of circulating pro-inflammatory cytokines, affecting sensitivity to glucocorticoids, brain functions and behaviour (Cohen & Khalaila, 2014).

To get a sense of the impact of stress stimulus on the individual, see the following flow of occurrences as examples: psychological responses will lead to increased emotions and more negative emotional states; physiological responses will lead to changes in the body's hormonal level, and may cause reactions such as increased heart rate or sweat production; behavioural responses will lead to changes in facial expression, body posture, or patterns of interaction with computers; and performance responses are related to changes in attention, logical reasoning, and productivity (Alberdi, Aztiria, & Basarab, 2016).

Usually, stress levels are measured on the basis of self-assessment questionnaires about psychological and physiological symptoms of the individual, where the information is subjective. Physiological symptoms provide the best results, analysing cortisol levels from blood and salivary samples (Centre for Studies on Human Stress, 2007).

Although these tests achieve the stress level of an individual, they do not allow continuous and real-time monitoring of stress, are not compatible with the worker daily routine and require the worker to move to health facilities. On the other hand, other methods of measuring stress levels, especially those dealing with behavioural symptoms, while they are those that can be used to continuously and non-intrusively measure the level of stress in the individual, they have not yet been extensively studied in the literature, and it is not yet known how to identify the level of precision they may have in the results. The continuous monitoring is a relevant aspect, as evidenced by the work of Oerlemans and Bakker (2014) and Derks and Bakker (2014).

Unlike other methods, the proposed method does not require the use of additional equipment and can be applied through devices that are perfectly adequate and adjusted to the user's current activity without requiring behavioural changes that could affect the measurement itself. Such property will allow an early diagnosis, making strategies defined for stress and anxiety control more effective.

Health Information Technology

We can find in the literature several studies demonstrating that information technologies have been fundamental in the various fields of research, including in health research, supporting the discovery of new diseases and therapeutics, and helping the evolution of health care (Gawande & Bates, 2000; Institute of Medicine, 2001). Another example, the adoption of information technology (IT) in the healthcare (Health IT) reduced the error rate of high-risk medication by 55% (Bates et al., 1998). However, until very recently, information systems have been seen only as a silent partner in the field of health care, that is not seen by patients, but have been critical in the health system (Eric Venn-Watson, 2014).

The future of health care is to place the patient as the main stakeholder of her/his own health care, providing the insider information, so he/she gets the ability to apply to herself/himself the most appropriate health care to his/her needs (Eric Venn-Watson, 2014).

mHealth was the label found for this new approach. mHealth or "mobile health" is a trend that has proven to be a valuable strategy for improved health systems by the closer proximity they allow among patients and health professionals. An example of the advantages of mHealth results from the use of mobile and monitoring devices, nowadays widespread worldwide, which ensure that professionals are able to access and share, in real time, patients' clinical data wherever they are, thus allowing faster access to information, obtaining knowledge of the patient's condition, defining the most appropriate treatment to be applied (Eric Venn-Watson, 2014), and reducing healthcare costs for both organizations and patients (doherty, 2013).

This component of medicine is still in its infancy, as no standards have yet been established. mHealth involves a many technologies such as SMS, GPRS, 3G and 4G mobile communications, georeferencing systems (GPS) and Bluetooth. These systems are a key part of mHealth's success as they enable the gathering of clinical data in real time, contribute to more accurate monitoring of health indicators of the world population, bring health professionals closer to patients and disseminate relevant information to diseases treatment. All these factors are fundamental for medical research evolution (WHO, 2011).

In a study conducted by the World Health Organization (WHO), the applicability of mHealth was tested in several countries (WHO, 2011). Continuous patient monitoring is an integral part of these tests. Continuous monitoring is performed using mobile technology in order to remotely manage, monitor and treat diseases, such as, diabetics or cardiac patients. The use of sensors installed in homes, including imaging devices such as smartphones, allow personal and more appropriated approach, perfectly adapted for each disease.

These devices were used to facilitate the gathering of patient data. This is a real reduction in the distance between patients and health facilities, where information collection is conventionally done. The WHO concluded that private institutions are the ones that most pursue to develop such solutions. In Switzerland, a company is using a tele-biometric mechanism for the gathering biometric information from diabetes disease patients. Through wearable biosensors they collect data on heart rhythms, blood pressure, blood sugar, etc., and consolidate this information on a server so that it can be analysed later (WHO, 2011).

Use of Mobile Technology in Healthcare

Nowadays, there is a focus on the development of technologies for promoting the evolution of medical action, as well as the proximity between health professionals and users. Besides this two goals, technology development also aims to provide tools for the users to have themselves an important role in their own health care (Eric Venn-Watson, 2014).

The COPD Navigator project, developed by LifeMap Solutions, aims to improve the way people with chronic lung diseases manage their health. This project was developed taking into account the guiding principles of evidence-based medicine, hoping to give doctors better results at a lower cost. This application measures, through multisensory devices, the activity of a patient in real-time. All data are stored and sent to the health care teams that follow these patients, so that in the next session they can take measures as to the course to follow by the patient in the treatment of her/his disease (LifeMap Solutions, n.d.). The COPD Navigator development team hopes in the future to develop algorithms that detect the worsening of symptoms so that both, patients and physicians, can be prevented and act immediately. Although COPD Navigator is nowadays targeting only pulmonary diseases, it was an important inspiration for this project by the use of emerging technologies that allow the continuous monitoring of chronic patients and because it adopted the approach of medicine based in the evidence.

Open mHealth is an open standard for mobile health data, developed by a non-profit organization with the same name, that aims provide the access to digital health data from disparate sources. Open mHealth envisions to create a global community of developers, product managers, health IT decision makers and clinical researchers to build an open framework for using digital health data accessible to all, thus developed a common language for standardize, store, integrate, share, visualize and process health data (Open mHealth, n.d.). A project developed by Open mHealth was dedicated to type 1 diabetics. This project aimed to solve a persistent problem in this type of patients: the need for continuous monitoring of their blood glucose and insulin levels and, consequently, their diet. Through the use of several mobile devices apps, Open mHealth gathered a panoply of data that, not only helped diabetics patients involved in the project to realize how to improve their quality of life, but also helped to raise among them awareness about hypoglycemic symptoms (Open mHealth, 2015). Open mHealth provides important information about research involving the use of technology for better health care, as well as several APIs that allow easily integrate health data.

Biovotion is another very relevant case study that motivated the development of this project. The company was created in 2011, in Switzerland, by specialists on monitoring by wearables and, nowadays, it is specialized in the continuous monitoring of physiological data. Biovotion provides services of data gathering and supply for health care providers, providing solutions for integration of monitoring data into healthcare systems. Biovotion mission is the improvement people's health and quality of life, the increasing of health care organizations outcomes and the costs reduction of health care systems in the medium to long term (Biovotion, n.d.). The solutions provided by Biovotion, as well as the device developed by them, was very relevant for the project described here, given the capabilities that allow to create an ecosystem conducive to the application of continuous monitoring through mobile devices and wearable technologies.

MAIN FOCUS OF THE CHAPTER

Stress-Off Mobile Application

For the mobile application development, we needed to know the use mobile devices in information systems context, the technology of smartphones, tablets and wearable devices, and how to build robust data-based apps. In this section, it is also presented the data set need for stress levels measurement, the requirements for the development of the mobile app and an overall view of the project ecosystem.

Mobile Devices and Information Systems

This project is part of a wider project that aims to create an ecosystem that involves different kind of technologies, such as Internet, mobile and wearable technologies. Smartphones are increasingly connected to the Internet and to wearable devices, as shown in Figure 1. The use of these devices is complementary. Smartphones, devices that are nowadays part of people's daily lives; smartbands biosensors are used for collecting physiological and environmental data. When paired with smartbands, smartphones allow the processing of collected data in order to provide feedback when it justified. The Internet provides the infrastructure for building a central

Figure 1. Evolution of pairing of wearable technologies with smartphones
Source: Cisco, 2014.

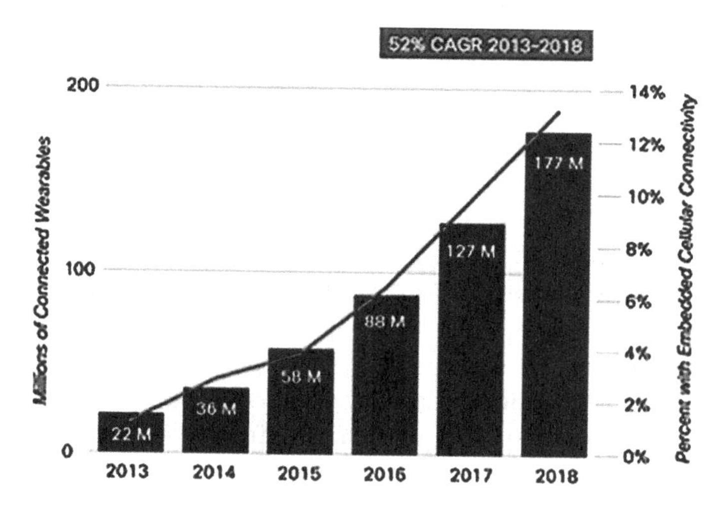

system where the data collected by the smartbands, duly identified, can be sent to a data repository, processed, analysed and used for knowledge extraction.

Mobile devices such as smartphones, tablets and wearables, have been widely and, for long, are being used in research for data collection and analysis. Mobile technology has thus contributed to the growth of economies and is regarded as the main driver for innovation (West, 2014)

The market of mobile technologies is very competitive and dynamic. Companies are constantly innovating, willing to adopt new technologies and release new products to market at a very fast rate. This constant innovation leads to a rapid evolution and, consequently, the expansion to new application fields that had not previously been considered. The pulmonology services of Viana do Castelo Hospital implemented a continuous monitoring system using a mobile application. This system allows the healthcare professionals to remotely monitor, in real time, chronic respiratory failure patients, what radically reduced hospital emergency visits for this reason (Noronha, 2016).

Wearable Devices

Wearables are computerized devices that, in addition to being portable computers, may be in physical contact with people for long periods of time (Picard & Healey, 1997). Wearable technologies provide appropriate platforms for quantifying, in the long term, the responses received regarding the physical and psychological state of the people. Thus, these technologies will provide valuable support for the improvement of individualized treatments for giving feedback for people and helping them to better know and manage their living habits (Sung, Marci, & Pentland, 2005).

Smartbands will be type of wearable that will be used within the scope of this project. These intelligent wristbands, also known as activity detectors, have biosensors in their structures to collect and store biometric data of people using them, based on their daily activities. Smartbands are able to collect many different data, such as kilometers traveled, calories consumed, sleep quality, among several others (Shih, Han, Poole, Rosson, & Carroll, 2015).

These devices can be easily synchronized with smartphones or computers, offering an improvement in the long-term individual follow-up and the opportunity of learning their habits of life for providing the user with tips on how to improve them (Sung et al., 2005). The wrist smartbands or wristbands include many sensors that allow, in real time, to collect the user biometric data and other relevant information about the activity that is being executed by the user. Microsoft Band 2, short name MSBand2, one of the market leaders' wristbands, has the following built-in sensors: optical heart rate sensor, accelerometer, gyrometer, GPS, brightness sensor, body temperature sensor, ultraviolet sensor, capacitive sensor, galvanic skin response, microphone and barometer. In the scope of this project, there will be used the following sensors:

- **Optical Sensor for Heart Rate:** It will allow to collect, in real time, data about their heart rhythm, an important data for evaluate the individuals stress level;
- **Accelerometer:** It will recognize what kind of movements are being performed by the user, so helping to understand what kind of job is being run and, in this project, to use as an input to assess possible symptoms of stress;
- **Body Temperature Sensor:** It will help to realize temperature fluctuations in the body of the individual, which can be a stress indicator;
- **Sensor of Ultraviolet Rays:** It allow to know the individual sun exposure to know e.g. whether the individual is in a room with natural lighting and realize the increased body temperature may result from the fact of being exposed to sunlight;
- **Light Sensor:** It will make it possible to recognize, for example, if the individual works in an environment with little light intensity, which is a factor that can stimulate the appearance of stress.

Most of these sensors are present in the vast offer of this type of devices available in the market. However, considering the need for high quality and the handling collected data, the project team has select MSBand2.

Smartphones, Tablets, and Mobile Apps

In the scope of this project, smartphones and tablets are a critical component that synchronizes with MSBand2 using a mobile application. They provide feedback to its user about the quality of his/her living habits and processes the data collected adequately for detecting and preventing chronic stress. According to the Oxford Dictionary, smartphone was first used during the 1980s where they referred to a phone with computer technology. These devices can perform several functions present on a computer and typically have a touch screen as a user interface, provide access to the Internet and use a proprietary operating system capable of running software. A tablet is a small laptop that accepts inputs directly on the screen without using a mouse or keyboard for the purpose (Oxford Dictionaries, n.d.).

Mobile applications, often referred to as "apps", are installed in the both types of devices smartphones and tablets. An app consists of software built specifically to run on an operating system present these devices: iOS, Android, Windows Phone, among others. These applications usually have the same characteristics of any other software designed for a computer. The usability, given the dimensions of these types of devices, is the most distinctive characteristics.

A study carried out by the Department of Information Systems of the University of Wisconsin demonstrates that the importance of the adoption of Information Technology (IT) in health care, emphasizing the use of smartphones (Park & Chen, 2007). The same study highlights that the current literature has raised many questions about how health care can benefit from the new developments in IT, in particular, how systems should be designed that allow the electronic registration of health data. Smartphones can offer several benefits to health professionals. Health professionals, using a smartphone with access to wireless networks, can be in permanent contact with their email and access their patients' electronic records. A concrete example provided in this study is the adoption of electronic prescription (Park & Chen, 2007).

These systems have the capacity to bring health professionals closer to their patients. The use of mobile devices permits to combine the continuous user monitoring with simultaneous supplying of relevant information about the measurements that are being made the user body. That feedback allows the healthcare professional to know, almost in real-time, the physical state of a patient and, simultaneously, allows the user to apply specific improvements considering some out-of-range indicator.

Feedback can have an impact on people's lives in many contexts. Haydon et al. (2012) used the mobile device with children with emotional disorders, who presented learning difficulties, especially in the calculation. iPad tablets were attributed to the children to promote active learning by giving immediate feedback of the child mistakes. The device also provided feedback on correct answers, emphasizing each correct answer, which naturally increased the likelihood that the student would respond correctly in subsequent responses. At the end of each session, a summarization of the results for a total performance result was shown.

Comparing the results obtained with this approach and the traditional approach, the use paper and pen, the study concluded that students who used the mobile device showed greater effectiveness and success in the responses (Haydon et al., 2012). It can then be concluded that feedback is a driver of the result obtained with the use of these devices, in a psychological context. Constant feedback for correct answers has an impact on how the child will address the following questions. It is in this perspective that the feedback will fit in this project, with the ambition that the user feels that the monitoring will deliver legible signs about his/her condition. The user will also want to know if is performing less appropriately in a given context and how can improve.

Since the mobile devices market is dominated by Android Operating System (OS), followed by iOS and Windows Phone, we decided to use a single mobile development platform able to deliver applications for these three OS. The Outsystems Platform, developed by OutSystems, was the selected mobile development platform. OutSystems has Portuguese origin, but the company moved its headquarters to the US in 2014. According to Forrester, OutSystems is a Low-Code Development Platforms. The greatest strengths of OutSystems Platform are its broad features and tools for database and process applications, mobile and web user experiences, integration, and collaboration (Richardson & Rymer, 2016). This decision allowed to speed up the development process and gain time for other equally important tasks in the project implementation.

The use of Outsystems Platform for the development of mobile applications brings several important benefits (OutSystems, n.d.): (1) the development of the application is based on visual objects, through component drag and drop; (2) the deployment of the application is done through a simple click, inside the platform; (3) it incorporates native behaviours of each platform into the applications, facilitating the implementation of functionalities that use the mobile devices components; (4) integration of web services without the need for hard code programming; (5) implementation of user experience standards embedded in the platform; (6) and the base architecture assures safety, scalability and performance.

Web Services and API

The data collected by the mobile application need to be available to health professionals, who also need to get data from other sources. Data repositories are usually used as the bridge among different systems that need to exchange information with each other. This project will implement a data repository for save the data collected by the end users in their devices and support the calculations needed by the health professionals to draw conclusions of the stress levels of their patients.

The concept of Web services is used quite frequently nowadays, but it is not always defined in the same way. Usually, Web services are viewed as accessible applications over the Web. A more complete definition for Web service defines it as an application that is accessed through the Web with a corresponding Application Programming Interface (API), published with an additional description in any directory service (Alonso, Casati, Kuno, & Machiraju, 2004).

The World Wide Web Consortium (W3C) redefined Web services as being a software application that is identified by a Uniform Resource Identifiers (URI), whose interfaces and bindings are capable of being defined, described and discovered as XML artifacts. A Web service directly supports interactions with other software agents based on XML messages exchanged via Internet-based protocols (Alonso et al., 2004), for example the HTTP protocol. There are some technologies that are considered the most appropriate for designing Web services, such as SOAP and RESTful.

SOAP, an acronym for Simple Object Access Protocol, is a protocol defined in XML format that aims to exchange text based messages, which must always contain a description that usually identifies the purpose of their invocation. This protocol implements Web Services standards, such as the description of the Web service and its functionalities (WSDL) and the Universal Description, Discovery and Integration (UDDI) standard that defines the rules of use of the Web service. SOAP is based on the HTTP protocol and its GET / POST headers, making the communication of messages independent of the environment in which they are made: different networks, different platforms, different operating systems (Hamad, Saad, & Abed, 2010).

REST is a software architecture modeled based in the way data is represented, accessed and modified on the Web. According REST architecture, data and functionalities are considered as resources, and can be accessed through links typically used on the Web, usually referred to as URI. Usually, to make these resources available, well defined operations are used. REST emphasizes a client-server architecture designed to be used through stateless communication protocols, where HTTP is the most commonly used protocol. RESTful Web services implement the REST architecture and aim to expose data and service functionality through URIs over HTTP. The HTTP headers are: GET, to get data; POST to create a new resource or data; UPDATE to change a resource or data; DELETE to remove a resource or data (Hamad et al., 2010).

The Application Programming Interface (API) was the mechanism selected to turn available the data gathered through the devices for external applications consumption. API incorporates one or more features, representing a mechanism that aims to reuse code. The API creates an abstraction level for the programmers, allowing them to execute functions without needing to realize how they were implemented, hiding the code, providing only one interface for it to be invoked (Stylos, 2009).

RESTful API is an application, based in the REST architecture, to turn available a set of functionalities, designed for a well-defined purpose, that can be called many times by many applications, avoiding the features rewriting.

Requested Data for Stress Levels Measurement

As already mentioned previously, physical, behavioural and environmental data influence the emergence of stress. In this sense, it is important to analyse, in particular, what may be the most relevant data within each of these factors, so that, in the context of continuous monitoring, that data could be analysed exhaustively.

Some studies show that work directly influences heart rate, which in cases of persistence leads to cardiovascular problems. A study by the American Heart Association found that the heart rate is higher during working hours and in the hours of leisure after work. However, against the expectations of the study team, stress at work does not significantly affect the individual at night and rest days (Vrijkotte, van Doornen, & de Geus, 2000). Therefore, it is important to analyse stress levels when the individual is under the pressure of work.

Blood pressure is also one of the factor that must be used to analyse the individual's stress levels. In the same American Heart Association study, systolic blood pressure increases significantly during working hours, when compared with hours of leisure after work hours. They also found higher blood pressure during work days than on rest days and that work stress affected during sleep hours, since blood pressure was higher (Vrijkotte et al., 2000). This information helps to understand the extent to which blood pressure reveals stress levels of the individual, making it an important data to collect and analyse.

In a study conducted at the Massachusetts Institute of Technology Hernandez et al. (2011) analysed the stress levels of call center workers through sweat gland alterations, using biosensors installed on people's skin, which collected data continuously. They defined this component of the organism as the best for stress analysis, since these glands are controlled by the sympathetic nervous system (Boucsein, 1992). The use of biosensors in the skin will give the possibility to monitor their activities. In this study, they also indicated that the data obtained differed according to age group, sex, ethnicity, and hormonal cycles. This study explores the use of biosensors for data collection, and praises the importance of analysing the influence of sweat glands to measure stress levels.

The level of cortisol released is one of the physical factors that best signals the level of stress in the individual. The scientific literature shows that high levels of cortisol directly influence learning and memory capacity and, among other problems, reduces the immune system capacity, decreases bone density, causes weight gain and heart problems, rises blood pressure (Bergland, 2013). It has been claimed that the cortisol system is activated when the individual feels that the attainment of their goals is threatened or that he/her cannot accomplish at all (Dickerson & Kemeny, 2004). Cortisol is easily collected through saliva (Kirschbaum & Hellhammer, 1994), so the adoption of this process facilitated the studies about people's daily activity. However, this process was not used in this project, given the difficulty of continuous and non-intrusive monitoring of stress level based on cortisol values.

The surrounding environment has also been considered one of the factors that can increase stress in people, for example, crowded places, noisy places, air pollution (Ulrich et al., 1991), and natural disasters. The coexistence of people with this type of environment can lead to increased levels of fatigue, frustration, tension, irritation or even panic. These factors must be considered in the context of continuous monitoring to combine the physical state of an individual with the environment in which he/she is involved, as it gives the possibility of perceiving the extent to which these can lead to situations of chronic stress.

The content of the data used in this project helps to demonstrate how, in a real scenario, the data can be best gathered, processed and supplied. The following biometric data were selected for the proof of concept artefact: ambient light, heart rate, heart rate interval, galvanic skin response and body temperature. In terms of the individual personal data, for proof of concept purposes, we are concerned to protect the individual identity, so we just selected the following attributes: weight, height, age, nationality, city, physical activity level, smoker/non-smoker flag.

After an analysis of the data collected by MSBand2, it was found that the incidences were recorded at a rate less than a second. Given that it is in the best interest save every reading, the transacted data volume would be massive. It was demonstrated a specific case of the analysis carried out for just one biometric attribute. The data from an ambient light sensor was gathered and records for 10 minutes. At the end of this period, MSBand2 recorded 1171 occurrences. Using this sample as a reference, the average number of records is around 117 records per minute, so for one day long, i.e., 1440 minutes, the average number of records would be 168,480 records for just this biometric attribute! Therefore, we were able to perceive the huge volume of occurrences that would be recorded for every individual for a short time period.

Data Repository Requirements

Considering the huge data to be recorded, at real time, and based on several cases, we decided to use the technique NoSQL.

The decision to build the Data Model using MongoDB was made according to the possibilities that MongoDB provides to manage data. Not just it stores huge amounts of data, but it's an elastic framework as well. This means that we can ask for information on MongoDB anyway it fits the purpose. Also, and this is the main reason why, due to the enormous amount of data collected for one biometric data for at least 10 minutes, we decided that a low weight DB engine would be required, and MongoDB was the one that we use because it fit our needs and is one of the most fastest framework when using NoSQL databases.

There are several very known systems that transact and process a high volume of data per second. Facebook and Amazon are a success cases in using this type of technology.

In 2011, Facebook already had more than 3.5 billion shared content per week (links, posts, etc.). They decided to develop their own technology that implements NoSQL notions, named Cassandra, to respond to replication, fault detection, caching, among other features. Amazon faced a challenge to improve the reliability of the large volume of data transacted among the several business applications it holds. The challenge was to change the way they distributed the data storage of their applications to deal with the increasing number of data transactions. To this end, Amazon developed Dynamo in 2007, a data storage technology based on the NoSQL specifications as well. As a result, Amazon's several services have remained available in 99.9995% of requests made (Lóscio, Oliveira, & Pontes, 2011).

MongoDB is another implementation of the NoSQL specifications oriented to documents. It is a very popular NoSQL database for storing collections of documents. Each document is an object that has a unique identifier associated with it and made by a set of fields or subdocuments. An important factor to highlight is that in a document-oriented NoSQL context, the model does not depend on a rigid schema, that is, it does not require a fixed structure as mandatory by conventional relational databases.

Mobile App Requirements

From the detailed analysis of the needs, carried out with the deep involvement of all stakeholders, the following mobile application requirements were considered in the development of a functional prototype:

- The mobile application must be available on iOS and Android platforms;
- Users must register, safeguarding their anonymity, therefore only generic data of the should be registered to enable the analyse of the selected stress indicators;
- Users must log in to view the available data analyses and track their health data;
- The application should provide a user-oriented dashboard, where the user can quickly observe his stress indicators status and drill down to detailed biometric data gathered from sensors;
- The user interfaces should be very user-friendly, without any technical requirement, so the user could intuitively access to information, rather displayed as charts;
- The dashboard will include a Key Performance Indicator (KPI), obtained through calculations over the gathered user biometric data, that will inform the user about their level of stress;
- A notification system that will recommend users for better health habits and advise the when they run the risk of chronic stress, keeping the users aware of the quality of their health habits.
- The application, based on the user gathered biometric data, should continuously acquire knowledge about individuals' habits and provide tips for improve them or build new habits.

Overall View of the Project Ecosystem

The development of the mobile app is part of a bigger project, under progress, that aims to create an ecosystem that enables the continuous monitoring of stress in individuals and supports the analysis by health professionals of the gathered biometric data. Another important project component has the goal implement mathematical calculations to determine the level of stress of an individual, so the calculation that was implemented in the mobile is for demonstration purposes only.

The continuous monitoring is done through wearable technology, the MSBand2 smartbands, which was the one chosen for implementing the proof of concept. The mobile app synchronizes and gathers the biometric data from the smartband, in real time, and sends that information for a central repository, were the data can be used for stress indicator calculation as well as for detailed analyse by health professionals. The development of the RESTful API allows the user and health professionals to analyse remotely biometric data and calculated indicators.

All collected data is stored in a high-performance database that will save, in real time, the gathered biometric data from mobile apps and, simultaneously, will quickly provide access to biometric data and calculated indicators to individual users and health professionals. For the implementation of the proof-of-concept, a virtual machine was created in Microsoft Azure Cloud Platform and configured with the MongoDB, the selected NoSQL database, and the API software. Figure 2 represents the global view of the ecosystem that has been implemented for the scope of this project.

The opinion of potential users of the system was solicited. The analysis of the answers tells us that there is a natural concern with security, privacy and the development of a set of laws that offer protection to the user, in particular through information that should be provided to them about the data collected and their use.

In the opinion of the respondents, the information system presented (which includes the use of the device) is safe, reliable and gives the necessary guarantees of privacy. In addition, they consider that the use of the device is not very complex and more practical than the questionnaire response on another platform.

It should be noted that the information about the system, in particular the device, is fundamental in the decision to participate in the project, while the need to use the device is not so decisive in that decision.

RESULTS AND FUTURE WORK

The Figure 3 present some screens of this prototype, which is in Portuguese language. The first 3 screens, from left to right, implement the login and personal data registry, required for future forecasting models. The screen at the right shows a menu for selecting the type of information to analyse.

Figure 2. Overall view of the project ecosystem

Figure 3. Example of 4 application screens

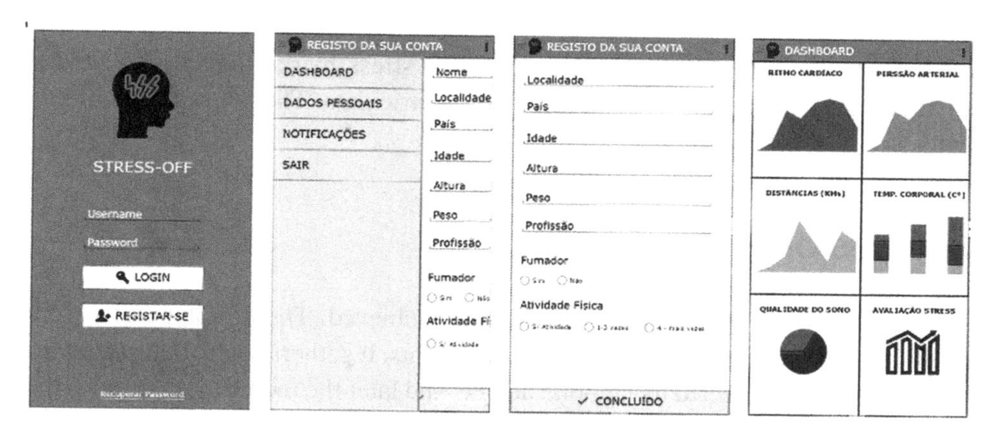

The Figure 4 shows the user notification center. The screen is divided into two tabs that correspond to "read notifications" and "unread notifications". When the user opens an "unread notification", the transition to the "read notifications" is automatic.

This first prototype should be used, in real context, by a sample of individuals that perform activities susceptible to create high levels of stress. It is an important landmark for the validation of the application, therefore, this prototype is planned to be used by medicine students of the University of Minho.

The following step will be the consumption of the data generated in the validation tests of the mobile application prototype. The large amount of data that will be generated requires the adoption of appropriate analytical models and techniques. For analysing and extracting knowledge from the data, we will perform data analytics, apply data mining concepts and explore machine learning algorithms.

Figure 4. Notifications center

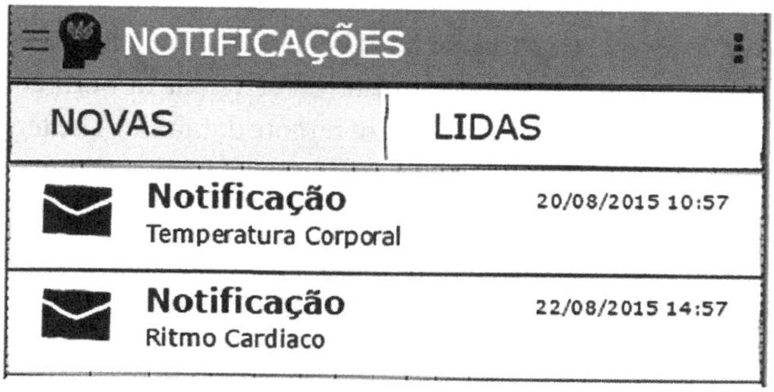

FUTURE RESEARCH DIRECTIONS

Building stricter models and determining more stress markers requires a device capable of providing greater variability of biometric data. The construction of this device is a goal for the future.

CONCLUSION

The goals proposed for this project were fully achieved. The prototype that were developed complies with the specified requirements. It gathers health data through a wearable device connected to a mobile device, and later the mobile device sends the data to the existing Quality of Life Information System (QoLIS). The participation and commitment of the end users was crucial for the successful application development; individuals to be monitored and health professionals created a very user-friendly environment.

The security and protection of data privacy has been guaranteed through the omission of any kind of information that denotes the identity of any individual and avoiding the publication of any data that could identify any individual. Although the mobile application requests several data attributes that could help to identify the individual who uses it, that data does not circulate to channels other than your personal smartphone.

The prototype of the mobile app accomplished successfully the purpose, gathering the end-user biometric data, proving the continuous monitoring through the stress indicator calculation and notifying user whenever a biometric indicator gets an abnormal value. The user experience feedback obtained from the prototype end-users proved to be widely accepted and complied with the requirements.

The wearable device used for continuous monitoring for stress measurement purposes proved to be reliable and accurate as a typical clinical device. The validation of the smartband truthfulness was an important result, because it ensures that the measurements made by the smartband could be used for the purpose of this project.

The Web service created demonstrated to be efficient in both communication and interoperability with the Microsoft Azure remote database. The technology used allowed to develop the service with a reasonable workload without requiring large scale implementations. The communication between que mobile app and the database are based in HTTP protocol and JSON data format, used by the Outsystems platform.

The MongoDb NoSQL database guarantees the scalability without the need to increase the processing load, space or even data structures. Despite the massive volume of data transactions, the database high availability assured an excellent behaviour.

REFERENCES

Alberdi, A., Aztiria, A., & Basarab, A. (2016). Towards an automatic early stress recognition system for office environments based on multimodal measurements: A review. *Journal of Biomedical Informatics*, *59*, 49–75. doi:10.1016/j.jbi.2015.11.007 PMID:26621099

Alonso, G., Casati, F., Kuno, H., & Machiraju, V. (2004). *Web Services*. Springer Berlin Heidelberg. doi:10.1007/978-3-662-10876-5_5

Bates, D. W., Leape, L. L., Cullen, D. J., Laird, N., Petersen, L. A., Teich, J. M., ... Seger, D. L. (1998). Effect of Computerized Physician Order Entry and a Team Intervention on Prevention of Serious Medication Errors. *Journal of the American Medical Association*, *280*(15), 1311. doi:10.1001/jama.280.15.1311 PMID:9794308

Bergland, C. (2013). *Cortisol: Why "The Stress Hormone" Is Public Enemy No. 1*. Retrieved September 24, 2016, from https://www.psychologytoday.com/blog/the-athletes-way/201301/cortisol-why-the-stress-hormone-is-public-enemy-no-1

Biovotion. (n.d.). *About Us*. Retrieved May 29, 2017, from http://www.biovotion.com/about-us/

Boucsein, W. (1992). *Electrodermal Activity*. Springer, US. doi:10.1007/978-1-4757-5093-5

Carvalho, P. (2011). *Estudo da fadiga por compaixão nos cuidados paliativos em Portugal: Tradução e adaptação cultural da escala "Professional Quality of Life 5."* Instituto de Ciências da Saúde, Universidade Católica Portuguesa, Porto. Retrieved from http://repositorio.ucp.pt/handle/10400.14/8918

Centre for Studies on Human Stress. (2007). *How to measure strees in humans?* Quebec, Canada: Author.

Cohen, M., & Khalaila, R. (2014). Saliva pH as a biomarker of exam stress and a predictor of exam performance. *Journal of Psychosomatic Research*, *77*(5), 420–425. doi:10.1016/j.jpsychores.2014.07.003 PMID:25439341

De Beeck, R. O., Hermans, V., & European Agency for Safety and Health at Work. (2000). *Research on work-related low back disorders*. Institute for Occupational Safety and Health.

Derks, D., & Bakker, A. (2014). Smartphone use, work–home interference, and burnout: A diary study on the role of recovery. *Applied Psychology*. Retrieved from http://onlinelibrary.wiley.com/doi/10.1111/j.1464-0597.2012.00530.x/full

Devereux, J. J., Buckle, P. W., & Vlachonikolis, I. G. (1999). Interactions between physical and psychosocial risk factors at work increase the risk of back disorders: An epidemiological approach. *Occupational and Environmental Medicine, 56*(5), 343–353. doi:10.1136/oem.56.5.343 PMID:10472310

Dickerson, S. S., & Kemeny, M. E. (2004). Acute Stressors and Cortisol Responses: A Theoretical Integration and Synthesis of Laboratory Research. *Psychological Bulletin, 130*(3), 355–391. doi:10.1037/0033-2909.130.3.355 PMID:15122924

Doherty, D. (2013). *Conferências saúde CUF - Mobile Health - Novas formas de olhar a saúde*. Retrieved July 13, 2017, from https://www.slideshare.net/3GDR/conferencias-saudecuf-david-doherty-23514751?ref=http://mhealthinsight.com/2013/05/18/join-us-in-portugal-for-mobile-health-new-ways-of-looking-at-health/

Eric Venn-Watson. (2014). *The evolution of health IT: Today, technology that supports clinical workflows is becoming the norm*. Retrieved July 13, 2017, from http://www.healthcareitnews.com/blog/evolution-health-it

European Agency for Safety and Health at Work. (2000). *Work-related low back disorders*. Luxembourg: Office for Official Publications of the European Communities. Retrieved from https://osha.europa.eu/en/tools-and-publications/publications/reports/204

European Agency for Safety and Health at Work. (2015). *Second European Survey of Enterprises on New and Emerging Risks (ESENER-2)*. Luxembourg: Publications Office of the European Union. Retrieved from https://osha.europa.eu/pt/node/7653/file_view

European Commission. (n.d.). *Health, Demographic Change and Wellbeing - European Commission*. Retrieved June 27, 2017, from https://ec.europa.eu/programmes/horizon2020/en/h2020-section/health-demographic-change-and-wellbeing

Gawande, A., & Bates, D. (2000). The use of information technology in improving medical performance. Part II. Physician-support tools. *MedGenMed: Medscape General*. Retrieved from http://europepmc.org/abstract/med/11104459

Hamad, H., Saad, M., & Abed, R. (2010). Performance Evaluation of RESTful Web Services for Mobile Devices. *Int. Arab J. E-Technol.* Retrieved from http://www.iajet.org/iajet_files/vol.1/no.3/Performance Evaluation of RESTful Web Services for Mobile Devices.pdf

Hassard, J., & Cox, T. (2015). *Work-related stress: Nature and management*. Retrieved June 29, 2017, from https://oshwiki.eu/wiki/Work-related_stress:_Nature_and_management

Haydon, T., Hawkins, R., & Denune, H. (2012). A comparison of iPads and worksheets on math skills of high school students with emotional disturbance. *Behavioral*. Retrieved from http://journals.sagepub.com/doi/abs/10.1177/019874291203700404

Hernandez, J., Morris, R. R., & Picard, R. W. (2011). *Call Center Stress Recognition with Person-Specific Models. Affective Computing and Intelligent Interaction*. Springer Berlin Heidelberg. doi:10.1007/978-3-642-24600-5_16

Hicks, T., & Caroline, M. (2007). *A guide to managing workplace stress*. Retrieved from https://www.google.com/books?hl=pt-PT&lr=&id=fcxUW9kyDukC&oi=fnd&pg=PA7&dq=A+Guide+to+Managing+Workplace+Stress+hicks&ots=l6AFkH1H6z&sig=Q2sIEc3SuUo_vUXanTlLlM2CkSk

Institute for Quality and Efficiency in Health Care. (2017). *Depression: What is burnout?* Retrieved from https://www.ncbi.nlm.nih.gov/pubmedhealth/PMH0072470/

Institute of Medicine. (2001). Crossing the quality chasm: A new health care system for the 21st century. Washington, DC: Author.

Kirschbaum, C., & Hellhammer, D. H. (1994). Salivary cortisol in psychoneuroendocrine research: Recent developments and applications. *Psychoneuroendocrinology*, *19*(4), 313–333. doi:10.1016/0306-4530(94)90013-2 PMID:8047637

Lazarus, R. S., & Folkman, S. (1984). *Stress*. New York: Appraisal, and Coping.

LifeMap Solutions. (n.d.). *COPD Navigator - LifeMap Solutions*. Retrieved February 24, 2017, from http://www.lifemap-solutions.com/products/copd-navigator/

Lóscio, B., Oliveira, H., & Pontes, J. (2011). NoSQL no desenvolvimento de aplicações Web colaborativas. *VIII Simpósio Brasileiro de*. Retrieved from http://www.addlabs.uff.br/sbsc_site/SBSC2011_NoSQL.pdf

Maslach, C., & Leiter, M. P. (2008). Early predictors of job burnout and engagement. *The Journal of Applied Psychology*, *93*(3), 498–512. doi:10.1037/0021-9010.93.3.498 PMID:18457483

Murcho, N., Jesus, S., & Pacheco, E. (2009). O mal-estar relacionado com o trabalho em enfermeiros: um estudo empírico. *Actas Do I Congresso Luso-Brasileiro de Psicologia*. Retrieved from https://scholar.google.pt/scholar?q=O+mal-estar+relacionado+com+o+trabalho+em+enfermeiros%3A+um+estudo+empírico&btnG=&hl=pt-PT&as_sdt=0%2C5

Noronha, N. (2016). *Doentes com insuficiência respiratória acompanhados via smartphone | SAPO Lifestyle*. Retrieved September 24, 2016, from http://lifestyle.sapo.pt/saude/noticias-saude/artigos/doentes-com-insuficiencia-respiratoria-acompanhados-via-smartphone?artigo-completo=sim

Oerlemans, W., & Bakker, A. (2014). Burnout and daily recovery: A day reconstruction study. *Journal of Occupational Health*. Retrieved from http://psycnet.apa.org/journals/ocp/19/3/303/

Open mHealth. (2015). *Case study: Type 1 diabetes - Open mHealth*. Retrieved May 24, 2017, from http://www.openmhealth.org/features/case-studies/case-study-type-1-diabetes/

Open mHealth. (n.d.). *About Us | Open mHealth*. Retrieved July 24, 2017, from http://www.openmhealth.org/organization/about/

OutSystems. (n.d.). *Low-Code Development Platform for Mobile and Web Apps | OutSystems*. Retrieved January 12, 2017, from https://www.outsystems.com/platform/

Oxford Dictionaries. (n.d.). *Smartphone - definition of smartphone in English*. Retrieved June 29, 2017, from https://en.oxforddictionaries.com/definition/smartphone

Park, Y., & Chen, J. V. (2007). Acceptance and adoption of the innovative use of smartphone. *Industrial Management & Data Systems*, *107*(9), 1349–1365. doi:10.1108/02635570710834009

Picard, R. W., & Healey, J. (1997). Affective wearables. *Personal Technologies*, *1*(4), 231–240. doi:10.1007/BF01682026

Richardson, C., & Rymer, J. R. (2016). The Forrester Wave. *Low-Code Development Platforms*, *Q2*, 18.

Shih, P. C., Han, K., Poole, E. S., Rosson, M. B., & Carroll, J. M. (2015). Use and Adoption Challenges of Wearable Activity Trackers. In *iConference 2015 Proceedings*. iSchools. Retrieved from https://www.ideals.illinois.edu/handle/2142/73649

Stansfeld, S. A., Fuhrer, R., Shipley, M. J., & Marmot, M. G. (1999). Work characteristics predict psychiatric disorder: Prospective results from the Whitehall II Study. *Occupational and Environmental Medicine*, *56*(5), 302–307. doi:10.1136/oem.56.5.302 PMID:10472303

Stylos, J. (2009). *Making APIs more usable with improved API designs, documentation and tools*. Pittsburgh: Carnegie Mellon University. Retrieved from http://search.proquest.com/openview/5d9b5ea97d780f089afab105565a902c/1?pq-origsite=gscholar&cbl=18750&diss=y

Sung, M., Marci, C., & Pentland, A. (2005). No Title. *Journal of Neuroengineering and Rehabilitation*, *2*(1), 17. doi:10.1186/1743-0003-2-17 PMID:15987514

Taormina-Weiss, W. (2012). *Workplace Stress - Symptoms and Solutions - Disabled World*. Retrieved July 20, 2016, from https://www.disabled-world.com/disability/types/psychological/workplacestress.php

Ulrich, R. S., Simons, R. F., Losito, B. D., Fiorito, E., Miles, M. A., & Zelson, M. (1991). Stress recovery during exposure to natural and urban environments. *Journal of Environmental Psychology*, *11*(3), 201–230. doi:10.1016/S0272-4944(05)80184-7

Vrijkotte, T. G. M., van Doornen, L. J. P., & de Geus, E. J. C. (2000). Effects of Work Stress on Ambulatory Blood Pressure, Heart Rate, and Heart Rate Variability. *Hypertension*, *35*(4), 880–886. doi:10.1161/01.HYP.35.4.880 PMID:10775555

West, D. M. (2014). *Going mobile: How wireless technology is reshaping our lives*. Brookings Institution Press. Retrieved from http://www.jstor.org/stable/10.7864/j.ctt7zsvqt

WHO. (2011). *mHealth: New horizons for health through mobile technologies: second global survey on eHealth*. Geneva, Switzerland: WHO Library Cataloguing-in-Publication Data. Retrieved from http://www.who.int/goe/publications/goe_mhealth_web.pdf

KEY TERMS AND DEFINITIONS

Burnout: State of chronic stress related with a profession.

Chronic Stress: Situation in which the autonomic nervous system is not able to adapt a response to a stressor.

Continuous Monitoring: Ability to monitor something as often as desired for as long as you want.

Data Visualization: Study of the form of presentation of data so that they become perceptible.

Health Information System: Organized set of health-related information.

Mobile Technologies: Set of technologies and software that are used in mobile devices.

Wearable Devices: Devices used in the body for collecting data relating to the individual.

Chapter 11
Mobile Technologies for Managing Non-Communicable Diseases in Developing Countries

Siddique Latif
Information Technology University (ITU), Pakistan & National University of Sciences and Technology, Pakistan

Muhammad Yasir Khan
Information Technology University (ITU), Pakistan

Adnan Qayyum
Information Technology University (ITU), Pakistan

Junaid Qadir
Information Technology University (ITU), Pakistan

Muhammad Usman
COMSATS Institute of Information Technology, Pakistan

Syed Mustafa Ali
Mercy Corps, Pakistan

Qammer Hussain Abbasi
University of Glasgow, UK

Muhammad Ali Imran
University of Glasgow, UK

ABSTRACT

Non-communicable diseases (NCDs) are the global leading cause of morbidity and mortality and disproportionately affect more in the less developed countries. Mobile technologies are being used for a variety of purposes in healthcare. Most importantly, they are enabling new ways for NCDs management by providing powerful tools to both doctors and patients for effective prevention and treatment. As the common

DOI: 10.4018/978-1-5225-5270-3.ch011

risk factors of NCDs are related to human behavior; therefore, mobile phone-based health solutions can be used to combat with rising burden of NCDs by focusing on behavioral change programs to promote a healthy lifestyle. This chapter discusses the common NCDs, their burden, and future estimated projections, and shows how mobile phone technologies can provide effective NCDs management in developing countries—which have a lot of issues in their healthcare systems.

1. INTRODUCTION

Non-communicable diseases (NCDs) are becoming the world's largest burden due to their chronic nature— which requires a long duration of care for their effective management. NCDs are causing 40 million annual deaths globally, equivalent to 70% of all deaths (see WHO (2017b)). The main types of NCDs are cardiovascular diseases, cancers, chronic respiratory diseases, and diabetes. Physical inactivity, unhealthy diets, and harmful use of tobacco smoke and alcohol are the major risk factors for NCDs. According to World Health Organization (WHO), excessive use of tobacco accounts for 7.2 million deaths every year; harmful use of alcohol caused over the half of 3.3 million annual deaths; excess intake of salt is attributed to 4.1 million annual deaths, and physical inactivity blamed for 1.6 million annual deaths. Similarly, rapid unplanned urbanization, globalization of unhealthy lifestyle and aging of the population are also causing the rise in NCDs. People of all age groups (i.e., children, adults and the elderly) from all the regions and countries are vulnerable to NCDs. Evidence shows that 15 million of premature deaths of people between the ages of 30 and 69 years are caused by NCDs, and 80% of these deaths occur in low and middle-income countries with an expected increase of 41.8 million by 2030 (Piot et al., 2016). In 2015, the burden of NCDs when compared with the statistics of 2000, it shows a remarkable increase in the burden of NCDs (as illustrated in Figure 1).

The epidemic of NCDs is a real threat to developing countries as it poses devastating challenges to their healthcare systems. Diabetes, cancer, hypertension, and cardiovascular disease (CVD) are the major cause of disability and deaths in low and middle-income areas (Slama et al., 2016). In 2015, WHO reported that 70% of death were caused by NCDs and among all those, diabetes mellitus considered as a major threat to life because it caused complications to many other diseases such as blindness, kidney failure, heart disease, diabetic foot (gangrene) (Pangaribuan & Junifer, 2014). In developing countries, the healthcare systems are already grappling with a large number of challenges. Lack of trained medical staff, unavailability of healthcare services and facilities, and poor health awareness among the population are the major problems in developing countries. Poverty is another cause for the

Figure 1. Percentage of total disease burden accounted for NCDs (total disability-adjusted life year (DALYs))
Source: Reproduced from GHDx, 2015.

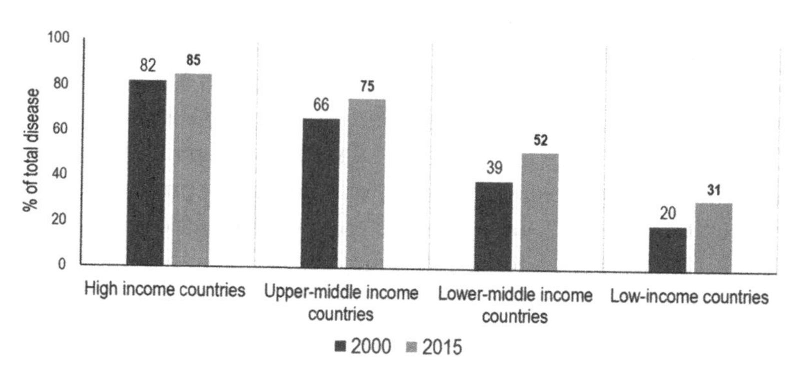

prevalence of NCDs in these countries because poor people are at the high risk to being exposed to the harmful products like tobacco, unhealthy diet practices due to low health awareness. Further, political instability in developing countries badly affects the formulation and implementation of policy for healthcare sectors. In these situations, management of NCDs is very problematic because of their long duration and high-cost of treatment. As a result, each year around 150 million people face the financial crisis, about 100 million people are living under the poverty line due to healthcare expenses, and unfortunately, above 90% of these people live in low-income countries (WHO, 2010b). According to United Nations Project, world population in 2030 will reach 8.5 billion and projected deaths due to NCDs will be 52 million which was 38 million in 2012 (WHO, 2014b). At the same time, the rapid increase in the older population is also accelerating the growth of NCDs. According to World Bank statistics, the total number of elderly people (65+) is 8% of total population in 2016 and this number is increasing drastically which is causing great economic cost (World, 2016).

The technology enabled solutions have incredible opportunities to combat these issues. Encouragingly, mobile phone technologies are the most powerful tools that can transform the current healthcare systems to a new ecosystem, where clinical staff can stay connected to remote areas with faster communication to enable efficient collaboration. According to International Telecommunication Union (ITU), there are 5 billion wireless subscribers and 70% of them living in low and middle-income countries (Kay, Santos, & Takane, 2011). Figure 2 highlights the rapid increase in the mobile phone subscriptions (in millions) in developed countries, developing countries and world during 2005 to 2017. In developing countries, mobile technologies can provide new ways to collaborate with the world best physicians. It allows patients to remotely communicate and consult with experts without any physical presence.

Health awareness and education, training of healthcare workers and epidemic outbreak tracing will be more easily tackled using mobile communication technologies. Despite the great opportunities of mobile phones based healthcare solutions, there are also some hurdles such as inadequate literacy skills, cultural and language barrier, lack of infrastructure, gender difference and societal norms that hinder the adoption of these innovative solutions in developing countries (Latif, Rana, Qadir, Imran, & Younis, 2017) which needs to be addressed by strategic planning to design human-centered solutions with the engagement of both public and private organizations.

The rest of the chapter is organized as follows. In section 3, we present an overview basic NCDs, their burden, and future projections. In section 4, a detailed discussion on the existing healthcare problems in developing countries is presented to motivate the provision of healthcare by using mobile technologies. Section 5 discusses the opportunities of mobile technologies for healthcare. In section 6, a case study on diabetes management using mobile phones is provided to validate the potential benefits of mobile phone technologies for NCDs management. Section 7 provides the future directions to boost the use and adoption of mobile technologies in developing countries followed by conclusions in section 8.

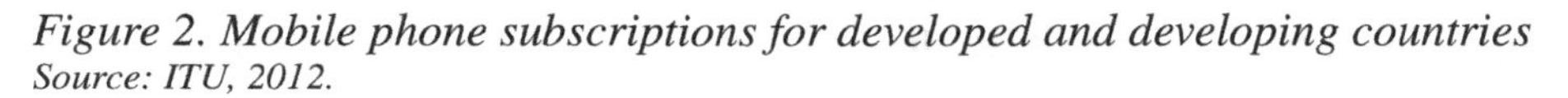

Figure 2. Mobile phone subscriptions for developed and developing countries
Source: ITU, 2012.

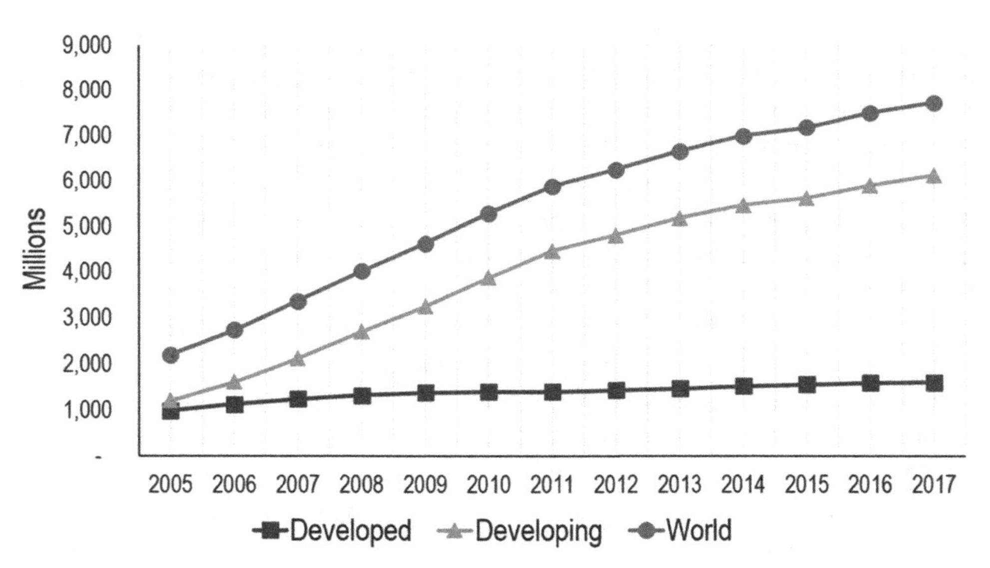

2. COMMON NON-COMMUNICABLE-DISEASES (NCDS)

Non-communicable diseases (NCDs) are not easily continuous from one person to another through direct contact. They are causing a number of premature deaths and disabilities throughout the globe. Table 1 shows the global NCDs burden and future projections.

It is noted that NCDs are increasingly becoming common in our society due to behavioral and environmental factors. The patients with NCDs require a long duration of systematic medical care due to their slow progression and chronic nature. The behavioral risk factors associated with different NCDs are listed in Table 2. Common NCDs along with their adverse effects on developing countries are discussed below.

2.1 Cancer

Cancer is the second leading cause of death worldwide and it is causing 8.8 million annual deaths globally (WHO, 2017a). Of these deaths, approximately 70% (i.e. 5.5 million) is occurring in the developing countries. It is also projected that cancer deaths will rise from 6.7 million in 2015 to 8.9 million in 2030 and deaths caused by cancer in developed countries is expected to remain stable over the next twenty

Table 1. Annually estimated global mortality, burden and projection (2015-2030)

Diseases	Casualties (Million)	Burden (Million)	Future Projection (Million)
Cancer	8.8	17.5	21.6
Diabetes	1.6	415.0	366.0
Cardiovascular	17.7	422.7	23.6
Chronic Respiratory	3.9	> 235.0	5.1

Source: WHO, 2017b; Wild et al., 2004; Roth et al., 2017; Fitzmaurice et al., 2017.

Table 2. Four shared risk factor for Non-communicable diseases

Disease	Tobacco Use	Unhealthy Diet	Physical Inactivity	Harmful Use of Alcohol
Cancer	✓	✓	✓	✓
Diabetes	✓	✓	✓	✓
Cardiovascular Disease	✓	✓	✓	✓
Chronic Respiratory	✓			

Source: Adapted from WHO, 2009.

years (Chan, 2010). These numbers of deaths are not enough alarming, it is worth mentioning that there will be 19 million new cancer cases to be diagnosed in 2025, solely based on the projected demographic changes (Atlas, 2012).

As per International Agency for Research on Cancer (IARC), the development of cancer also depends upon aging factor. The other important factors are unhealthy lifestyles and rapid unplanned urbanization. The frequency of cancer grows dramatically with age, due to a buildup of risks for some cancers that rise with age. Moreover, the tendency for cellular repair mechanisms is also less effective in older people. In developing countries, there are not enough financial resources, advanced infrastructure, technology and trained staff to cope with cancer patients. Because it requires screening and early detection which involve surgery, radiotherapy, or chemotherapy. Such treatments are very costly and mostly reserved for rich people. Poor patients have very little chance of getting proper cancer treatment due to the shortage of cancer hospitals or without financial support. Many of patients remain uncounted and never reach to cancer treatment center for appropriate care. This catastrophe will soon become a crisis due to the rapid rise in global cancer burden in developing countries where populations continue to grow. To reduce the significant suffering, disability, and deaths by cancer, cost-effective programs are needed for screening, early diagnosis, palliative care and treatment.

2.2 Cardiovascular Diseases (CVDs)

The global burden of CVDs is escalating expeditiously due to health transition in the developing countries. CVDs refer to the disorders related to blood vessels and heart that include coronary heart disease, rheumatic heart disease, cerebrovascular disease, and other conditions. Disorders of blood vessels cause heart attack, chest pain (angina) or stroke. While heart disorders affect heart's muscles, valves or rhythm. CVDs are mostly caused by unhealthy diet, harmful use of tobacco and alcohol, and physical inactivity. Cessation of tobacco use, avoiding alcohol, salt reduction in diet, consuming fruits and vegetables along with regular physical activity reduce the risk of cardiovascular disease.

According to WHO, CVDs are the leading cause of global deaths. CVDs are causing 31% (approximately 17.7 million) of annual deaths and more than 75% of this number is contributed by the developing world. In 2015, out of the 17 million premature deaths by NCDs, 82% were in low and middle-income countries, and 37% were caused by CVDs. In developing countries, people do not have an integrated healthcare system for early detection and treatment of CVDs' risk factors as compared to the developed nations. They have less or limited access to equitable and cost-effective care services for their actual needs. As a result, many people in these countries are detected late and die a premature death from CVD, often in their most productive age. Similarly, poor people are affected most due to out-of-pocket spending and catastrophic health expenditure. In the developing countries, most of the CVDs can be prevented by reducing behavioral risk using population-wide as well as individual level strategies to provide very cheap and effective integrated health interventions. These interventions should be feasible to implement in low-resource settings.

2.3 Diabetes

Diabetes is a growing health challenge that poses a threat to the economies of both developing and developed nations. The rise of diabetes is fueled by nutrition transition, sedentary lifestyles, and increase in obesity. Asia's large population tend to have diabetes at younger ages due to high usage of smoking and alcohol, high intake of carbohydrates and decreased physical activities (Hu, 2011). Further, the development of diabetes in children is another major global threatening challenge as well that is propelling the upsurge in diabetes patients (Hossain, Kawar, & El Nahas, 2007). Diabetes is a chronic metabolic disease that is characterized by the raised levels of blood glucose. This can lead to a severe damage to various body organs such as kidneys, heart, eyes, and nerves. There are three types of diabetes that are type 1, type 2 and gestational diabetes. The most common diabetes is type 2, usually in adults. This is caused when the body does not make sufficient insulin or become resistant to insulin. Type 2 diabetes starts slowly and patients suffer from overweight, blur vision and slow healing. In type 1 diabetes, the pancreas produces very small amount or no insulin by itself. Patients with type 1 diabetes suffer from rapid weight loss, increase in thirst and appetite, fatigue and increased urination. The sufferers of gestational diabetes are pregnant women and blood sugar returns to normal after the delivery.

The burden of diabetes is substantial, at least one in 20 worldwide deaths are caused by diabetes and it is projected that diabetic patients could double in developing countries from 115 million diagnosed in 2000 to 284 million in 2030 (GENEVA, 2003). According to International Diabetes Federation (IDF), in 2015, 415 million adults were living with diabetes globally and the estimated figure in 2040 will be 642 million people between the age 20-79 years. Diabetes and other NCDs are imposing a double burden of diseases on the developing countries as these countries are still struggling to deal with the infectious diseases like HIV/AIDS, tuberculosis and malaria. In developing countries, approximately 9.1% people of the entire population are living with diabetes and over 40.6% people are undiagnosed (Ingelheim, n.d.). Early diagnosis and regular physical activities with proper diet can help to manage diabetes. Similarly, cessation of tobacco is also crucial to avoid complications. In order to overcome diabetes in developing countries, there is a need for proper programs and facilities that can provide a cost-effective screening for early signs of diabetes-related disorders like retinopathy and kidney disease, etc.

2.4 Chronic Respiratory Diseases (CRDs)

Chronic respiratory diseases (CRDs) include asthma, pulmonary hypertension, lung diseases and chronic obstructive pulmonary disease (COPD). These diseases are disorders of airways and other lungs structures. CRDs are considered as the major cause of premature deaths in adults globally (Chuchalin et al., 2014). The spread of CRDs is increasing worldwide among children and elderly and it is projected that the prevalence of CRDs will increase both in developed and developing countries (Fairall et al., 2005). The risk factors for CRDs are tobacco smoke, air population, occupational dust and chemicals, and frequent respiratory infections in childhood. According to the recent survey of Center of Disease Control (CDC), one in 13 people (about 24 million people) had asthma in 2016. This situation is worse among children i.e. about 1 in 10 children (8.6%) had asthma in 2016. Further, WHO reported that above 3 million people die due to COPD every year which is an estimated 6% of global deaths and more than 90% these deaths are in low and middle-income countries. CRDs represent a major healthcare challenge in developing countries due to their frequency, severity, and economic impact. There are various forms of treatment that can assist to control symptoms and improve the quality of life for the patients with CRDs.

3. HEALTHCARE PROBLEM IN DEVELOPING COUNTRIES

For centuries, communicable diseases were considered the main causes of death globally. The advancements in medical research in terms of antibiotics, vaccination, and improvement in healthy lifestyle have increased life expectancy. These things are causing the rise of NCDs with disproportionately higher impact in developing countries. Between 2008 and 2030, deaths caused by NCDs are projected to rise by 45% in low-income countries while 1% and 12% in highand middle-income countries respectively (as shown in Figure 3). In Africa, deaths from NCDs are growing faster than anywhere else in the globe (WHO, 2010a). Combined with the persistent problem of managing infectious diseases, developing countries are now confronted with a double disease burden. Therefore, developing countries need new ways to ensure early access to affordable medications with limited resources to cope with the increasing burden of NCDs.

Figure 3. NCDs attributed to a growing share of global deaths, especially in developing regions
Source: Adapted from Nikolic et al., 2011.

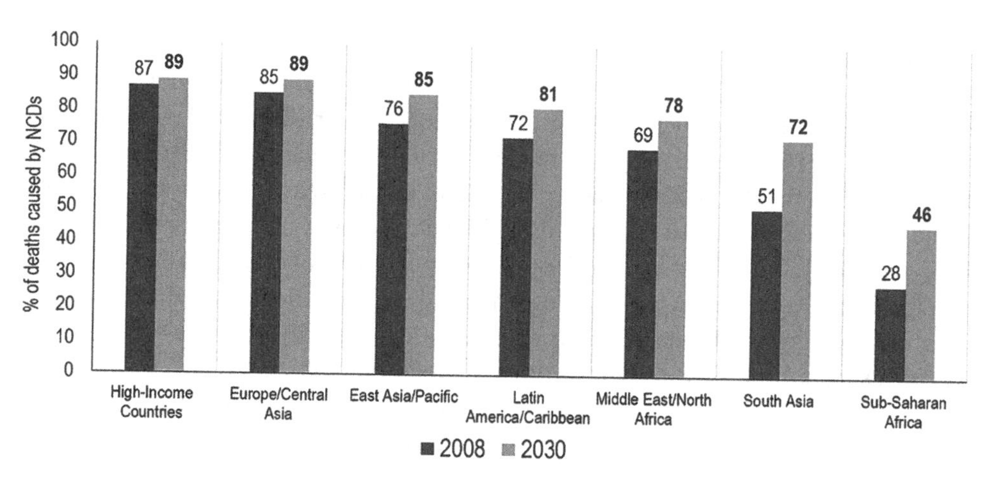

3.1 Lack of Monitoring Facilities

Healthcare systems provide health services to facilitate and maintain people's health. These actions are highly influenced by political, social, cultural factors of a country. Developing countries have lower budgets and poor policies for the improvements in healthcare systems. People are less likely to receive proper care due to the geographically uneven distribution of health services and the quality of care is very poor when they are available. For example, healthcare facilities in some developing countries lack water, proper sanitation, and hygiene which are causing risk within institutions to which patients expect healing (UNICEF & WHO, 2015). Therefore, poor people in developing countries tend to have less access to quality of care, and millions of people suffer and die from the disease for which effective treatments and interventions exist. In poor countries, economic resources are not sufficient to provide and support the provision of essential healthcare facilities (O'Donnell, 2007). Patients suffering from NCDs needs proper monitoring for effective treatment but in developing countries, due to poor health monitoring and early detection services, the prevalence of NCDs is rising and demanding an immediate attention to overcome this burden.

3.2 Lack of Healthcare Professionals

Access to the health care services is a basic necessity that should be available to every person without any discrimination. People in developing countries are unable to avail proper treatments due to limited access to basic healthcare facilities with the additional problem of unavailability of resources. The shortage arises due to limited availability of physicians, services, and lack of trained medical staff (Latif, Qadir, Farooq, & Imran, 2017). The accessibility and availability of healthcare services are quite challenging in developing countries especially in rural areas where residents have to travel long distances to visit some specialist or doctor. For example, in Bangladesh, 70% of the population resides in rural areas (Islam & Biswas, 2014) but most of the health facilities concentrated in urban areas ignoring the majority of the population. Because of limited healthcare professionals, many people with diseases often left undiagnosed until outcomes become more fatal. These situations are creating an unnecessary burden of suffering on the patients and their families. In future, situation will become more challenging as highlighted by WHO– there will be a shortfall of 12.9 million healthcare workers in 2035 (WHO, 2013). Many countries have raised their healthcare workforce with the threshold of 23 skilled professionals per 10,000 people, but there are still 83 countries especially in Asia and sub-Saharan Africa with insufficient healthcare professionals.

3.3 Cost Constraints

NCDs are presenting an unprecedented challenge to global economies and hurdle their social development. In developing countries, the economic burden of NCDs is affecting people at all levels especially to the households exposed to the poverty. Due to expensive treatment for NCDs, millions of people face the financial crisis in low-income countries (WHO, 2010b). According to IDF, one in 11 adults has diabetes and 12% of the global health expenditure is spent on treatment of diabetes. Cost of medicine and regular hospital visits are the largest components that inhibit many NCDs patients from seeking the treatment. A study on rural Ghana has shown that the cost of insulin alone is 60% of monthly income of those on minimum daily wages (Aikins, 2005). While in Pakistan, the direct cost of co-morbidity is 45% greater than the direct cost of patients without co-morbidity (Khowaja, Khuwaja, & Cosgrove, 2007). Most importantly, in these poor countries, there is no financial support for the treatment of NCDs and all financial burden is solely borne by household (Kankeu, Saksena, Xu, & Evans, 2013). According to Assistant Secretary Planning and Evaluation (ASPE) brief 2016, 43.4% of the rural population reported that they did not have usual care facilities and 26.5% got delayed treatment or did not receive care in the last year due to high costs. Therefore, there is a crucial need for the cost-effective solutions to enable the population to avail equal health facilities without any financial discrimination. In this regard, mobile technology can be a useful platform to provide a cost-effective solution.

3.4 Problems With Infrastructure

The purpose of infrastructure is to provide timely and effectively the essential public health care services. Healthcare infrastructure provides the facilities to prevent disease, promote health and healthy activities, and be prepared for both acute and chronic challenges. A public health infrastructure is based on three key factors: competent and qualified workforce, up-to-date data and health providers able to access and respond timely to public health needs. Generally, healthcare facilities in developing countries have evolved to cope with the burden of communicable infectious diseases. The new pressing challenge is the prevention and management of NCDs. India is the second largest populated country in the world with inadequate health care infrastructure (Kumar & Gupta, 2012). Now, there is a need of thinking new strategies on how such medical services in developing countries are financed. These services include the provision of appropriate and effective treatments for timely and cost-effective screening and diagnostic of people.

4. OPPORTUNITIES FOR MOBILE TECHNOLOGIES IN NON-COMMUNICABLE-DISEASES (NCDS)

Mobile technologies have great potential to be used as economical, robust, secure and real-time diagnostic tool or health monitoring device. Figure 4 shows a general model for mobile phone-based NCDs management system which enables a two-way communication among patients and doctors. In this way, patients can easily share their health data with doctors to seek treatments from experts.

These mobile phone-based systems can provide various powerful methods for advancing NCDs healthcare such as awareness, promotion, prevention, early diagnosis, surveillance, monitoring, treatment adherence and de-addiction as highlighted in Figure 5 and these opportunities are discussed below.

4.1 Health Education and Awareness

The prevalence of NCDs is rising and these diseases usually require a strong behavioral modification for preventive and strategic care. It is noted that more than half of disease burden can be prevented by behavioral and life style changes, and increased health awareness among the population (mHealth Alliance, 2013). Similarly, lack of awareness about the disease can also lead to a large number of deaths in a society when a disease outbreak happens. Therefore, for effective control on the rise of NCDs health education and awareness are very crucial. NCDs can be largely avoided and managed effectively by reducing four common risk factors, i.e., unhealthy diet,

Figure 4. General architecture for mobile phone based NCDs management system

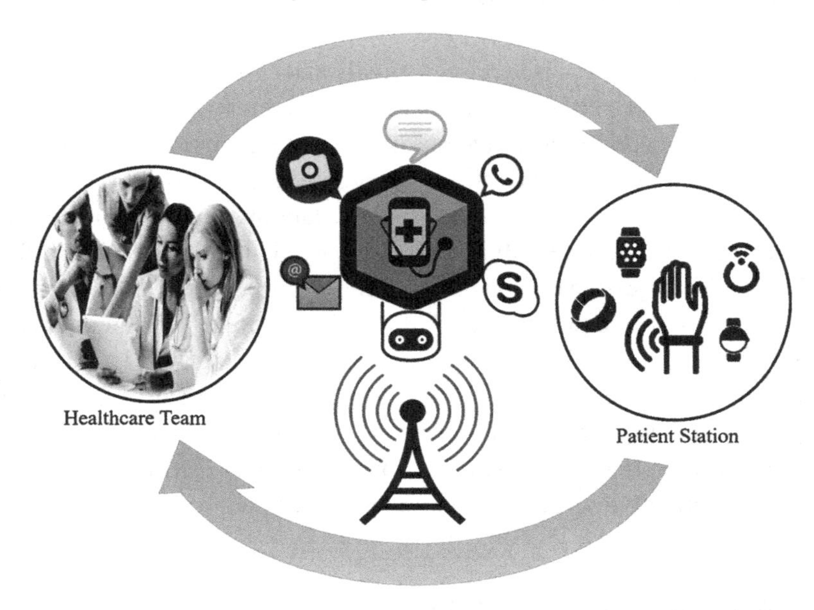

Figure 5. Mobile technologies opportunities for non-communicable-diseases (NCDs)

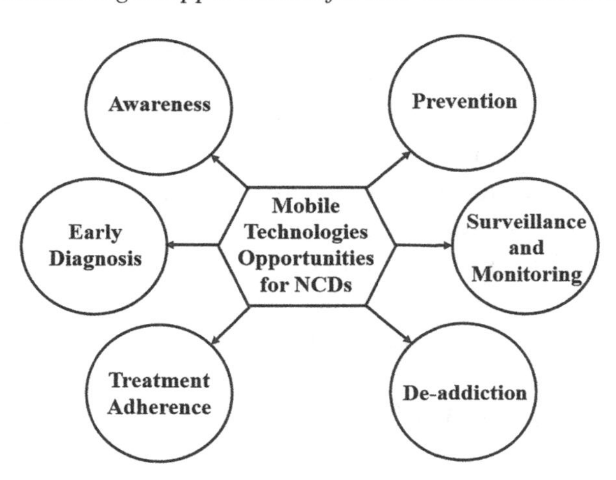

tobacco use, alcohol consumption, and physical inactivity. It has been shown that 80% of heart disease and stroke, 80% diabetes, and 40% cancer disease can be prevented by avoiding these risk factors (Kumar & Gupta, 2012). Mobile phone based services like audio and textual feedback have potential to provide health interventions and education in resource-constrained environments, and they have high penetration among patients with NCDs including adults, elderly, with less education, and other with socioeconomic risk factors. In developing countries, mobile phones have already been successfully utilized in the infectious diseases and communicable diseases to train health workers, to ensure treatment adherence, and to enhance health education among the population. In the case of NCDs, there is good evidence on mobile phone technologies to increase health awareness among people (Majumdar, Kar, Kumar, Palanivel, & Misra, 2015). A joint initiative, between WHO and ITU, aimed to use mobile phones as a channel to promote healthy habits related to the prevention and reduction of NCDs. This project provides toolkits and technical stuff to encourage mobile based NCD prevention programs (ITU & WHO, 2012).

4.2 Prevention and Early Diagnosis

In many developing countries, healthcare systems are hospital-centered and people used to admit in these hospitals after acute conditions, long-term complications and disabilities. The treatment of these patients in the later stages of NCDs is very expensive and requires high technology-intensive operations. Due to large population per hospital, it becomes very difficult to accumulate every patient and a large number of people with the risk of NCDs remain undiagnosed and those who are diagnosed have limited access to treatment. To halt the rising trend of NCDs, integrated approaches for prevention and early detection of NCDs are required.

Mobile phone-based prevention and early diagnostic systems are the key strategic tools that can fill this gap by providing cost-effective and sustainable healthcare interventions with limited resources. These solutions increase the involvement and interest of people in their health and empower them to monitor their own health. Further, these interventions improve people health literacy which result in the early detection of disease and avoid costly high technology treatments. It had been proved that assessment of preventive care can reduce cardiovascular risk such as heart attacks, strokes. Similarly, early treatment of diabetes patients can avoid diabetic nephropathy and expensive renal dialysis (WHO et al., 2010). An important aspect of the persuasive technique is that it can be leveraged for human behavior development and social good (Harris, Qadir, Khan, & Ahmad, 2017). Mobile phone sensors such as GPS or proximity sensors can be used to nudge the individuals towards making the right choice for their health on the basis of their routine activities. These solutions can cut healthcare costs by reducing illness and hospital admissions by keeping patients physically active and adhere them to their medications. Further, smartphone based applications can be integrated with wearable body sensors to provide effective health monitoring and early diagnosis.

4.3 Surveillance and Monitoring

Surveillance is a process of continuous, systematic collection, analysis, and interpretation of public health data to design, improve, implement, and evaluate the public healthcare practices. The purpose of surveillance and monitoring system is to track the health-related activities of people to take proper actions against diseases. For NCDs, surveillance and monitoring are very crucial that keep doctors to be updated about patients' health status. The classical techniques for monitoring the patients are not sufficient to provide enough information regarding patients' health. Wireless sensors and smartphones are being used for sending patients data to physicians (Dobkin & Dorsch, 2011). This enables the clinicians to monitor compliance and risk factor during the medication. Feedback from doctors motivate the patients with NCDs for self-management, which can avoid hospitalization and also help in detecting medical complications that need re-hospitalization. Disease surveillance based on home monitoring using mobile phones are more sophisticated and cost-effective solutions for the elderly population with NCDs in the community. This can avoid the physical presence and reduce the cost of visits. Patients do not have to wait for the appointment, they can consult with their doctors remotely based on the health records. Further, the supplies of medicines can also be tracked using mobile technologies that can prevent from stockouts of essential medicines and can also be used to produce insights for drug development procedure.

4.4 Treatment Adherence

Treatment adherence is of significant importance and it is the measure of how correctly a patient follows the treatment advice e.g. medication and drug compliance. It can also refer to other conditions as well such as self-care, correct use of a medical device and self-directed activities like exercises and diet plan, etc. In developed countries, average treatment adherence to chronic diseases is 50% and it is considerably low in developing countries (Mwangi & Mukanya, 2017). According to BMC (BioMed Central) health services research (Gadkari & McHorney, 2012), nearly two-thirds examiner reported that the patients usually forget to take their medication and consequently failed to adhere to their treatment prescription. Various factors such as negligence in self-medication, complex medical regimen, forgetfulness and lack of interaction with healthcare providers are blamed for poor treatment adherence. Treatment adherence in NCDs is very important, as these diseases require self-care and management, hence poor adherence can pose patients to various potential life-threatening risks. Mobile phone-based behavior change interventions (i.e., using text messaging) to encourage patients to adhere their medication, increase exercise, improve their diets plans, and to reduce harmful use of tobacco are very successful

(Cole-Lewis & Kershaw, 2010). In a study (Beaglehole, Ebrahim, Reddy, Voute, & Leeder, 2008), it is demonstrated that compliance improved significantly due to behavioral interventions, which reduced the demand for monitoring facilities and feedbacks from doctors. Mobile phone can be used for behavioral changes programs by sending tailored messages or interactive voice response (IVR) based on the individuals' beliefs, traits, and abilities.

5. CASE STUDY: THE SCOPE OF MOBILE PHONES FOR DIABETES MANAGEMENT IN DEVELOPING COUNTRIES

Mobile phones are proved to be the most suited and prosperous technology that is being harnessed in developing countries settings to improve NCDs management and care. They can be used as a tool for the collecting health data, surveillance of service delivery, and improvement of primary care by providing evidence-based care. As a diagnostic tool, mobile phones can help healthcare workers in screening and diagnostic trails for NCDs in the community. In this section, we perform a case study on diabetics' management using mobile phone-based solutions to validate their prosperity for NCDs management. Mobile technologies offer almost similar healthcare solutions to NCDs, therefore, we present some common solutions for diabetic that can have similar applicability to other NCDs. Following are some potential use cases, where mobile phones are playing a major role in diabetes management.

5.1 Screening and Diagnosis of Diabetes

Over the last few decade, the prevalence of diabetes is rapidly increasing especially type 2 diabetes. The alarming fact is that half of the people do not have knowledge about the condition. It seems logical that earlier detection of diabetes may help. For example, early detection and treatment of patients with diabetes can reduce morbidity (Association et al., 2013). The implication of such early detection system needs to screen the people regularly. Mobile phones have great potential to be used as the tool to screen the people on regular basis. Diabetic retinopathy (DR) is one of the major causes of blindness in adults with diabetes and its impact is high in most parts of the world. For example, the prevalence of DR in India is 18% (i.e., over 65 million people) (Rema et al., 2005). Therefore, such large number of patients need to be screened every year for DR. An ideal screening method require trained ophthalmologist. In developing countries, insufficient number and non-availability of ophthalmologists, and a high cost of the conventional screening procedures are the limiting factors for screening such a large numbers of people with diabetes. Mobile phone based solution can be effectively utilized in such situations. For example, in

(Rajalakshmi et al., 2015) smartphone-based retinal color photography was used for screening DR in the clinic and for teleophthalmology. They have screened 301 patients with type 2 diabetes and proved that smartphone-based retinal photography is effective for screening and diagnosis of DR with high sensitivity and specificity. Similarly, iBGStar is an innovative blood glucose reading device that can be connected to any Apple iOS device. Readings of blood glucose can be immediately displayed using iBGStar diabetes manager application (app). Hence, such mobile phone-based solutions can be harnessed in developing countries to screen the population at the individual as well as community level. For example, Bluetooth-based glucometers can be used by health workers in developing country for screening and detecting un-diagnosed diabetes patients at the community level by reducing the laboratory.

Machine learning models have abilities to be used for the diagnostic of diabetes using data gathered by mobile phones. A mobile phone-based system was proposed by (Sridar & Shanthi, 2017) proposed that can diagnose diabetes based on the real-time data from glucometer and retinal image features. In this study, the authors used feature selection technique to get the optimal features. These features are given to neural network to generate a dataset with minimum errors. This system was able to effectively diagnose type I and type II diagnosis mellitus with improved accuracy. Bourouis et al. (2014) proposed an innovative and low-cost smartphone based intelligent system that use a microscopic lens to perform eye examinations in remote areas. In this technique, they used Artificial Neural Networks to analyze the retinal images of patients for the identification of retinal diseases caused by diabetes.

These solutions show the potential of mobile phone technologies for screening and diagnosis of diabetics. In most of these solutions, mobile phones are used to collect health data in real-time that is transmitted to the operating system for diagnosis or screening of patients. Such solutions are very suitable for developing countries, where monitoring facilities are very limited. Another important aspect of mobile phone-based systems is that they can also be exploited for screening and diagnostic of other NCDs such as cancer, asthma, etc.

5.2 Evidence-Based Treatment

Evidence-based treatments or interventions based on the routine activities of patients, their diet plans, and medical history are very crucial for effective management and treatments of NCDs. Mobile phone can be used to capture patients' health data using wireless sensing devices or built-in sensors like camera, microphone, accelerometer, and GPS. This data is transmitted or shared in the form of health records to the doctors in many ways such as SMS, MMS, Bluetooth, Internet, and physical memory cards. When the mobile connection is not available, this data can be stored on the mobile phones' internal memory that can be transmitted later. In this way, patients' health data is available to the medical practitioners for better treatments based on patients' medical records. For example, diabetes is a chronic disease and its effective management requires optimal control over blood pressure, glucose and lipids to a determined level. In a current setting of developing countries, patients are unable to get tailored treatments based on their medical history. It is also evident that there is a large gap between actual recommendations and effective use of drugs for diabetes (Pan, Yang, Jia, Weng, & Tian, 2009). These facts highlight the importance of evidence-based treatment that can support and improve the quality of care.

Mobile phones are being utilized to record the patient's routine activities and vital signs, necessary for effective care. In a pilot project, cell phone-based diabetes management system proved to be a very encouraging solution, as it can facilitate treatment decisions based on organized data and reduces logbook time for making trends from patient history (Quinn et al., 2008). A smartphone-based gestational diabetes management system is proved to be very effective by providing an improved glucose control with a reduced number of clinic visits (Mackillop & team, 2014). This system uploads blood glucose data in real-time by reducing manual transcription errors. Nurses can respond quickly to the patient by analyzing this data without the frequent time-intensive clinic visits. These solutions highlight the potential benefits of evidence-based diabetes treatment systems that can enable a two-way communication between healthcare professionals and patient. Such mobile-phone-based systems can also help non-physician health workers to manage uncomplicated cases of diabetes in remote areas where the trained physicians are the major constraint.

5.3 Remote Diabetes Care

Remote monitoring using mobile phones can make NCDs' patients to be more self-aware and more accountable for their health knowing that their data is being monitored by their healthcare provider. Patients engagement to their medical records in real-time provide an increased efficiency with a reduced healthcare costs by avoiding patients' physical presence. The advancements in mobile communication technologies offer numerous ways of monitoring patients' health remotely. It provides myriad of software and hardware applications that can be used for remote monitoring of NCDs. For, example, mobile phones can be used collects oxygen concentration in blood, glucose level, blood pressure readings, and other patients' activities and their emotions. This remotely captured health data can help physicians in various ways to gain deep insights into patients' life. In the case of diabetes, mobile phone-based remote monitoring systems can help diabetes patients to control hypertension and their blood pressure, etc. Smartphone applications can help the diabetes patients in different ways such as long-term tracking and surveillance of their health, prescription management to improve diabetes control and quality of life. They can also be used to register individuals eating behaviors regularly that is very helpful to track the blood glucose level.

Diabetes Interactive Diary is a mobile phone-based program to enable a direct communication between a physician and patients. Both parties can interact through Short Message Service (SMS) so glucose levels can be monitored and insulin dose can be directly suggested (Daskalaki, Prountzou, Diem, & Mougiakakou, 2012). In a study by (Logan et al., 2007), a home blood-pressure management system was developed as a pilot-test to actively engage patients. This system consisted of a Bluetooth-enabled blood-pressure monitor, mobile phone, and other related components. This home-based BP management system proved to be very effective among the patients having uncontrolled hypertension. Similarly, mobile phones enable remote interventions and coaching of complex skills that can help diabetes patients to control blood sugar in severe condition. Using mobile phones, patients with diabetes condition are now able to collect their daily routine data, and by sharing this data with experts, they receive concrete and tailored feedback manage their illness (Kollmann, Riedl, Kastner, Schreier, & Ludvik, 2007).

Such mobile phone-based solutions are very suitable for the people of remote areas, where there is a scarcity of health professional. Mobile phone-based monitoring systems can be used to alert the healthcare providers in any serious condition that need immediate treatment. Such timely detection of the severe condition is the key to prevention from serious deterioration. Therefore, many of the interventions of NCDs can be provided by capturing vital indicators or sign with mobile phones.

5.4 Lifestyle Management

Mobile phones are the most common tool to manage the daily routine life of diabetes patients by developing existing practices and interventions. There are a number and variety of mobile applications that have been proved to be very effective in managing patients' lifestyle with improved health outcomes. Lack of proper diagnosis and health management of patients with diabetes can lead to various serious consequences such as heart disease, blindness, kidney failure, stroke and severe foot sores. Therefore, health management of diabetes patients is very helpful in living a managed life.

Mobile phones are being used in helping people with diabetes to manage their lifestyle for during Ramadan [1]. Throughout the month of Ramadan, health management is very challenging for people with diabetes. Mobile phone enables people to receive gaudiness through text messages on daytime fasting and evening feasts. mDiabetes is a project that has been launched in 2013, in Senegal for the month of Ramadan to help people with diabetes and increase awareness to prevent complications due to fasting and feasting (WHO, 2014a). This project was proved to be very effective and most people found it very helpful. Weight management of diabetes patient is a very crucial factor. Mobile phone applications with Integrating Body Mass Index (BMI) can bring an optimal balance between the body weight and required calorie for the body of a patient. Similarly, automated data capture using the mobile phone can also be used to keep the physical record (i.e., step count) of the patient. This data is very helpful for patients to understand the need for increasing or decreasing the intensity of physical work. In rural areas of developing countries, people do not have knowledge about the causes and symptoms of NCDs particularly, and often have very limited access to basic healthcare facilities (Latif et al., 2017). Mobile phone-based life management tools such as blood glucose monitor, diet management, weight management, education and awareness using automated data capturing of daily routine activity are very helpful in improving their daily life.

6. FUTURE DIRECTIONS TO PROMOTE MOBILE TECHNOLOGIES FOR NCDS MANAGEMENT

As highlighted in this chapter, NCDs are increasing rapidly worldwide and are a leading cause of morbidity and mortality. In this context, use of mobile technologies to halt the epidemic of NCDs is crucial. Particularly, to meet the goal of adopting mobile technologies for NCDs management for a longer term, the promotion of such interventions to the public, healthcare service providers and stakeholders is required. Various directions to promote the use of mobile technologies for NCDs management in developing countries are discussed as follows.

6.1 Behavioral Change for Technology Adoption

Technology adoption is influenced by various factors such as age, gender, socioeconomic norms and privacy concerns. These factors hinder the individuals of developing countries in adopting technology for healthcare, e.g., people are vulnerable to their privacy and misuse of their personal data. Therefore, to overcome such issues, the establishment of ethical committees is required at the national level, policies should be made to ensure that individuals data must be used securely. Such actions will increase the sense of satisfaction among the people and will leverage them for behavioral change.

Any modification or transfiguration in human behavior is known as behavioral change, there are various factors that influence the human behavior. Various behavioral theories attempt to describe different views on the change in human health behavior. Changing behavior is a central objective for technology adoption (Organization,

2002) and human behavior plays a vital role in this regard. In developing countries, immoderate human behaviors such as tobacco use and unhealthy diet plan, etc., affect directly in raising risks for NCDs. In this context, mobile applications can be used to play an empowering role in changing behavior (Hartin et al., 2016).

To promote technology adopting behavior for a longer term, it requires a multifaceted approach to advocate communities for changing behavior for better management of NCDs. In this context, various strategies can be followed, for example, local community leaders, clergy, community activists and religious leaders can be involved in changing social norms of peoples. Various communal gatherings, public seminars, meetings and advertisement at governmental level can be deemed as ideal places for this purpose. To accomplish the goal of reducing morbidity and mortality associated with NCDs in developing countries, acceptance and adoption of new mobile technologies is required by both healthcare service providers and patients.

6.2 Public and Private Partnership

The partnership between public and private sectors can act as the foundation for establishing an adaptable, sustainable and results-oriented framework for NCDs management. To stop the increasing epidemic of NCDs in developing countries, development of advanced infrastructure by utilizing mobile technologies for the betterment of diagnosis and treatment processes is required. Therefore, public and private sectors should focus on strategic planning and partnership to meet the objectives of developing mobile-based health ecosystem by utilizing their resources, core competencies and knowledge. There are many opportunities for mobile technologies in healthcare delivery services, but implementing them on a larger scale is more challenging in developing countries. While joint ventures of public and private organizations with common vision and targets have promising potential to overcome this challenge. In this context, the government, non-profit/for-profit private sector and non-government organizations can be engaged to stimulate maximum cooperation in halting NCDs epidemic. For example, in Pakistan, a public-private partnership led by NGO Heartfile with Ministry of Health and regional WHO office was undertaken for developing a national action plan for strategic management of NCDs (Heartfile, 2004).

6.3 Mobile Applications and Their Interoperability

Mobile applications can aid the individuals in empowering self-healthcare and there are millions of application available for different platforms such as Apple and Android. According to report on mobile health (mHealth) economics, there are 259,000 health apps available for iOS and Android. Despite the various benefits of these applications, their authenticity is questioned due to the lack of regulation and there is a limited number of authentic health applications. Authenticity problem leads to the necessity of imposing a condition to mHealth market to ensure weather, these applications are secure and reliable.

Mobile applications interoperability means the ability of mobile apps to work together within or across clinical practices to improve the healthcare delivery services. Today, mobile applications are not interoperable across different platforms and it is very challenging to deliver patient-centric healthcare solutions. Therefore, research is needed on the development of interoperable applications to provide improved accessibility, better opportunities for diagnosis and treatments of NCDs.

6.4 Usability and Human Centered Solutions

Usability testing of mobile-based solutions is crucial in analyzing how end users interact with the technology to make a tangible impact on rising epidemic of NCDs. The evaluation outcomes can be utilized to enhance patient-centered healthcare delivery services. Also, knowing usability trends will assist in the assessment that technology is effectively and efficiently used by users or not. However, there are various challenges such as age-related factors, user's attitude, inadequate availability of learning support and other technology-related issues encountered in measuring usability. These factors directly affect the outcomes of analysis and should be considered in the design process. Hence, while designing a human-centered system or application, to be deployed in a developing country, characteristics and social behaviors of targeted users should be considered. To provide effective human-centered solutions, it is very important to understand users' needs and requirements as well as how they would interact with the system.

7. CONCLUSION

The overall burden of NCDs is rising due to environmental and behavioral risk factors, and their management is very challenging in developing countries because of various persistent healthcare problems such as lack of trained staff, limited monitoring facilities, and cost constraints. Despite tremendous opportunities of mobile interventions for the cessation of NCDs epidemic, there are many factors hampering their implementation in developing countries. Keeping in view on these critical aspects, this chapter gave in-depth insights, opportunities, and challenges for the successful implementation of mobile technologies based innovative solutions for NCDs management in developing countries. Also, major health issues, possible ways to overcome encountered challenges and various directions to promote mobile interventions are highlighted. A case study is presented, to validate the potentials of mobile technology for diabetes management that are also helpful for designing solutions for other NCDs. In developing countries, mobile technologies based innovative healthcare solutions would be more effective if noteworthy problems faced in deploying such solutions are given adequate deliberation.

REFERENCES

Aikins, A. G. (2005). Healer shopping in africa: New evidence from rural-urban qualitative study of ghanaian diabetes experiences. *BMJ (Clinical Research Ed.)*, *331*(7519), 737. doi:10.1136/bmj.331.7519.737 PMID:16195290

Association, A. D. (2013). Standards of medical care in diabetes—2013. *Diabetes Care*, *36*(Supplement 1), S11–S66. doi:10.2337/dc13-S011 PMID:23264422

Atlas, C. (2012). *Human development index ranking*. Retrieved from https://goo.gl/kqGSVS

Beaglehole, R., Ebrahim, S., Reddy, S., Voute, J., & Leeder, S. (2008). Prevention of chronic diseases: A call to action. *Lancet*, *370*(9605), 2152–2157. doi:10.1016/S0140-6736(07)61700-0 PMID:18063026

Bourouis, A., Feham, M., Hossain, M. A., & Zhang, L. (2014). An intelligent mobile based decision support system for retinal disease diagnosis. *Decision Support Systems*, *59*, 341–350. doi:10.1016/j.dss.2014.01.005

Chan, M. (2010). *Cancer in developing countries: Facing the challenge*. Retrieved from https://goo.gl/YeYsQL

Chuchalin, A. G., Khaltaev, N., Antonov, N. S., Galkin, D. V., Manakov, L. G., Antonini, P., ... Demko, I. (2014). Chronic respiratory diseases and risk factors in 12 regions of the russian federation. *International Journal of Chronic Obstructive Pulmonary Disease*, *9*, 963. doi:10.2147/COPD.S67283 PMID:25246783

Cole-Lewis, H., & Kershaw, T. (2010). Text messaging as a tool for behavior change in disease prevention and management. *Epidemiologic Reviews*, *32*(1), 56–69. doi:10.1093/epirev/mxq004 PMID:20354039

Daskalaki, E., Prountzou, A., Diem, P., & Mougiakakou, S. G. (2012). Realtime adaptive models for the personalized prediction of glycemic profile in type 1 diabetes patients. *Diabetes Technology & Therapeutics*, *14*(2), 168–174. doi:10.1089/dia.2011.0093 PMID:21992270

Dobkin, B. H., & Dorsch, A. (2011). The promise of mhealth: Daily activity monitoring and outcome assessments by wearable sensors. *Neurorehabilitation and Neural Repair, 25*(9), 788–798. doi:10.1177/1545968311425908 PMID:21989632

Fairall, L. R., Zwarenstein, M., Bateman, E. D., Bachmann, M., Lombard, C., Majara, B. P., & … . (2005). Effect of educational outreach to nurses on tuberculosis case detection and primary care of respiratory illness: Pragmatic cluster randomised controlled trial. *BMJ (Clinical Research Ed.), 331*(7519), 750–754. doi:10.1136/bmj.331.7519.750 PMID:16195293

Fitzmaurice, C., Allen, C., Barber, R. M., Barregard, L., Bhutta, Z. A., Brenner, H., & ... others. (2017). Global, regional, and national cancer incidence, mortality, years of life lost, years lived with disability, and disability-adjusted life-years for 32 cancer groups, 1990 to 2015: A systematic analysis for the global burden of disease study. *JAMA Oncology, 3*(4), 524–548. doi:10.1001/jamaoncol.2016.5688 PMID:27918777

Gadkari, A. S., & McHorney, C. A. (2012). Unintentional non-adherence to chronic prescription medications: How unintentional is it really? *BMC Health Services Research, 12*(1), 98. doi:10.1186/1472-6963-12-98 PMID:22510235

GENEVA. (2003). *Diabetes cases could double in developing countries in next 30 years.* Retrieved from https://goo.gl/5b7kBf

GHDx. (2015). *Global burden of disease study 2015 (gbd 2015) data resources.* Retrieved from http://ghdx.healthdata.org/gbd-2015

Harris, A., Qadir, J. K., & Ahmad, U. (2017). *Persuasive technology for human development: Review and case study.* arXiv preprint arXiv:1708.08758

Hartin, P. J., Nugent, C. D., McClean, S. I., Cleland, I., Tschanz, J. T., Clark, C. J., & Norton, M. C. (2016). The empowering role of mobile apps in behavior change interventions: The gray matters randomized controlled trial. *JMIR mHealth and uHealth, 4*(3), e93. doi:10.2196/mhealth.4878 PMID:27485822

Heartfile. (2004). *National action plan for prevention and control of noncommunicable diseases and health promotion in Pakistan.* Retrieved from http://heartfile.org/pdf/NAPmain.pdf

Hossain, P., Kawar, B., & El Nahas, M. (2007). Obesity and diabetes in the developing world—a growing challenge. *The New England Journal of Medicine, 356*(3), 213–215. doi:10.1056/NEJMp068177 PMID:17229948

Hu, F. B. (2011). Globalization of diabetes. *Diabetes Care*, *34*(6), 1249–1257. doi:10.2337/dc11-0442 PMID:21617109

Ingelheim, B. (n.d.). *The worldwide occurrence of type 2 diabetes*. Retrieved from https://goo.gl/wCNwsr

Islam, A., & Biswas, T. (2014). Health system in bangladesh: Challenges and opportunities. *American Journal of Health Research*, *2*(6), 366–374. doi:10.11648/j.ajhr.20140206.18

ITU & WHO. (2012). *Mhealth for NCDs (WHO-ITU joint work-plan)*. Retrieved from https://goo.gl/QjEopz

Kankeu, H. T., Saksena, P., Xu, K., & Evans, D. B. (2013). The financial burden from non-communicable diseases in low-and middle-income countries: A literature review. *Health Research Policy and Systems*, *11*(1), 31. doi:10.1186/1478-4505-11-31 PMID:23947294

Kay, M., Santos, J., & Takane, M. (2011). mhealth: New horizons for health through mobile technologies. *World Health Organization, 64*(7), 66–71.

Khowaja, L. A., Khuwaja, A. K., & Cosgrove, P. (2007). Cost of diabetes care in out-patient clinics of karachi, pakistan. *BMC Health Services Research*, *7*(1), 189. doi:10.1186/1472-6963-7-189 PMID:18028552

Kollmann, A., Riedl, M., Kastner, P., Schreier, G., & Ludvik, B. (2007). Feasibility of a mobile phone–based data service for functional insulin treatment of type 1 diabetes mellitus patients. *Journal of Medical Internet Research*, *9*(5), e36. doi:10.2196/jmir.9.5.e36 PMID:18166525

Kumar, A., & Gupta, S. (2012). *Health infrastructure in India: Critical analysis of policy gaps in the indian healthcare delivery*. Vivekanand International Foundation.

Latif, S., Qadir, J., Farooq, S., & Imran, M. A. (2017). How 5g wireless (and concomitant technologies) will revolutionize healthcare? *Future Internet*, *9*(4), 93. doi:10.3390/fi9040093

Latif, S., Rana, R., Qadir, J., Imran, M., & Younis, S. (2017). Mobile health in the developing world: Review of literature and lessons from a case study. *IEEE Access: Practical Innovations, Open Solutions*, *5*, 11540–11556. doi:10.1109/ACCESS.2017.2710800

Logan, A. G., McIsaac, W. J., Tisler, A., Irvine, M. J., Saunders, A., Dunai, A., ... Trudel, M. (2007). Mobile phone–based remote patient monitoring system for management of hypertension in diabetic patients. *American Journal of Hypertension, 20*(9), 942–948. doi:10.1016/j.amjhyper.2007.03.020 PMID:17765133

Mackillop, L. (2014). *Smartphone based management of gestational diabetes.* Retrieved from https://goo.gl/Pq4N9c

Majumdar, A., Kar, S. S., Kumar, G., Palanivel, C., & Misra, P. (2015). mhealth in the prevention and control of non-communicable diseases in India: Current possibilities and the way forward. *Journal of Clinical and Diagnostic Research : JCDR, 9*(2), LE06.

mHealth Alliance. (2013). *Mhealth opportunities for non-communicable diseases among the elderly*. Retrieved from https://goo.gl/3Pk1Ks

Nikolic, I. A., Stanciole, A. E., & Zaydman, M. (2011). *Chronic emergency: Why NCDs matter*. Academic Press.

O'Donnell, O. (2007). Access to health care in developing countries: Breaking down demand side barriers. *Cadernos de Saude Publica, 23*(12), 2820–2834. doi:10.1590/S0102-311X2007001200003 PMID:18157324

Pan, C., Yang, W., Jia, W., Weng, J., & Tian, H. (2009). Management of chinese patients with type 2 diabetes, 1998–2006: The diabcare-china surveys. *Current Medical Research and Opinion, 25*(1), 39–45. doi:10.1185/03007990802586079 PMID:19210137

Pangaribuan, & Junifer, J. (2014). Diagnosis of diabetes mellitus using extreme learning machine. In *Information technology systems and innovation (ICITSI), 2014 international conference on* (pp. 33–38). Academic Press.

Piot, P., Caldwell, A., Lamptey, P., Nyrirenda, M., Mehra, S., Cahill, K., & Aerts, A. (2016). Addressing the growing burden of non–communicable disease by leveraging lessons from infectious disease management. *Journal of Global Health, 6*(1), 010304. doi:10.7189/jogh.06.010304 PMID:26955469

Quinn, C. C., Clough, S. S., Minor, J. M., Lender, D., Okafor, M. C., & Gruber-Baldini, A. (2008). Welldoc™ mobile diabetes management randomized controlled trial: Change in clinical and behavioral outcomes and patient and physician satisfaction. *Diabetes Technology & Therapeutics, 10*(3), 160–168. doi:10.1089/dia.2008.0283 PMID:18473689

Rajalakshmi, R., Arulmalar, S., Usha, M., Prathiba, V., Kareemuddin, K. S., Anjana, R. M., & Mohan, V. (2015). Validation of smartphone based retinal photography for diabetic retinopathy screening. *PLoS One*, *10*(9), e0138285. doi:10.1371/journal.pone.0138285 PMID:26401839

Rema, M., Premkumar, S., Anitha, B., Deepa, R., Pradeepa, R., & Mohan, V. (2005). Prevalence of diabetic retinopathy in urban india: The chennai urban rural epidemiology study (cures) eye study, i. *Investigative Ophthalmology & Visual Science*, *46*(7), 2328–2333. doi:10.1167/iovs.05-0019 PMID:15980218

Roth, G. A., Johnson, C., Abajobir, A., Abd-Allah, F., Abera, S. F., & Abyu, G. (2017). Global, regional, and national burden of cardiovascular diseases for 10 causes, 1990 to 2015. *Journal of the American College of Cardiology*.

Slama, S., Kim, H.-J., Roglic, G., Boulle, P., Hering, H., Varghese, C., & Tonelli, M. (2016). *Care of non-communicable diseases in emergencies*. London: Lancet.

Sridar, K., & Shanthi, D. (2017). Design and development of mobile phone based diabetes mellitus diagnosis system by using ann-fp-growth techniques. *Biomedical Research*.

UNICEF & WHO. (2015). *Water, sanitation and hygiene in health care facilities: status in low and middle income countries and way forward*. Author.

WHO. (2009). *Four noncommunicable diseases, four shared risk factors*. Retrieved from https://goo.gl/ixGLRJ

WHO. (2010). *Package of essential noncommunicable (pen) disease interventions for primary health care in low-resource settings*. WHO.

WHO. (2010a). *Back ground paper: Non communicable diseases in low and middle income countries*. Geneva: WHO.

WHO. (2010b). *Global status report on noncommunicable diseases 2010*. Retrieved from https://goo.gl/NOucW

WHO. (2013). *Global health workforce shortage to reach 12.9 million in coming decades*. Geneva, Switzerland: World Health Organization.

WHO. (2014a). *Mobile phones help people with diabetes to manage fasting and feasting during Ramadan*. Retrieved from https://goo.gl/spN2E4

WHO. (2014b). *World health statistics*. Retrieved from https://goo.gl/9q2sSA

WHO. (2017a). *Cancer by world health organization*. Retrieved from http://www.who.int/cancer/en/

WHO. (2017b). *Noncommunicable diseases and their risk factors.* Retrieved from http://www.who.int/ncds/introduction/en/

Wild, S. H., Roglic, G., Green, A., Sicree, R., & King, H. (2004). Global prevalence of diabetes: estimates for the year 2000 and projections for 2030: response to Rathman and Giani. *Diabetes Care, 27*(10), 2569–2569. doi:10.2337/diacare.27.10.2569-a PMID:15451948

World, B. (2016). *Population ages 65 and above.* Retrieved from https://goo.gl/wPzRdX

World Health Organization. (2002). *Reducing risks, promoting healthy life*. Author.

ENDNOTE

[1] A holy month of Islamic Calendar in which Muslims observe fasting from dawn till dusk.

Chapter 12
UX Design Guideline for Health Mobile Application to Improve Accessibility for the Visually Impaired:
Focusing on Disease Retinitis Pigmentosa

Woo Jin Kim
Kyungpook National University, South Korea

Il Kon Kim
Kyungpook National University, South Korea

Jongoh Kim
Universität Heidelberg, Germany

Minji Kim
Kyungpook National University, South Korea

ABSTRACT

A health mobile application (app) has enabled users to access personal health records at any time and place. As an app provides health service to users, it is crucial for an app to be accessible to every user. However, often an app does not provide proper visual aid for users who are visually impaired. The authors restrict the range of visually impaired to retinitis pigmentosa (RP) patients in this chapter. RP is a rare type of progressive retinal disease that is hard to cure. Unfortunately, there are no established guidelines to assist RP patients in using their degrading sense of vision. In this chapter, the authors review WCAG (web content accessibility guidelines) specified by W3C, analyze the UX designs of 140 popular health apps chosen based on the number of download counts in app stores, and propose a set of standard-compliant UX design guidelines to assist the visually impaired (RP) in accessing visual data and evaluate its compliance compared to WCAG.

DOI: 10.4018/978-1-5225-5270-3.ch012

INTRODUCTION

According to the annual Internet World Stats, there are over 3.27 billion internet users people have access to the Internet as of June 2015 (Miniwatts Marketing Group, 2015). According to Mobility Report November 2015 issue by Ericsson, the number of smartphone users in the world exceeded 3.4 billion, amounting to 45.9% of the number of mobile phone users (Ericsson, 2015). One significant reason why the web has become a prominent media form is that it has offered easy access to information to everyone. Tim O'Reilly, the founder of the O'Reilly Media Inc., proposed Web 2.0 in October 2004 and it has empowered people to create, save, and share data at any place as long as they are connected to the Internet. At the same time, introduction of mobile devices capable of web browsing provided the physical structure for access to data by any individual at any time and place. As Web 2.0 has matured, Mobile Web 2.0 was introduced and it enabled the visually impaired to communicate with physically healthy people and to participate in social activities (Martin & Chuck, 2015, pp40-44). The trend of empowering people to access data regardless of physical conditions has continued to date in 2015, and apps in 'Medical' and 'Health & Fitness' categories are receiving increased attention on the market, which are expected to increase in value from 2.4 billion dollars in 2014 to 26 billion dollars in 2017 in total at both Apple's AppStore and Google's PlayStore according to Research2Guidance (Research 2Guidance, 2013). Such an explosive growth of health apps is partly due to the availability of more various types of health devices matched with the general consumer trend that actively seeks to stay healthy.

Paying attention to health on individual level is a positive trend for aging societies because public health policy and budget for public health care cannot cover all aspects of public health in the face of increasing life expectancy. In addition, if an individual proactively participate in health management by oneself and the management results in decrease in one's overall medical cost, the management can be an effective way to improve one's quality of life. From the viewpoint of public service administrator, this approach would certainly help reduce the public health cost in aging societies. However, to date in 2015, Apple and Google does not provide proper Design Guideline to planners, developers, and designers enabling them to develop an App which provides easy access to RP patients. As a result, visually impaired users could not utilize an App as effective as other people who have healthy eyesight. This is a violation of the "Twenty-First Century Communications and Video Accessibility Act" enacted by the Obama Administration in October 2010 (Federal Communications Commission, 2010). In this paper, we seek to shed light on factors that hinder easy access to data for the visually impaired in health apps. We propose a set of standard-compliant UX design guidelines to assist the visually impaired(RP) in accessing visual data and evaluate its compliance compared to WCAG.

Thus, this paper is organized as follows. In background section, we discuss behavioral aspects of retinitis pigmentosa patients as they are the target of our discussion. "Solutions and recommendations" section describes how we chose the 140 apps in the categories of 'Medical', 'Health & Fitness' in US, Canada, Japan, Australia, Republic of Korea, Germany, and UK, followed by our analysis of UX design in these apps. We also describe a set of WCAG (Web Content Accessibility Guidelines)-based UX design guidelines that could improve data accessibility for RP patients. Finally, in "future research directions" section, we describe how our guidelines can be applied to the MyTh Platform.

BACKGROUND

Definition of UX Design

'UX (User eXperience) is commonly referred to 'User-centric Product', and 'Service Design'. These definitions are not inaccurate. However, they could have various interpretations due to combination of words not used together conventionally. Therefore, specific objective and exercise of this definition is ambiguous. To eliminate the obscurity, I define UX Design as 'Object Oriented Design' from Cooper Design Journal. The key characteristic of 'Object Oriented Design' is that it covers various aspects of a product: mechanism, visual form, physical form, etc.

Definition of Retinitis Pigmentosa

With retinitis pigmentosa, the retina progressively degenerates due to malfunction of photoreceptors at the retina. Symptoms arise due to malfunctions of rod and cone cells and both eyes are affected. RP is one of the most common type of inherited retinal disorders affecting one person out of every four thousand people. It is estimated that around two million people suffer from RP globally. In European countries, RP is responsible for 10% of legal blindness for the general population, and above 30% if only population over 65 years of age is considered.

Description of RP Patients' Visual Perception

Symptoms of RP may not appear initially, but as time progresses, symptoms gradually appear such as night blindness, declining peripheral vision and visual motion detection, and inability to discern color. Total blindness may result in with varying speed to each patient. A healthy person has a 10mm angle of view while an RP patient starts from 10mm and deteriorates to 200mm over time, i.e., side vision is gradually lost except for central vision. Symptoms of an RP patient are listed in Table 1 (RP Research, 2014).

Other than these characters, patients with RP frequently have higher rates of other visual impairments including amblyopia, myopia, hyperopia, strabismus. Observation of behavioral and symptomatic characteristics of persons with visual impairments are listed in Table 2 (Kim Woo Jin, 2012).

Table 1. Symptoms of RP

Night Blindness	Seeing becomes hard to impossible in dimly light environments.
Photophobia	Light sources such as sunlight or fluorescent light can be the cause. Reading black characters with white background becomes challenging.
Peripheral Vision Reduction	Side vision is reduced from outer edges towards the central vision. Cognition of still and moving objects becomes a challenge.
Central Vision Reduction	Central vision is affected at a later stage of RP. Objects appear distorted and reading becomes difficult. Visual cognition degenerates.
Color Blindness	Colors may initially be difficult to discern, and shapes of objects in specific color combinations are not recognized.

Source: RP Research, 2014.

Table 2. Observation of RP patients

In the presence of amblyopia, the patient moves very close to the target object to see.
Narrowed visual field causes elongated letters or images hard to recognize.
Presence of diplopia hampers visual perception when eye movement changes direction.
Pointers such as a finger is used when following text.
Searching through visually scattered data pieces is challenging.
Visual perception degrades if brightness difference is subtle.
Colors of similar hue appear identical for a color-blind patient.
Text reading fatigues significantly more than image viewing.
Visual perception is greatly affected by lighting in the environment.
Fast visual change on computer screen causes difficulty in perception and the patient relies more on sound cues than visual.
Focus on the target frequently changes or blurs.
Presence of astigmatism causes double vision and visual distortion.
Shuffling arrangement of items of a familiar diagram causes difficulty in perception.
Trust is put to information obtained through senses other than vision.

Source: Kim Woo Jin, 2012.

Based on this table, we classify visual perception patterns of an RP patient as 'low vision', 'color blindness', and 'visual field defect' like represented in Figure 1, which are applied to our analysis of WCAG and the 140 health apps that we will be presenting later in this chapter.

CONTEXTUALIZATING THE PROBLEM

In this paper, the key is to find ways to improve the health app accessibility of RP patients with limited physical abilities. The reason for focusing on the health app here is that mobile apps are the closest way to care their health of RP patients. Of course, the RP patients classified as information-poor are inevitably subject to differences in the amount and quality of information that can be acquired by non-disabled people for any reason. However, if they do not have the means to support them in spite of their willingness to actively manage their health, then it is nothing more than an ideal. One example is the fact that only 60% of the registered population of people with visual disabilities in Korea, including RP patients, are engaged in economic activities. Furthermore, 60% of those who are engaged in economic activities are reported to have lower income levels than non-disabled people. By the way, how about 40% of people who are not doing economic activities are living their lives? There are differences in the welfare policies between countries, but in Korea, there is a tendency to focus on the welfare benefits facing the eyes rather than the ones that can be self-reliant. A typical example of this is the reduction of the medical expenses to visually impaired. This is not a dimension of prevention, but the perspective of post-treatment aimed at treating diseases and supporting them. In other words, it is difficult to expect support from the preventive level. After all, they also have to do health care themselves. However, the amount of information they can acquire is insufficient, and there is no way to take any measures because there is no way

Figure 1. Behavioral and symptomatic characteristics of RP patients

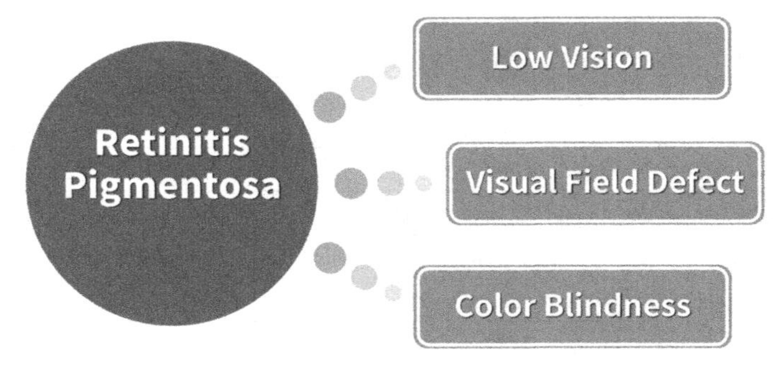

to accommodate a variety of services for non-disabled people. This fragmentary problem is linked to the problem of an aging society that has already begun. More diseases will occur during the lifetime and more costs will have to go into welfare benefits to solve them. The cost should be provided from the people who makes vigorous economic activities, but the birth rate, which is declining with the passage of time, is throwing away that expectation. It may sound somewhat magnificent, but this study began with the hope that by preventing these social problems and providing more opportunities to RP patients, they would be able to play their roles and obligations as members of society. This researcher, RP patient like them, also wants to change the Korean society from an active point of view. In the following chapters, UX design guidelines are proposed as a means to identify and supplement the factors impeding access to their health apps from the perspective of RP patients. Of course, these guidelines are not designed to hinder the aesthetics of formwork in the visual design arena. I hope that you are aware of the fact that it is a guide to things to consider while maintaining current visual satisfaction. In addition, UX Design guidelines derived through a series of procedures aim to improve the accessibility of patients with RP, but if you expand your vision a little more, you will see limited range of physical abilities such as low vision, color blindness, I would like to emphasize that securing the accessibility of the environment makes it possible to acquire more consumers from a business perspective.

TESTS AND RESULTS

We analyzed 140 most downloaded health apps on the AppStore and PlayStore in USA, Canada, Japan, Australia, Republic of Korea, Germany, and United Kingdom as of November 2015 (Tables from 10 to 16 included in APPENDIX) based on the guideline like Figure 2. Our analysis charts the 140 apps based on the visual perception patterns of RP patients as well as W3C WCAG.WCAG is a guideline provided by W3C (World Wide Web) for mobile contents. Table 3 shows the specifics of WCAG and Table 4 describes the criteria for selecting the 140 health apps.

W3C WCAG

Table 4 describes the criteria for selecting health apps for our analysis.

Analysis of compliance for the 140 health apps was conducted from November 1 till 20 in 2015, and Tables from 5 to 8 show the results of our analysis for the 16 guidelines in WCAG. Each cell shows the number of apps that comply with the 16 guidelines of WCAG out of the top five apps, and the compliance ratio is calculated based on the numbers.

Figure 2. Process of establishment validating UX Design Guideline

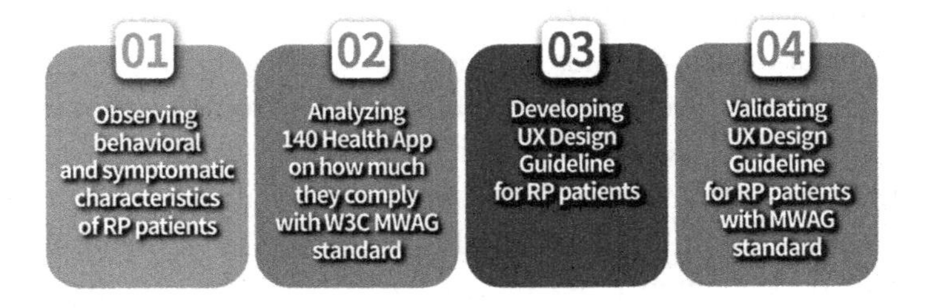

Table 3. WCAG (Web Content Accessibility Guidelines)

1. Perceivable	1. Information conveyed solely with color
	2. Large pages or large images
	3. Multimedia with no captions: caption
	4. Audio-onlyprompts(beeps) for important information (warnings, errors)
	5. Non-text objects(Images, sound, video) with no text alter native
	6. Text entry
	7. Content for matted using tables or CSS, and reading order not correct when linearized (for example when CSS or tables not rendered)
2. Operable	8. Mouse required for interaction and navigation
	9. Scripting required to operate content Special plug-in required
	10. Missing or inappropriate page title
	11. Inconsistency between focus (tab) order and logical document content sequence: focusing
	12. Non descriptive link label
3. Understandable	13. Long words, long and Complex sentences, jargon
	14. Content spawning new windows without warning user
	15. Blinking, moving, scrolling or auto-updating content
4. Robust	16. Invalid or unsupported markup Scripting required to generate content

This analysis of the 140 apps revealed widely varying WCAG compliance ratio between 60% to 90% among different countries like above chart (Figure 3), but the overall average was at 83.2%. Unfortunately, the most important guideline No. 5, "non-text objects (images, sound, video) with no text alternative" was observed by only 12.8% of the analyzed apps. For a user with low vision, color blindness, or visual field defect, access to data is not supported with alternative methods by most apps, which results in lowered data accessibility for the visually impaired. We observed that there is a room for improvement in UX, UI, and visual communication

Table 4. Criteria for health app selection

1. Countries (Seven)	USA / Canada / Japan / Australia / Republic of Korea / Germany / United Kingdom
	Based on World Internet Usage and Population Statistics published in June 30, 2015 by Internet World Stats, three continents with highest Internet engagement activities were selected.
2. App Stores (Two)	AppStore(Apple)/ PlayStore(Google)
	Largest app markets by number of mobile apps in 2015
3. Categories (Two)	Medical/ Health & Fitness
	AppStore (Apple), PlayStore (Google) Largest app markets by number of mobile apps in 2015
4. Ranking (Top Five)	Top 5 most downloaded apps in each category
	Top 5 most downloaded free apps in real time in AppStore and PlayStore each as ranked by AppAnnie for November 2015

Table 5. Medical apps in app store (ios)

App Store (iOS): Medical																	
Country	**WCAG Compliance Ratio**	**WCAG Items (5 Apps per Country)**															
		1	**2**	**3**	**4**	**5**	**6**	**7**	**8**	**9**	**10**	**11**	**12**	**13**	**14**	**15**	**16**
USA	88.8%	5	5	5	5	0	4	5	5	5	2	5	5	5	5	5	5
Canada	88.8%	5	5	5	5	0	4	5	5	5	4	4	4	5	5	5	5
Japan	82.5%	5	4	5	5	0	3	5	5	5	2	3	4	5	5	5	5
Australia	91.3%	5	5	5	5	0	4	5	5	5	5	5	4	5	5	5	5
Republic of Korea	78.8%	5	3	5	5	1	3	5	5	5	2	2	2	5	5	5	5
Germany	91.3%	5	5	5	5	1	4	5	5	5	5	5	4	4	5	5	5
United Kingdom	92.5%	5	5	5	5	1	5	5	5	5	5	4	4	5	5	5	5

design for the visually impaired from the list of obstacles to RP patients in accessing visual data (BACKGROUND) along with our WCAG compliance analysis result. We propose detailed UX design guidelines based on our observations and it is summarized in Table 9.

The UX design guidelines as suggested above were checked with WCAG for compliance and the result is shown in Figure 4.

Figure 5 illustrates that our UX Design Guidelines for RP Patients meets 81% of the specifications in WCAG. Considering that the data accessibility for aurally and mentally challenged people was considered when WCAG was drafted, our UX

Table 6. Health and fitness apps in app store (ios)

App Store (iOS): Health and Fitness																	
Country	**WCAG Compliance Ratio**	**WCAG Items (5 Apps per Country)**															
		1	**2**	**3**	**4**	**5**	**6**	**7**	**8**	**9**	**10**	**11**	**12**	**13**	**14**	**15**	**16**
USA	90.0%	5	4	5	5	1	5	5	5	5	5	4	4	5	4	5	5
Canada	90.0%	5	5	5	5	1	3	5	5	5	5	5	4	5	4	5	5
Japan	87.5%	5	5	5	5	1	4	5	5	5	5	5	2	5	3	5	5
Australia	86.3%	5	5	4	5	1	2	5	5	5	4	5	5	5	3	5	5
Republic of Korea	78.8%	5	4	5	5	0	4	5	5	5	3	3	3	3	3	5	5
Germany	92.5%	5	5	5	5	1	5	5	5	5	5	5	3	5	5	5	5
United Kingdom	88.8%	5	4	3	5	0	4	5	5	5	5	5	5	5	5	5	5

Table 7. Medical apps in PlayStore (Android)

PlayStore (Android): Medical																	
Country	**WCAG Compliance Ratio**	**WCAG Items (5 Apps per Country)**															
		1	**2**	**3**	**4**	**5**	**6**	**7**	**8**	**9**	**10**	**11**	**12**	**13**	**14**	**15**	**16**
USA	70.0%	5	4	4	4	0	2	4	5	4	2	4	1	5	4	4	4
Canada	80.0%	4	4	4	5	1	2	5	5	5	3	4	2	5	5	5	5
Japan	63.8%	5	5	5	5	0	1	4	5	5	0	1	1	1	4	4	5
Australia	81.3%	5	4	4	5	0	3	3	5	5	4	4	4	5	4	5	5
Republic of Korea	68.8%	5	3	5	5	1	4	5	5	5	1	1	2	4	3	4	3
Germany	91.3%	5	4	5	5	2	4	5	5	5	4	5	4	5	5	5	5
United Kingdom	82.5%	5	4	5	5	0	2	4	5	4	5	5	3	4	5	5	5

design guidelines is a meaningful improvement as our UX design provides visual aid for people with low vision, color blindness and visual field defect.

FUTURE RESEARCH DIRECTIONS

DIN, the German Institute for Standardization, asserts that 'Standards ensure innovation'. I believe DIN's contention is true. Then, it is critical for all nations

Table 8. Health and fitness apps in PlayStore (Android)

PlayStore (Android): Health and Fitness																	
Country	**WCAG Compliance Ratio**	**WCAG Items (5 Apps per Country)**															
		1	**2**	**3**	**4**	**5**	**6**	**7**	**8**	**9**	**10**	**11**	**12**	**13**	**14**	**15**	**16**
USA	86.3%	5	4	5	5	1	5	5	5	5	3	5	1	5	5	5	5
Canada	78.8%	5	5	5	5	0	4	5	5	5	2	2	1	5	4	5	5
Japan	85.0%	5	4	5	5	0	4	5	5	5	3	4	4	4	5	5	5
Australia	85.0%	5	5	5	5	1	3	5	5	5	2	4	5	3	5	5	5
Republic of Korea	73.8%	5	3	5	5	1	4	5	5	5	2	2	3	2	4	5	3
Germany	80.0%	5	5	4	5	1	2	5	5	5	3	3	3	4	4	5	5
United Kingdom	76.3%	5	4	4	5	2	3	5	5	4	1	4	1	5	4	4	5

Figure 3. Compliance rate of 140 health applications between countries and operating systems

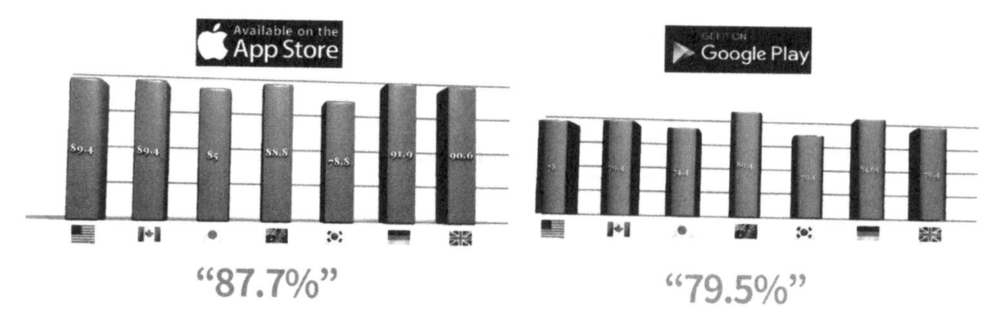

to understand the definitions of health and standardization in Health Informatics area. The definition of health is well described in one World Health Organization document (International Organization for Standardization, 2011) that suggests requirements for Electronic Health Records (ISO/TC 215 Health informatics, 2015). World Health Organization claims that the ideal vision of health (and consequently of health information) is "A state of complete physical, mental and social well-being and not merely the absence of disease or infirmity"

Compliance of health apps to international standards is essential to make the apps more accessible to people while health apps gain popularity in fast pace in the app market. As the health app market is driven by individual consumers rather than hospitals, UX of end-users is vital as much as UX of medical professionals.

Table 9. UX Design guidelines for the visually impaired (RP)

Category	Guideline / Discovered issues
1. Increased Data Exposure	A. Support for zoom in/out for main content
	Discovered issues: • Most mobile apps do not support zooming features. • Even if zooming is supported, the zoom configuration is initialized when page is turned.
	B. Support for alternative color schemes
	Discovered issues: • White background triggers photophobia symptoms. • Low contrast / gradation makes perception of visual data difficult.
2. Data Accessibility	C. Intuitive navigation
	Discovered issues: • Sub-items are difficult to find during navigation. • Area for touch input is too narrow and incorrect input is recognized. • Geolocation finding is not supported.
	D. Menu intuitively responding to user's intention
	Discovered issues: • It is not possible to verify if the app is functioning as intended. • Current status of app functioning is not available.
	E. Support for input method other than touch
	Discovered issues: • No input method is supported other than touching screen. • Touch area is limited due to small screen size.
3. Information Acquisition	F. Use highly legible fonts.
	Discovered issues: • Visually fancy fonts are used resulting in low legibility for some screen resolutions. • Zooming-in causes pixelation of fonts. • Serif and Sans-serif fonts are mixed and causes difficulty in reading.
	G. Highlighting main images that user can access
	Discovered issues: • Distinguishing ads and content is difficult due to their placement design. • Selective manipulation of text/image is difficult on the content.
	H. Highlighting action-triggering media
	Discovered issues: • Locating media content within web-page is difficult. • Action-triggering media are not obvious in the interface. • Only page errors and pop-ups are visually alerted.
4. Data Search	I. Put text input box always in the same location.
	Discovered issues: • During text input, touching a point outside the keyboard changes page layout. • Link menu is accessible during text input. • Input window does not support zoom in/out.
	J. Show search result immediately for the search text box.
	Discovered issues: • Content is searched only through navigation. • Quick search does not exist and user has to heavily rely on visual information on screen.
	K. Speech recognition for text input
	Discovered issues: • Text input or menu selection is done only through touching screen.

Figure 4. Compliance of our proposed UX design guidelines to WCAG

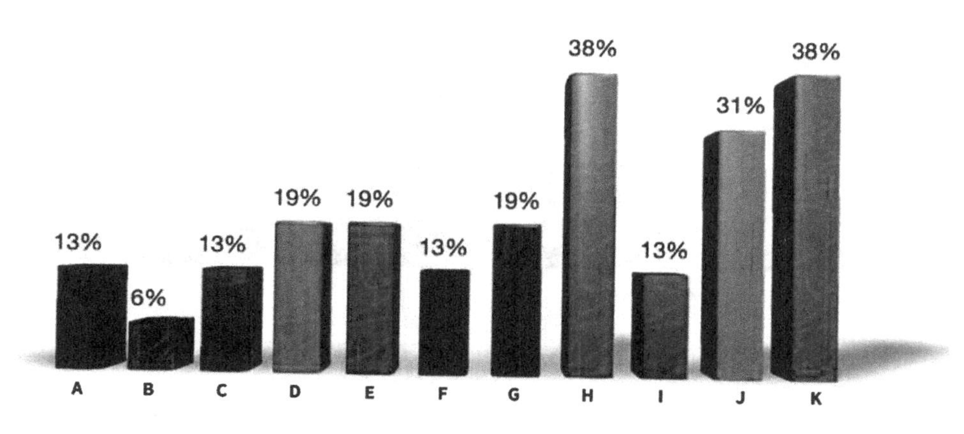

Figure 5. Visual data accessibility compliance ratio for different types of visual impairment

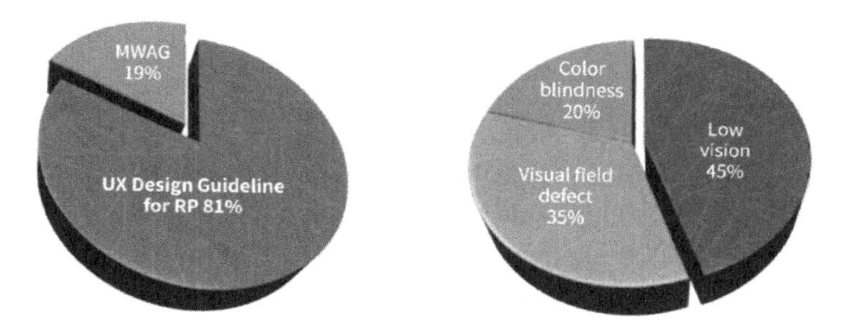

Thus, health app developers should do their best to meet their users' needs. From this perspective, our next project is worth to mention. Our lab is planning to develop a practical health information standard-based platform called MyTh. Compliance evaluation could have being processed through our MyTh Platform.

We named our platform MyTh as every man desire both healthy body and longevity which is so difficult to achieve that seems like a myth. RP patients also dream that their remaining visualability would persist throughout their lives. This paper was started to contribute to achieving that dream through the MyTh Platform. Summary characteristics of MyThPlatform are shown in Figure 6.

Figure 6. MyTh platform concept

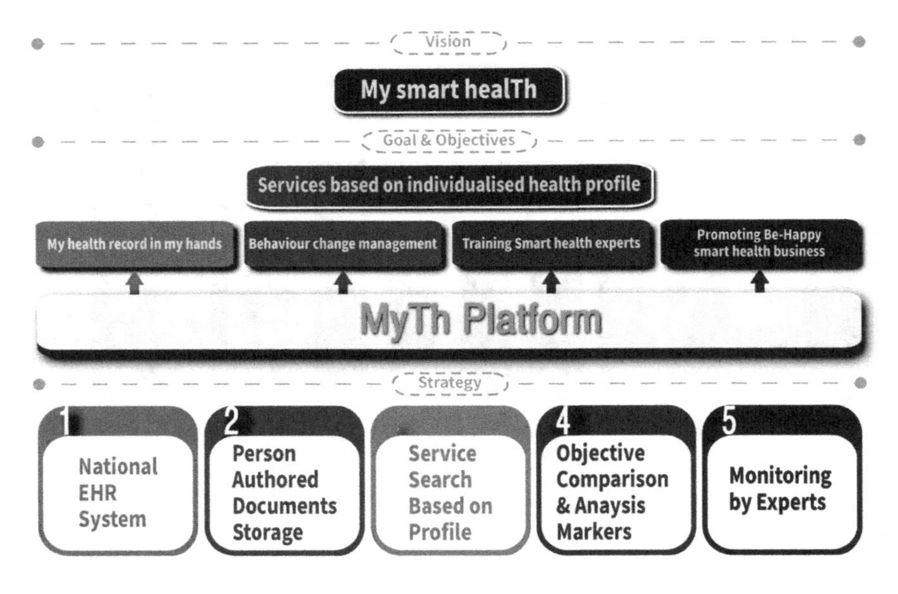

CONCLUSION

In this paper, we reviewed the physical challenges that RP patients face and analyzed the requirements of WCAG considering visual impairment. Then we analyzed 140 apps to draft a set of UX design guidelines that also complies with WCAG to assist the visually impaired. In future research, an analysis could be conducted on how RP patients acquire information as RP symptoms progress and another set of UX design guidelines could be proposed to ensure data accessibility for them. As a side note, one of the authors of this paper is an RP patient and working together in a team offered a tremendous opportunity to learn about the everyday challenges of a visually challenged person. We hope health apps in the future could offer improved data accessibility for the visually impaired by following the design guidelines suggested in this paper.

ACKNOWLEDGMENT

This research was supported by the Institute for Information & communications Technology Promotion(IITP) grant funded by the Korea government(MSIP) [grant number 10041145, Self-Organized Software platform (SoSp) for Welfare Devices].

REFERENCES

Ericsson. (2015). Mobile Subscriptions Q3 2015. *Ericsson Mobility Report: on the pulse of the networked society, No. EAB-15:037849 Uen*, 4.

Federal Communications Commission. (2010). *Twenty-First Century Communications and Video Accessibility Act* [Pub. L. 111-260]. Author.

International Organization for Standardization. (2011). *Health informatics - Requirements for an electronic health record.* Author.

ISO/TC 215 Health Informatics. (2015, Apr 15). Geneva: International Organization for Standardization.

Kim, W. J. (2013). *Study on design for improving the mobile-web accessibility of users who have blind* (Master's Thesis). Available from National Assembly Library. (KDMT1201317130)

Martin & Chuck. (2011). *The Third Screen: Marketing to Your Customers in a World Gone Mobile*. London: Nicholas Brealey Publishing.

Miniwatts Marketing Group. (2015, Nov. 20). *World internet usage and population statistics*. Author.

Research2Guidance. (2013, Mar 4). *Mobile-Health-Market-Report-2013-2017*. Author.

Research, R. P. (2014). *Retinitis Pigmentosa Q&A*. Seoul, South Korea: Elsevier Korea.

KEY TERMS AND DEFINITIONS

MyTh Platform: A platform under development in Korea as a concept to implement PHR (personal health record).

RP (Retinitis Pigmentosa): An inherited, degenerative eye disease that causes severe vision impairment. In addition, the first author had the same disease, so this study could be started.

Service Design: The activity of planning and organizing people, infrastructure, communication, and material components of a service in order to improve its quality and the interaction between the service provider and its customers.

UX Design (User eXperience Design): A process of enhancing user satisfaction with a product by improving the usability, accessibility, and pleasure provided in the interaction with the product.

WCAG (Web Content Accessibility Guidelines): A part of a series of web accessibility guidelines published by the web accessibility initiative (WAI) of the World Wide Web Consortium (W3C), the main international standards organization for the internet.

APPENDIX

The following data was updated on October 27, 2017.

Table 10. 20 most downloaded health apps in USA

OS	App Name	Brief Explanation	OS	App Name	Brief Explanation
iOS	Amwell: Doctor Visits 24/7	Remote health app user can contact with doctor at any time.	iOS	Healow	Show data of doctor, date, and medical record.
iOS /Android	CareZone	Manage medication and prescriptions for medical institutions.	iOS /Android	iTriage	Possible to access medical DB, medical contents.
iOS /Android	CVS/pharmacy	Prescription history access and finds CVS/pharmacy store.	iOS	MyChart	Manage user's health information and support to communicate with doctor.
iOS /Android	Ear Spy: Super Hearig	Helps improve hearing.	iOS	My Diet Coach – Weight Loss	Provides food information and BMI calculator.
Android	Fabulous: Motivate Me!	Improve health and productivity through step-by-step programs.	iOS	MyFitnessPal	Quick and easy calorie reduction based on lots of data.
iOS /Android	Figure 1 – Medical Images	App for healthcare professionals.	Android	Period and Ovulation Tracker, Ovulation calculator	Calculate ovulation and infertility days and informs in advance.
iOS /Android	Fitbit	Collecting tracking activity data of daily life.	iOS /Android	Period Tracker Lite	Women's Life Cycle Healthcare.
iOS	FollowMyHealth	Take control of health information.	iOS /Android	Runtastic Results Fitness App	Provide 12-week training plan.
iOS	GoodRx – Save On Prescriptions	Compare price of medication between hospitals.	iOS /Android	ScriptSaveWellRx Rx Discounts	Manages medicine prescription, searches and compares the price of drug.
Android	Google Fit – Fitness Tracking	Records speed, path and altitude to give a customized coaching.	iOS	SleepCyclealarmclock	Helping user get up and provides detailed sleep statistics and graph.

Table 11. 20 most downloaded health apps in Canada

OS	App Name	Brief Explanation	OS	App Name	Brief Explanation
Android	30 Day Fit Challenge	Helps improve health and fitness with 30-day challenge.	Android	Medscape	Provides major medical resources used by medical experts.
Android	Anatomy Learning-3D Atlas	Provides 3D image of anatomy.	Android	mySugr: the blood sugar tracker made just for you	Diabetes logbook app.
iOS	Baby Tracker(Feed timer, sleep, diaper log)	Track all of child's important information for parents, doctors and caregivers.	iOS /Android	Nike+ Run Club	Tracks and stores all user's and gives personalized coaching plans.
iOS /Android	Clue Period Tracker	Women's Life Cycle Healthcare.	iOS /Android	Pregnancy +	Supports user at every stage of being a parent.
iOS /Android	Deep Sleep and Relax Hypnosis	Sleep assistant through sound, hypnotic technique	iOS /Android	Pregnancy & Baby App Glow, Countdown Calendar	Support mother while pregnancy.
iOS /Android	Epocrates Plus	Enable better patient care by delivering the right information.	iOS /Android	Relax Melodies: Sleep Sounds, White Noise & Fan	Helps to sleep comfortably naturally.
iOS /Android	GiveBlood	Find blood donation clinic and make a blood donation appointment.	iOS /Android	Shoppers Drug Mart	Possible to access personalized offers and load them directly to digital Shoppers Optimum Card.
iOS /Android	Leafly Marijuana Strain and Dispensary Reviews	Provides marijuana strains and news of cannabis and clinics.	iOS	Sworkit: Workouts & Plans	Workout program for everyone.
iOS /Android	Lose It! – Calorie Counter	Calculates the daily calorie budget and track calorie from food, weight and activity.	Android	Twilight	Helps resolve sleep problems.
iOS /Android	Medavie Blue Cross Mobile	Ensures user to access to Blue Cross benefits.	iOS /Android	Weight Watchers Mobile	Helps user lead a healthier, more active life.

Table 12. 19 most downloaded health apps in Japan

OS	App Name	Brief Explanation	OS	App Name	Brief Explanation
iOS /Android	3 minute workout	Provides adequate exercise for 3 minutes daily.	iOS /Android	Lab clock	Records and measures the gap of contraction.
iOS /Android	Calorie calculation & diet recording Free support application: calorie size	Diet management application to support the right balance of diet and exercise.	Android	Lifelog	Tracking important information like sleep time, walking distance, and more.
iOS /Android	Casual dieting-weight manager-	Helps manage weight by recording weight, body fat and meals.	iOS /Android	Lunarna: physiology / ovulation day prediction menstrual day management application	Women's Life Cycle Healthcare.
iOS /Android	COCOLOLO - Stress check with camera & AI Kimochi forecast -	Measures stress tensions and provide expert advice on the stress tensions value.	iOS /Android	"Medicine notebook plus" Japanese medicine notebook electronic medication notebook application	Possible to record and verify the medication.
iOS	Dieting log of mumps	Guides the appropriate calorie intake according to user's profile.	iOS /Android	Ninaru Deliver information for pregnant women until pregnancy ~ childbirth!	Informs pregnant women with necessary information.
iOS	Ear age check! iPhone version	Ear age check game.	iOS /Android	Pediatric surgeon recommended! Vaccination scheduler Baby · Child-raising for mothers and children	Support child's vaccination.
iOS /Android	Everlasting egg club - a popular app for pregnancy and childbirth	Pregnancy app for mom and dad.	iOS	Sleep Meister - Sleep Cycle Alarm Lite	Alarm application using a device's built-in accelerometer to sense body motion.
iOS/ Android	Eye Ticket - Do not wait for reservation	Makes an appointment without standing in line.	iOS /Android	Stress Scan	Stress measurement by analyzing the waveform of the heart.
iOS	Eye Training Cocololo	Good game for near or primitive, presbyopia, etc.	iOS /Android	Tsutsukitoka: Can be shared by a couple	Manages fetus growth progress, health and weight.
iOS /Android	Instant Heart Rate	Heart rate measurement and monitoring.			

Table 13. 19 most downloaded health apps in Australia

OS	App Name	Brief Explanation	OS	App Name	Brief Explanation
iOS /Android	Appointuit	Manage appointment with doctor.	iOS /Android	MedicineList+	Provides a comprehensive medication reminder system.
Android	Doctor Mole – Skin cancer	Assist people in learning about the skin cancer.	Android	MIMS For Android	Provide comprehensive and up to date medicines information.
iOS /Android	Emily Skye's Workouts	Guides through workouts designed to get maximum results with instructional video.	Android /iOS	MINDBODY – Fitness & Wellness	Find fitness classes near user and possible to book proper classes.
iOS /Android	Express Plus Medicare	Interacting with Medicare and accessing user's Medicare information.	iOS /Android	My OSHC Assistant	User can manage health cover and access to information about user's OSHC.
Android /iOS	First Aid – Australian Red Cross	Conduct training to cope with emergencies.	iOS /Android	Period Diary	Women's Life Cycle Healthcare.
Android /iOS	Garmin Connect™ Mobile	Helps to improve physical fitness and health care.	iOS /Android	Seven – 7 Minute Workout	Provide guides toward user's fitness goal adapted for user's fitness level.
Android /iOS	GP4Me	Hospital reservation app	iOS /Android	SkinVision Prevent Skin Cancer	Checks skin to see if it has skin cancer.
iOS /Android	Kindara: Fertility Tracker	Calculate cycles for pregnancy.	iOS /Android	Smiling Mind	Meditation app
Android /iOS	MadBarz – Bodyweight Workouts	Provides categorized exercise depending on the weights.	Android /iOS	Strava Running and Cycling	Support to record activity, compare performance over time.
Android /iOS	MedAdvisor	Helps user stay in control of medication.	Android /iOS	Walk with Map My Walk	Track walking through maps and get feedback to improve performance.

Table 14. 15 most downloaded health apps in Republic of Korea

OS	App Name	Brief Explanation	OS	App Name	Brief Explanation
iOS /Android	All that baby	Manage the growth of the child.	iOS /Android	Moabebe, all about childcare	Provides all information about childcare.
Android	Blue light blocking screen filter	Protect eyes from blue light	iOS /Android	MyAngel2	Helping pregnant, childbirth and parenting.
iOS /Android	Disease	Search for disease classifiers.	iOS /Android	Pink diary	Women's Life Cycle Healthcare
iOS /Android	Druginfo	Search for disease classifiers.	iOS /Android	Samsung Seoul Hospital	Easy to check contents related to medical care and health information contents.
iOS /Android	Emergency Medical Information	Providing emergency medical information.	iOS /Android	Saybebe	Provide embryo ultrasound video taken in hospital.
iOS /Android	Hwahae	Analysis ingredients of cosmetic.	iOS	Sunny ~ Calm wave & ocean sounds to Sleep Relax Meditate on the beach with rain and sea birds	Provide sounds helpful for sleep.
iOS	Labor Pains Timer	Contraction timer.	iOS	ubiHeartrate	Measure heart rate.
iOS /Android	Mi Fit	Track, and follow user's health and fitness data day and night.			

Table 15. 20 most downloaded health apps in Germany

OS	App Name	Brief Explanation	OS	App Name	Brief Explanation
iOS /Android	30 Day Ab Challenge FREE	Provides a simple 30 day workout plan.	Android	Lose weight without dieting	Helps lose weight in healthy way.
iOS /Android	All diseases	Provides detailed information about disease vocabulary and major disease	iOS /Android	myPill® Birth Control Reminder	Notify time to take medicine and never miss a pill.
Android	BMI Calculator	Daily calorie intake, consumption and BMI index calculation	Android	Ovulation calendar from urbia.de	Helps user get pregnant by using recorded user's cycle data.
iOS /Android	Calorie Counter by FatSecret	Tracks food, weight and exercise then calculate caloric intake and consumption.	iOS /Android	Pill Reminder	Notify time to take medicine and calculate remaining medicine.
iOS /Android	Calorie counter - FDDB Extender	Provides a detailed view of diet statistics to help get a motivation of diet.	iOS /Android	PillReminder - Think about it	Notify time to take medicine.
iOS /Android	Current Drugs	Provides huge medication DB	iOS /Android	Runtastic Running & Fitness	Tracking in real time using GPS and help user to jogging, running, etc.
iOS /Android	Freeletics Running	Provides effective fitness training program adapted to user.	iOS /Android	Search for doctors jameda	Makes sure user find a good doctor anytime, anywhere.
iOS /Android	iMamaiPapa	Health andpractical tips forpregnant women.	iOS	Stepz-Pedometer	Convenient step counter.
Android	Lady Pill Reminder	Keep track of taking birth control pills and notify to take pill.	iOS /Android	UP by Jawbone	Provides customized services by tracking sleep, activity, and diet.
iOS /Android	Lifesum: Diet & Health Plan	Provides personalized guide to healthy living using user's data.	Android	WC-Finder Deutschland	Provides a toilet database and displays it on the map.

Table 16. 19 most downloaded health apps in Germany

OS	App Name	Brief Explanation	OS	App Name	Brief Explanation
iOS /Android	Baby and Child First Aid by British Red Cross	First aid app with useful videos for child and baby.	iOS /Android	NICE BNF	Provides information on prescribing, dispensing and administering medicines.
iOS /Android	Babylon – pocket doctor	Possible to access and monitor healthcare and speak to expert online doctor.	iOS /Android	Ovia Fertility Period Tracker	Women's Life Cycle Healthcare.
Android	Bluelight Filter for Eye Care	Reduces fatigue of eyes caused by blue light.	iOS	Pregnancy Sonic	Fetal Baby Heartbeat App.
iOS /Android	Calm: Meditation	Guides to less anxiety and better sleep.	iOS /Android	SnoreLab: Record your Snoring	Detects the snoring sound and recommend a correction method.
iOS /Android	Headspace: Guided Meditation	Meditation app.	iOS /Android	Stop, Breathe & Think	Meditation guidance app.
Android	ICE – in case of emergency	Store information needed by rescuers and doctors effective in case of emergency.	iOS /Android	Superdrug	Provides benefits and rewards of being a digital Health &Beautycard member.
iOS /Android	Map My Run – GPS Running	Track and map every run and gives feedback and provides stats.	iOS /Android	SystemOnline	Patient health management app.
iOS /Android	My Pregnancy & Baby Today	Gives expert advice and guidance from the time of pregnancy until baby's first year	iOS /Android	ViewRanger GPS – Trail	Discovers trail guides and navigates.
iOS /Android	my PureGym	Finds, books and manages class.	iOS /Android	Visual Anatomy Free	Shows all body anatomy systems in 3D.
iOS /Android	NHSGiveBlood	App managing overall data of blood donation..			

Compilation of References

Abowd, G. D., Dey, A. K., Brown, P. J., Davies, N., Smith, M., & Steggles, P. (1999, January). Towards a better understanding of context and context-awareness. In *Handheld and ubiquitous computing* (pp. 304–307). Springer Berlin Heidelberg. doi:10.1007/3-540-48157-5_29

Abreu, J., Rebelo, S., Paredes, H., Barroso, J., Martins, P., Reis, A., ... Filipe, V. (2017). *Assessment of Microsoft Kinect in the Monitoring and Rehabilitation of Stroke Patients. In Recent Advances in Information Systems and Technologies* (pp. 167–174). Cham: Springer; doi:10.1007/978-3-319-56538-5_18

Acioli, C. G. (2014). *A educação na sociedade de informação e o dever fundamental estatal de inclusão digital* (Doctoral dissertation). Available from BDTD/IBICT Theses database: http://tede2.pucrs.br/tede2/handle/tede/4283

Ahmad, A., Farid, M. S., Ismail, Y., & Kadir, B. (2016). Review of customer adoption on mobile payment. *Journal Postgraduate, 1*(2).

Ahson, S. A., & Ilyas, M. (2008). *RFID handbook: applications, technology, security, and privacy*. CRC Press. doi:10.1201/9781420055009

Aikins, A. G. (2005). Healer shopping in africa: New evidence from rural-urban qualitative study of ghanaian diabetes experiences. *BMJ (Clinical Research Ed.), 331*(7519), 737. doi:10.1136/bmj.331.7519.737 PMID:16195290

Akhlaq, A., & Ahmed, E. (2013). The effect of motivation on trust in the acceptance of internet banking in a low income country. *International Journal of Bank Marketing, 31*(2), 115–125. doi:10.1108/02652321311298690

Alberdi, A., Aztiria, A., & Basarab, A. (2016). Towards an automatic early stress recognition system for office environments based on multimodal measurements: A review. *Journal of Biomedical Informatics, 59*, 49–75. doi:10.1016/j.jbi.2015.11.007 PMID:26621099

Ali, S., & Khusro, S. (2016). Mobile phone sensing: A new application paradigm. *Indian Journal of Science and Technology, 9*(19). doi:10.17485/ijst/2016/v9i19/53088

Alliance, S. C. (2011). The mobile payments and NFC landscape: A US perspective. *Smart Card Alliance*, 1–53.

Allums, S. (2014). *Designing mobile payment experiences: Principles and best practices for mobile commerce*. O'Reilly Media, Inc.

Ally, M., Balaji, V., Abdelbaki, A., & Cheng, R. (2017). Use of Tablet Computers to Improve Access to Education in a Remote Location. *Journal Of Learning For Development*, *4*(2), 221–228.

Alonso, G., Casati, F., Kuno, H., & Machiraju, V. (2004). *Web Services*. Springer Berlin Heidelberg. doi:10.1007/978-3-662-10876-5_5

Anliker, U., Ward, J., Lukowicz, P., Trster, G., Dolveck, F., Baer, M., ... Vuskovic, M. (2004). AMON: A wearable multiparameter medical monitoring and alert system. *IEEE Transactions on Information Technology in Biomedicine*, *8*(4), 415–427. doi:10.1109/TITB.2004.837888 PMID:15615032

Anon. (2010). *Plan- och bygglag (2010:900)*. Available at: http://www.notisum.se/rnp/sls/lag/20100900.HTM

Anon. (2014). *Construction Law*, Poland. Available at: http://www.architektura.info/index.php/prawo

Apple Inc. (2017). *Apple Watch Series 1 - Apple*. Retrieved March 21, 2017, from http://www.apple.com/apple-watch-series-1/

Aragall, F. (2003). *Technical Assistance Manual*. Available at: http://www.eca.lu/index.php/documents/eucan-documents/13-2003-european-concept-for-accessibility-2003/file

Aragall, F., Neumann, P., & Sagramola, S. (2013). *Design for All in progress, from theory to practice*. Available at: http://www.eca.lu/index.php/documents/eucan-documents/29-eca-2013-design-for-all-in-progress-from-theory-to-practice/file

Arocena, R., & Senker, P. (2003). Technology, inequality, and underdevelopment: The case of Latin America. *Science, Technology & Human Values*, *28*(1), 15–33. doi:10.1177/0162243902238493

Association, A. D. (2013). Standards of medical care in diabetes—2013. *Diabetes Care*, *36*(Supplement 1), S11–S66. doi:10.2337/dc13-S011 PMID:23264422

Atlas, C. (2012). *Human development index ranking*. Retrieved from https://goo.gl/kqGSVS

Baptista, G., & Oliveira, T. (2015). Understanding mobile banking: The unified theory of acceptance and use of technology combined with cultural moderators. *Computers in Human Behavior*, *50*, 418–430. doi:10.1016/j.chb.2015.04.024

Barnes, S., & Corbitt, B. (2003). Mobile banking: Concept and potential. *International Journal of Mobile Communications*, *1*(3), 273–288. doi:10.1504/IJMC.2003.003494

Barrera, J. F., Mira, A., & Torroba, R. (2013). Optical encryption and QR codes: Secure and noise-free information retrieval. *Optics Express*, *21*(5), 5373–5378. doi:10.1364/OE.21.005373 PMID:23482108

Bastos, R., & Dias, M. S. (2008). Automatic Camera Pose Initialization, using Scale, Rotation and Luminance Invariant Natural Feature Tracking. *Journal of WSCG, 16*(1–3), 97–104.

Bates, D. W., Leape, L. L., Cullen, D. J., Laird, N., Petersen, L. A., Teich, J. M., ... Seger, D. L. (1998). Effect of Computerized Physician Order Entry and a Team Intervention on Prevention of Serious Medication Errors. *Journal of the American Medical Association, 280*(15), 1311. doi:10.1001/jama.280.15.1311 PMID:9794308

Beaglehole, R., Ebrahim, S., Reddy, S., Voute, J., & Leeder, S. (2008). Prevention of chronic diseases: A call to action. *Lancet, 370*(9605), 2152–2157. doi:10.1016/S0140-6736(07)61700-0 PMID:18063026

Bergland, C. (2013). *Cortisol: Why "The Stress Hormone" Is Public Enemy No. 1*. Retrieved September 24, 2016, from https://www.psychologytoday.com/blog/the-athletes-way/201301/cortisol-why-the-stress-hormone-is-public-enemy-no-1

Biamino, G. (2011, March). Modeling social contexts for pervasive computing environments. In *Pervasive Computing and Communications Workshops (PERCOM Workshops), 2011 IEEE International Conference on* (pp. 415-420). IEEE. 10.1109/PERCOMW.2011.5766925

Biovotion. (n.d.). *About Us*. Retrieved May 29, 2017, from http://www.biovotion.com/about-us/

Bissoli, M. Â. M. A. (2001). *Planejamento turístico municipal com suporte em sistemas de informação*. Futura.

Blokdyk, G. (2017). *Mobile payment complete self-assessment Guide*. CreateSpace Independent Publishing Platform.

Bonilla, M. H. S. (2010). Políticas públicas para inclusão digital nas escolas. *Revista Motrivivência*, 40-60.

Boucsein, W. (1992). *Electrodermal Activity*. Springer, US. doi:10.1007/978-1-4757-5093-5

Bourouis, A., Feham, M., Hossain, M. A., & Zhang, L. (2014). An intelligent mobile based decision support system for retinal disease diagnosis. *Decision Support Systems, 59*, 341–350. doi:10.1016/j.dss.2014.01.005

Bozdougan, D. (2015). MALL revisited: Current trends and pedagogical implications. *Procedia: Social and Behavioral Sciences, 195*, 932–939. doi:10.1016/j.sbspro.2015.06.373

Bressler, D. M., & Bodzin, A. M. (2016). Investigating Flow Experience and Scientific Practices During a Mobile Serious Educational Game. *Journal of Science Education and Technology, 25*(5), 795–805. doi:10.100710956-016-9639-z

Brito, D., Viana, T., Lourenço, A., Sousa, D., & Paiva, S. (2017). A Mobile Solution to Help Visually Impaired People in Public Transports and in Pedestrian Walks. *International Journal of Sustainable Development and Planning, 13*(2).

Bronfenbrenner, U. (1944). A constant frame of reference for sociometric research: Part II. Experiment and inference. *Sociometry*, *7*(1), 40–75. doi:10.2307/2785536

Brown, B., & Chalmers, M. (2003). *Tourism and mobile technology*. ECSCW.

Bruce, E. (2016). *Apple Pay essentials*. Packt Publishing.

Brunschwiler, T., Straessle, R., Weiss, J., Michel, B., Van Kessel, T., & Ko, B. J. … Muehlner, U. (2017). CAir : Mobile-Health Intervention for COPD Patients. In *IEEE 19th International Conference on e-Health Networking, Applications and Services (Healthcom)* (pp. 17–19). IEEE.

Buhalis, D., & Law, R. (2008). Progress in information technology and tourism management: 20 years on and 10 years after the Internet - The state of eTourism research. *Tourism Management*, *29*(4), 609–623. doi:10.1016/j.tourman.2008.01.005

Burchardt, T., Le Grand, J., & Piachaud, D. (2002). Degrees of exclusion: developing a dynamic, multidimensional measure. In J. Hills, J. Le Grand, & D. Piachaud (Eds.), *Understanding Social Exclusion* (pp. 30–43). Oxford, UK: Oxford University Press.

Burston, J. (2015). Twenty years of MALL project implementation: A meta-analysis of learning outcomes. *ReCALL*, *27*(1), 4–20. doi:10.1017/S0958344014000159

Buttussi, F., & Chittaro, L. (2008). MOPET: A context-aware and user-adaptive wearable system for fitness training. *Artificial Intelligence in Medicine*, *42*(2), 153–163. doi:10.1016/j.artmed.2007.11.004 PMID:18234481

Carter, I. (2013). *Human behavior in the social environment*. AldineTransaction.

Carvalho, P. (2011). *Estudo da fadiga por compaixão nos cuidados paliativos em Portugal: Tradução e adaptação cultural da escala "Professional Quality of Life 5."* Instituto de Ciências da Saúde, Universidade Católica Portuguesa, Porto. Retrieved from http://repositorio.ucp.pt/handle/10400.14/8918

Castledine, E., Eftos, M., & Wheeler, M. (2011). *Build Mobile Websites and Apps for Smart Devices*. SitePoint.

CEN-CENELEC. (2014). *Guide for addressing accessibility in standards*. Available at: www.cen.eu

Centre for Studies on Human Stress. (2007). *How to measure strees in humans?* Quebec, Canada: Author.

Chan, M. (2010). *Cancer in developing countries: Facing the challenge*. Retrieved from https://goo.gl/YeYsQL

Christensen, C. M. (2013). *The innovator's dilemma: when new technologies cause great firms to fail*. Harvard Business Review Press.

Christensen, C., & Raynor, M. (2013). *The innovator's solution: Creating and sustaining successful growth*. Harvard Business Review Press.

Chuchalin, A. G., Khaltaev, N., Antonov, N. S., Galkin, D. V., Manakov, L. G., Antonini, P., ... Demko, I. (2014). Chronic respiratory diseases and risk factors in 12 regions of the russian federation. *International Journal of Chronic Obstructive Pulmonary Disease*, *9*, 963. doi:10.2147/COPD.S67283 PMID:25246783

Ciampa, K. (2014). Learning in a mobile age: An investigation of student motivation. *Journal of Computer Assisted Learning*, *30*(1), 82–96. doi:10.1111/jcal.12036

Clark, R. C., & Mayer, R. E. (2016). *E-learning and the science of instruction: Proven guidelines for consumers and designers of multimedia learning*. John Wiley & Sons. doi:10.1002/9781119239086

Clayton, K., & Murphy, A. (n.d.). Smartphone Apps in Education: Students Create Videos to Teach Smartphone Use as Tool for Learning. *The Journal of Media Literacy Education*, *8*(2), 99–109.

Cofield-Poole, B. (2016). *A touch with technology: Creating a foundation for meaningful digital inclusion through local government* (Order No. 10585321). Available from ProQuest Dissertations & Theses Global. (1880188513). Retrieved from https://search.proquest.com/docview/1880188513?accountid=26642

Cogo, D., Brignol, L.D., & Fragoso, S. (2014). Práticas cotidianas de acceso às TICs: outro modo de compreender la inclusão digital. *Palabra Clave - Revista de Comunicación, 18*(1), 156-183.

Cohen, M., & Khalaila, R. (2014). Saliva pH as a biomarker of exam stress and a predictor of exam performance. *Journal of Psychosomatic Research*, *77*(5), 420–425. doi:10.1016/j.jpsychores.2014.07.003 PMID:25439341

Cole-Lewis, H., & Kershaw, T. (2010). Text messaging as a tool for behavior change in disease prevention and management. *Epidemiologic Reviews*, *32*(1), 56–69. doi:10.1093/epirev/mxq004 PMID:20354039

Coll, C. (2013). El currículo escolar en el marco de la nueva ecología del aprendizaje. *Aula de innovación educativa, 219*, 31-36.

Comninos, A., Esselaar, S., Ndiwalana, A., & Stork, C. (2008). *M-banking the Unbanked.* Academic Press.

Coorevits, L., Schuurman, D., Oelbrandt, K., & Logghe, S. (2016). Bringing Personas to Life: User experience Design through Interactive Coupled Open Innovation. *Persona Studies*, *2*(1), 97–114. doi:10.21153/ps2016vol2no1art534

Correia, R. C., Duarte, J. P., & Leitão, A. M. (2010). MALAG: a discursive grammar interpreter for the online generation of mass customized housing. *Proceedings of the workshop in 4th Conference Design Computing and Cognition.*

Coskun, V., Ok, K., & Ozdenizci, B. (2011). *Near field communication (NFC): From theory to practice*. John Wiley & Sons.

Coursera. (n.d.). *Gamification*. Retrieved December 3, 2017, from https://www.coursera.org/learn/gamification

Crickx, A. (2014). *Recognito: Text Independent Speaker Recognition in Java*. Available in: <https://github.com/amaurycrickx/recognito>

Cristino, P. C. N. V. (2012). *Mobile Banking: fatores determinantes na adesão* (Doctoral dissertation). Escola Superior de Comunicação Social.

Cruz, P., Barretto Filgueiras Neto, L., Munoz-Gallego, P., & Laukkanen, T. (2010). Mobile banking rollout in emerging markets: Evidence from Brazil. *International Journal of Bank Marketing*, *28*(5), 342–371. doi:10.1108/02652321011064881

Cunningham, P., & Fröschl, F. (2013). *Electronic business revolution: opportunities and challenges in the 21st century*. Springer Science & Business Media.

Cysneiros, L. M. (1997). Integrando requisitos nao funcionais ao processo de desenvolvimento de software. Rio de Janeiro: Academic Press.

Dahlberg, T., Mallat, N., Ondrus, J., & Zmijewska, A. (2008). Past, present and future of mobile payments research: A literature review. *Electronic Commerce Research and Applications*, *7*(2), 165–181. doi:10.1016/j.elerap.2007.02.001

Dalton, N., Dalton, R., & Höelscher, C. (2015). People Watcher: an app to record and analyzing spatial behavior of ubiquitous interaction technologies. *Proceedings of the 4th International Symposium on Pervasive Displays (PerDis '15)*, 1–6. Available at: http://oro.open.ac.uk/42643/

Daskalaki, E., Prountzou, A., Diem, P., & Mougiakakou, S. G. (2012). Realtime adaptive models for the personalized prediction of glycemic profile in type 1 diabetes patients. *Diabetes Technology & Therapeutics*, *14*(2), 168–174. doi:10.1089/dia.2011.0093 PMID:21992270

David Baszucki. (n.d.). *Where are people playing? - Roblox Blog*. Retrieved October 31, 2017, from https://blog.roblox.com/2012/02/where-are-people-playing/

de Andrade, J. V. (2006). *Turismo: fundamentos e dimensões*. Ática.

De Araújo, R. B. (2003, May). Computação ubíqua: Princípios, tecnologias e desafios. In XXI Simpósio Brasileiro de Redes de Computadores (Vol. 8, pp. 11-13). Academic Press.

De Beeck, R. O., Hermans, V., & European Agency for Safety and Health at Work. (2000). *Research on work-related low back disorders*. Institute for Occupational Safety and Health.

de Carvalho, R. L. (2004). *Gestão de relacionamento com o cliente via Internet para grupos de pesquisa*. Goiânia, Goiás, Brasil: Universidade Federal de Goiás.

De Cheveigné, A., & Kawahara, H. (2002). YIN, a fundamental frequency estimator for speech and music. *The Journal of the Acoustical Society of America*, *111*(4), 1917–1930. doi:10.1121/1.1458024 PMID:12002874

De Mello, E. D. F. F., & Teixeira, A. C. (2011, November). A interação social descrita por Vigotski e a sua possível ligação com a aprendizagem colaborativa através das tecnologias em rede. In Anais do Workshop de Informática na Escola (Vol. 1, No. 1, pp. 1362-1365). Academic Press.

Deloitte. (2015). *Contactless mobile payments (finally) gain momentum*. Available at https://www2.deloitte.com/content/dam/Deloitte/fpc/Documents/secteurs/services-financiers/deloitte_tmt-predictions-2015-contactless-mobile-payments_en.pdf

Dennehy, D., & Sammon, D. (2015). Trends in mobile payments research: A literature review. *Journal of Innovation Management*, *3*(1), 49–61.

Denzin, N. K., & Lincoln, Y. S. (2008). *Collecting and interpreting qualitative materials*. Los Angeles, CA: Sage.

Derks, D., & Bakker, A. (2014). Smartphone use, work–home interference, and burnout: A diary study on the role of recovery. *Applied Psychology*. Retrieved from http://onlinelibrary.wiley.com/doi/10.1111/j.1464-0597.2012.00530.x/full

DESA. (2016). *Leaving no one behind: the imperative of inclusive development. Report on the world social situation*. DESA.

Deterding, S., Dixon, D., Khaled, R., & Nacke, L. (2011). From game design elements to gamefulness: defining gamification. In *Proceedings of the 15th international academic MindTrek conference: Envisioning future media environments* (pp. 9–15). Academic Press. 10.1145/2181037.2181040

Devereux, J. J., Buckle, P. W., & Vlachonikolis, I. G. (1999). Interactions between physical and psychosocial risk factors at work increase the risk of back disorders: An epidemiological approach. *Occupational and Environmental Medicine*, *56*(5), 343–353. doi:10.1136/oem.56.5.343 PMID:10472310

Dey, A. K. (2001). Understanding and using context. *Personal and Ubiquitous Computing*, *5*(1), 4–7. doi:10.1007007790170019

Diário da República 1ª série. (2006). *Regime da acessibilidade aos edifícios e estabelecimentos que recebem público, via pública e edifícios habitacionais*. Available at: https://dre.pt/application/dir/pdf1s/2006/08/15200/56705689.pdf

Dicheva, D., Dichev, C., Agre, G., & Angelova, G. (2015). Gamification in education: A systematic mapping study. *Journal of Educational Technology & Society*, *18*(3), 75.

Dickerson, S. S., & Kemeny, M. E. (2004). Acute Stressors and Cortisol Responses: A Theoretical Integration and Synthesis of Laboratory Research. *Psychological Bulletin*, *130*(3), 355–391. doi:10.1037/0033-2909.130.3.355 PMID:15122924

Dineshwar, R., & Steven, M. (2013, February). An investigation on mobile banking adoption and usage: A case study of Mauritius. In *Proceedings of 3rd Asia-Pacific Business Research Conference* (pp. 25-26). Academic Press.

Dobkin, B. H., & Dorsch, A. (2011). The promise of mhealth: Daily activity monitoring and outcome assessments by wearable sensors. *Neurorehabilitation and Neural Repair*, *25*(9), 788–798. doi:10.1177/1545968311425908 PMID:21989632

Doherty, D. (2013). *Conferências saúde CUF - Mobile Health - Novas formas de olhar a saúde.* Retrieved July 13, 2017, from https://www.slideshare.net/3GDR/conferencias-saudecuf-david-doherty-23514751?ref=http://mhealthinsight.com/2013/05/18/join-us-in-portugal-for-mobile-health-new-ways-of-looking-at-health/

Domingos, C. R. R. (2012). *Mobile Banking: factores influenciadores da utilização de APPS bancárias* (Doctoral dissertation). Instituto Superior de Economia e Gestão.

Dörnyei, Z. (1998). Motivation in second and foreign language learning. *Language Teaching, 31*(3), 117–135. doi:10.1017/S026144480001315X

dos Anjos, E. S., de Paula Souza, F., & Vieira Ramos, K. (2006). Novas tecnologias e turismo: Um estudo do site Vai Brasil. *Caderno Virtual de Turismo, 6*(4).

Duggal, P. (2013). *Mobile payments & mobile law.* Saakshar Law Publications.

Duretti, S., Marchioro, C. E., Marasso, L., Vicari, C., Fiorano, L., Papas, E. G., … Falda, S. (2015). ALL4ALL: IoT and telecare project for social inclusion. *2015 IEEE 1st International Forum on Research and Technologies for Society and Industry, RTSI 2015 - Proceedings, 80*, 17–22. 10.1109/RTSI.2015.7325065

Eagle, N., & Pentland, A. (2005). Social serendipity: Mobilizing social software. *IEEE Pervasive Computing, 4*(2), 28–34. doi:10.1109/MPRV.2005.37

Editorial. (2017). Alipay vs WeChat Pay – who is winning the battle? *ASEAN Today*. Available at https://www.aseantoday.com/2017/02/alipay-vs-wechat-pay-who-is-winning-the-battle/

Elaish, M. M., Shuib, L., Ghani, N. A., Yadegaridehkordi, E., & Alaa, M. (2017). Mobile learning for English Language Acquisition: Taxonomy, Challenges, and Recommendations. *IEEE Access: Practical Innovations, Open Solutions, 5*, 19033–19047. doi:10.1109/ACCESS.2017.2749541

Eloy, S. (2014). "Rabo-de-bacalhau" Building Type Morphology : Data to a Transformation Grammar-based Methodology for Housing Rehabilitation. *Arq.Urb, 11*, 31–47.

Emamian, P., & Li, J. (2017). SimpleHealth – a Mobile Cloud Platform to Support Lightweight Mobile Health Applications for Low-end Cellphones. In *IEEE 19th International Conference on e-Health Networking, Applications and Services (Healthcom)* (pp. 1–6). IEEE.

Eric Venn-Watson. (2014). *The evolution of health IT: Today, technology that supports clinical workflows is becoming the norm.* Retrieved July 13, 2017, from http://www.healthcareitnews.com/blog/evolution-health-it

Ericsson .(2015). Mobile Subscriptions Q3 2015. *Ericsson Mobility Report: on the pulse of the networked society, No. EAB-15:037849 Uen*, 4.

European Agency for Safety and Health at Work. (2000). *Work-related low back disorders.* Luxembourg: Office for Official Publications of the European Communities. Retrieved from https://osha.europa.eu/en/tools-and-publications/publications/reports/204

European Agency for Safety and Health at Work. (2015). *Second European Survey of Enterprises on New and Emerging Risks (ESENER-2)*. Luxembourg: Publications Office of the European Union. Retrieved from https://osha.europa.eu/pt/node/7653/file_view

European Commission. (n.d.). *Health, Demographic Change and Wellbeing - European Commission*. Retrieved June 27, 2017, from https://ec.europa.eu/programmes/horizon2020/en/h2020-section/health-demographic-change-and-wellbeing

Fairall, L. R., Zwarenstein, M., Bateman, E. D., Bachmann, M., Lombard, C., Majara, B. P., & … . (2005). Effect of educational outreach to nurses on tuberculosis case detection and primary care of respiratory illness: Pragmatic cluster randomised controlled trial. *BMJ (Clinical Research Ed.)*, *331*(7519), 750–754. doi:10.1136/bmj.331.7519.750 PMID:16195293

Fakomogbon, M.A., & Bolaji, H.O. (n.d.). Effects of Collaborative Learning Styles on Performance of Students in a Ubiquitous Collaborative Mobile Learning Environment. *Contemporary Educational Technology, 8*, 268-279.

Fallata, S. M. (2016). *Information and communications technology integrating at tatweer schools: Understanding experiences of Saudi female English as foreign language teachers* (Order No. 10141361). Available from ProQuest Dissertations & Theses Global. (1822226200). Retrieved from https://search.proquest.com/docview/1822226200?accountid=26642

Febraban. (n.d.). Retrieved from https://portal.febraban.org.br/

Federal Communications Commission. (2010). *Twenty-First Century Communications and Video Accessibility Act* [Pub. L. 111-260]. Author.

Ferguson, N. (2008). *The ascent of money: A financial history of the world*. Penguin.

Ferraiolo, D. F., Sandhu, R., Gavrila, S., Kuhn, D. R., & Chandramouli, R. (2001). Proposed NIST standard for role-based access control. *ACM Transactions on Information and System Security*, *4*(3), 224–274. doi:10.1145/501978.501980

Ferreira, C. E. C. (2010). *Os bancos brasileiros na internet: um estudo de caso sobre os motivadores para a sua forma de atuação neste ambiente* (Doctoral dissertation).

Fielding, R. (2015). *Representational State Transfer (REST)*. Retrieved June 23, 2017, from https://www.ics.uci.edu/~fielding/pubs/dissertation/rest_arch_style.htm

Finžgar, L., & Trebar, M. (2011). Use of NFC and QR code identification in an electronic ticket system for public transport. In *The 19th International Conference on Software, Telecommunications and Computer Networks* (pp. 1–6). IEEE.

Fisher, K. D. (2010). Technology-enabled active learning environments: An appraisal. CELE Exchange. Centre for Effective Learning Environments, 2010(6–10), 1–8.

Fitzmaurice, C., Allen, C., Barber, R. M., Barregard, L., Bhutta, Z. A., Brenner, H., & ... others. (2017). Global, regional, and national cancer incidence, mortality, years of life lost, years lived with disability, and disability-adjusted life-years for 32 cancer groups, 1990 to 2015: A systematic analysis for the global burden of disease study. *JAMA Oncology*, *3*(4), 524–548. doi:10.1001/jamaoncol.2016.5688 PMID:27918777

Flick, U. (2014). An introduction to qualitative research. *Sage (Atlanta, Ga.)*.

Fling, B. (2009). *Mobile Design and Development: Practical concepts and techniques for creating mobile sites and web apps (Animal Guide)*. O'Reilly Media.

Fogg, B. J. (2002, December). Persuasive technology: Using computers to change what we think and do. *Ubiquity*, *2002*, 5. doi:10.1145/764008.763957

Fokkenrood, H., Verhofstad, N., van den Houten, M., Lauret, G., Wittens, C., Scheltinga, M., & Teijink, J. (2014). Physical activity monitoring in patients with peripheral arterial disease: Validation of an activity monitor. *European Journal of Vascular and Endovascular Surgery*, *48*(2), 194–200. doi:10.1016/j.ejvs.2014.04.003 PMID:24880631

Forrester. (2017). *European mobile payments will almost triple by 2021*. Available at https://www.forrester.com/report/European+Mobile+Payments+Will+Almost+Triple+By+2021/-/E-RES137528

Friedman, M. (1994). *Money mischief: Episodes in monetary history*. Houghton Mifflin Harcourt.

Fundação Lemann e Meritt. (2016). *Censo Escolar*. Retrieved September 14, 2017, from http://qedu.org.br/

Furió, D., González-Gancedo, S., Juan, M.-C., Seguí, I., & Costa, M. (2013). The effects of the size and weight of a mobile device on an educational game. *Computers & Education*, *64*, 24–41. doi:10.1016/j.compedu.2012.12.015

Gadkari, A. S., & McHorney, C. A. (2012). Unintentional non-adherence to chronic prescription medications: How unintentional is it really? *BMC Health Services Research*, *12*(1), 98. doi:10.1186/1472-6963-12-98 PMID:22510235

Garg, G. (2015). *QR Code Payments: Everything you need to know*. Scanova. Available at https://scanova.io/blog/blog/2015/04/08/qr-code-payment/

Garg, P., Tian, L., Criqui, M., Liu, K., Ferrucci, L., Guralnik, J., ... McDermott, M. (2006). Physical activity during daily life and mortality in patients with peripheral arterial disease. *Circulation*, *114*(3), 242–248. doi:10.1161/CIRCULATIONAHA.105.605246 PMID:16818814

Garmin. (2017). *vívoactive® HR | Garmin*. Retrieved March 21, 2017, from https://buy.garmin.com/en-US/US/p/538374

Garris, R., Ahlers, R., & Driskell, J. E. (2002). Games, motivation, and learning: A research and practice model. *Simulation & Gaming*, *33*(4), 441–467. doi:10.1177/1046878102238607

Gaspar, F. (2016). ARch4maps : a mobile augmented reality tool to enrich paper maps. In *Proceedings of the Symposium on Simulation for Architecture and Urban Design* (pp. 179–182). London: SIMAUD. Available at http://www.scs.org/simaud

Gawande, A., & Bates, D. (2000). The use of information technology in improving medical performance. Part II. Physician-support tools. *MedGenMed: Medscape General.* Retrieved from http://europepmc.org/abstract/med/11104459

GENEVA. (2003). *Diabetes cases could double in developing countries in next 30 years.* Retrieved from https://goo.gl/5b7kBf

Ghag, O., & Hegde, S. (2012). A comprehensive study of google wallet as an NFC application. *International Journal of Computers and Applications, 58*(16).

GHDx. (2015). *Global burden of disease study 2015 (gbd 2015) data resources.* Retrieved from http://ghdx.healthdata.org/gbd-2015

Glauber, N. (2015). *Dominando o Android Studio, do básico ao avançado.* São Paulo: Novatec.

Gomes, S. (2015). *Requalificação do Bairro S. Nicolau e Realidade Aumenta Aplicada nos Mapas de Arquitetura.* Instituto Universitário de Lisboa.

Gonçalves, C., Rocha, T., Reis, A., & Barroso, J. (2017). *AppVox: An Application to Assist People with Speech Impairments in Their Speech Therapy Sessions.* Recent Advances in Information Systems and Technologies. doi:10.1007/978-3-319-56538-5_59

Google Inc. (n.d.). *Android Apps on Google Play.* Retrieved October 31, 2017, from https://play.google.com/store/apps

Google. (2017). *Fit – Google.* Retrieved March 21, 2017, from https://www.google.com/fit/

Gornik, H., & Beckman, J. (2005). Peripheral Arterial Disease. *Circulation, 111*(13), 169–172. doi:10.1161/01.CIR.0000160581.58633.8B PMID:15811861

Goundar, S. (2011, December). What is the Potential Impact of Using Mobile Devices in Education? *SIG GlobDev, 4,* 1–30.

Government, H. M. (2014). *The Building Regulations 2010. Access to and use of Buildings. Document M.* Available at: http://webarchive.nationalarchives.gov.uk/20151113141044/http://www.planningportal.gov.uk/uploads/br/br_pdf_ad_m_2013.pdf

Greengard, S. (2015). *The internet of things.* Cambridge, MA: MIT Press.

Groves, R. M. (2004). *Survey errors and survey costs* (Vol. 536). John Wiley & Sons.

Guerreiro, R., Eloy, S., & Lopes, P. F. (2012). *Lisbon Pedestrian Network. 2ª Conferência do PNUM Morfologia Urbana nos Países Lusófonos.*

Guimarães, A. S., & Johnson, G. F. (2007). *Sistemas de informações: administração em tempo real.* Qualitymark Editora Ltda.

Gup, B. E. (2003). *The future of banking.* Greenwood Publishing Group.

Haig, M. (2002). *Mobile marketing: The message revolution.* Kogan Page Publishers.

Hamad, H., Saad, M., & Abed, R. (2010). Performance Evaluation of RESTful Web Services for Mobile Devices. *Int. Arab J. E-Technol.* Retrieved from http://www.iajet.org/iajet_files/vol.1/no.3/Performance Evaluation of RESTful Web Services for Mobile Devices.pdf

Hari, R., & Kujala, M. V. (2009). Brain basis of human social interaction: From concepts to brain imaging. *Physiological Reviews, 89*(2), 453–479. doi:10.1152/physrev.00041.2007 PMID:19342612

Harris, A., Qadir, J. K., & Ahmad, U. (2017). *Persuasive technology for human development: Review and case study.* arXiv preprint arXiv:1708.08758

Harris, J. R. (2011). *The nurture assumption: Why children turn out the way they do.* Simon and Schuster.

Hartin, P. J., Nugent, C. D., McClean, S. I., Cleland, I., Tschanz, J. T., Clark, C. J., & Norton, M. C. (2016). The empowering role of mobile apps in behavior change interventions: The gray matters randomized controlled trial. *JMIR mHealth and uHealth, 4*(3), e93. doi:10.2196/mhealth.4878 PMID:27485822

Hassard, J., & Cox, T. (2015). *Work-related stress: Nature and management.* Retrieved June 29, 2017, from https://oshwiki.eu/wiki/Work-related_stress:_Nature_and_management

Hawthorn, N. (2015). 10 things you need to know about the new EU data protection regulation. *ComputerWorld UK Daily Digest.* Retrieved June 10, 2017, from: http://www.computerworlduk.com/security/10-things-you-need-know-about-new-eu-data-protection-regulation-3610851/4/

Haydon, T., Hawkins, R., & Denune, H. (2012). A comparison of iPads and worksheets on math skills of high school students with emotional disturbance. *Behavioral.* Retrieved from http://journals.sagepub.com/doi/abs/10.1177/019874291203700404

Heartfile. (2004). *National action plan for prevention and control of noncommunicable diseases and health promotion in Pakistan.* Retrieved from http://heartfile.org/pdf/NAPmain.pdf

Hernandez, J., Morris, R. R., & Picard, R. W. (2011). *Call Center Stress Recognition with Person-Specific Models. Affective Computing and Intelligent Interaction.* Springer Berlin Heidelberg. doi:10.1007/978-3-642-24600-5_16

Hicks, T., & Caroline, M. (2007). *A guide to managing workplace stress.* Retrieved from https://www.google.com/books?hl=pt-PT&lr=&id=fcxUW9kyDukC&oi=fnd&pg=PA7&dq=A+Guide+to+Managing+Workplace+Stress+hicks&ots=l6AFkH1H6z&sig=Q2sIEc3SuUo_vUXanTlLlM2CkSk

Holden, C. L., & Sykes, J. M. (2012). Leveraging mobile games for place-based language learning. *Developments in Current Game-Based Learning Design and Deployment, 27.*

Horwitz, E. K., Horwitz, M. B., & Cope, J. (1986). Foreign language classroom anxiety. *Modern Language Journal*, *70*(2), 125–132. doi:10.1111/j.1540-4781.1986.tb05256.x

Hoseinitabatabaei, S. A., Gluhak, A., Tafazolli, R., & Headley, W. (2014). Design, realization, and evaluation of uDirect-An approach for pervasive observation of user facing direction on mobile phones. *IEEE Transactions on Mobile Computing*, *13*(9), 1981–1994. doi:10.1109/TMC.2013.53

Hossain, P., Kawar, B., & El Nahas, M. (2007). Obesity and diabetes in the developing world—a growing challenge. *The New England Journal of Medicine*, *356*(3), 213–215. doi:10.1056/NEJMp068177 PMID:17229948

Howard, S. K., & Thompson, K. (2015). Seeing the system: Dynamics and complexity of technology integration in secondary schools. *Education and Information Technologies*, *21*(6), 1877–1894. doi:10.100710639-015-9424-2

Hsu, Y.-C., & Ching, Y.-H. (2015). A review of models and frameworks for designing mobile learning experiences and environments. *Canadian Journal of Learning and Technology*, *41*(3). doi:10.21432/T2V616

Hu, F. B. (2011). Globalization of diabetes. *Diabetes Care*, *34*(6), 1249–1257. doi:10.2337/dc11-0442 PMID:21617109

Hussain, M., Al-Haiqi, A., Zaidan, A. A., Zaidan, B. B., Kiah, M. L. M., Anuar, N. B., & Abdulnabi, M. (2015). The landscape of research on smartphone medical apps: Coherent taxonomy, motivations, open challenges and recommendations. *Computer Methods and Programs in Biomedicine*, *122*(3), 393–408. doi:10.1016/j.cmpb.2015.08.015 PMID:26412009

IBGE. (2016). *Estatísticas*. Retrieved September 12, 2017, from http://downloads.ibge.gov.br/downloads_estatisticas.htm

IBGE. (n.d.). *Resultados do Universo do Censo Demográfico 2010*. Retrieved from http://www.ibge.gov.br/home/

Ibrahim, N., Ahmad, W. F. W., & Shafe, A. (2016). Practitioners' validation on effectiveness of multimedia Mobile Learning Application for children. *International Conference on Computer And Information Sciences*, *3*, 103-108. 10.1109/ICCOINS.2016.7783197

Imaja, I. M., & Ndayizigamiye, P. (2017). A design of a mobile health intervention for the prevention and treatment of Cholera in South Kivu in the Democratic Republic of Congo. In *IEEE Global Humanitarian Technology Conference (GHTC)* (pp. 1–5). IEEE. 10.1109/GHTC.2017.8239251

INE. (2015). *Envelhecimento da população residente em Portugal e na União Europeia*. Retrieved July 3, 2017, from https://www.ine.pt/xportal/xmain?xpid=INE&xpgid=ine_destaques&DESTAQUESdest_boui=224679354&DESTAQUESmodo=2&xlang=pt

Ingelheim, B. (n.d.). *The worldwide occurrence of type 2 diabetes*. Retrieved from https://goo.gl/wCNwsr

Institute for Quality and Efficiency in Health Care. (2017). *Depression: What is burnout?* Retrieved from https://www.ncbi.nlm.nih.gov/pubmedhealth/PMH0072470/

Institute of Medicine. (2001). Crossing the quality chasm: A new health care system for the 21st century. Washington, DC: Author.

International Organization for Standardization. (2011). *Health informatics - Requirements for an electronic health record.* Author.

Islam, A., & Biswas, T. (2014). Health system in bangladesh: Challenges and opportunities. *American Journal of Health Research*, 2(6), 366–374. doi:10.11648/j.ajhr.20140206.18

ISO. (1991). *ISO/IEC 9126: Information Technology-Software Product Evaluation-Quality Characteristics and Guidelines for Their Use.* ISO.

ISO/IEC. (2009). *ISO/IEC 2nd WD 27002 (revision) - Information technology - Security techniques – Code of practice for information security management.* ISO Copyright Office.

ISO/TC 215 Health Informatics. (2015, Apr 15). Geneva: International Organization for Standardization.

ITU & WHO. (2012). *Mhealth for NCDs (WHO-ITU joint work-plan).* Retrieved from https://goo.gl/QjEopz

Jackson, E. M. (2004). *The PayPal wars: Battles with eBay, the media, the mafia, and the rest of planet Earth.* World Ahead Publishing.

Jean, G., & Simard, D. (2011). Grammar teaching and learning in L2: Necessary, but boring? *Foreign Language Annals*, *44*(3), 467–494. doi:10.1111/j.1944-9720.2011.01143.x

Jecan, S., Rusu, L., Arba, R., & Mican, D. (2017). Mobile application for elders with cognitive impairments. In *Internet Technologies and Applications* (pp. 155–160). ITA. doi:10.1109/ITECHA.2017.8101928

Jennings, H. H. (1959). *Sociometry in group relations: A manual for teachers.* American Council on Education.

Jovanov, E., Milenkovic, A., Otto, C., & de Groen, P. C. (2005). A wireless body area network of intelligent motion sensors for computer assisted physical rehabilitation. *Journal of Neuroengineering and Rehabilitation*, 2(1), 16–23. doi:10.1186/1743-0003-2-6 PMID:15740621

Kachru, B. B. (2006). The English language in the outer circle. *World Englishes*, *3*, 241–255.

Kankeu, H. T., Saksena, P., Xu, K., & Evans, D. B. (2013). The financial burden from non-communicable diseases in low-and middle-income countries: A literature review. *Health Research Policy and Systems*, *11*(1), 31. doi:10.1186/1478-4505-11-31 PMID:23947294

Kay, M., Santos, J., & Takane, M. (2011). mhealth: New horizons for health through mobile technologies. *World Health Organization, 64*(7), 66–71.

Kelly, K. M. (2015). *Apps for Behavior Analysts*. Available at: https://batechsig.com/2015/03/09/apps-for-behavior-analysts/

Khalilzadeh, J., Ozturk, A. B., & Bilgihan, A. (2017). Security-related factors in extended UTAUT model for NFC based mobile payment in the restaurant industry. *Computers in Human Behavior*, *70*(C), 460–474. doi:10.1016/j.chb.2017.01.001

Khan, W. Z., Xiang, Y., Aalsalem, M. Y., & Arshad, Q. (2013). Mobile phone sensing systems: A survey. *IEEE Communications Surveys and Tutorials*, *15*(1), 402–427. doi:10.1109/SURV.2012.031412.00077

Kharif, O. (2015). Fraudulent smartphone payments are becoming a pricey problem. *Bloomberg Businessweek*. Available at https://www.bloomberg.com/news/articles/2015-02-13/mobile-payment-fraud-is-becoming-a-pricey-problem

Khowaja, L. A., Khuwaja, A. K., & Cosgrove, P. (2007). Cost of diabetes care in out-patient clinics of karachi, pakistan. *BMC Health Services Research*, *7*(1), 189. doi:10.1186/1472-6963-7-189 PMID:18028552

Kim, C. P. J. L. J., & Lee, S. J. L. D. (2017). *"Don't Bother Me. I'm Socializing!": A Breakpoint-Based Smartphone Notification System*. Academic Press.

Kim, W. J. (2013). *Study on design for improving the mobile-web accessibility of users who have blind* (Master's Thesis). Available from National Assembly Library. (KDMT1201317130)

Kim, D. J., Ferrin, D. L., & Rao, H. R. (2008). A trust-based consumer decision-making model in electronic commerce: The role of trust, perceived risk, and their antecedents. *Decision Support Systems*, *44*(2), 544–564. doi:10.1016/j.dss.2007.07.001

Kim, D., & Solomon, M. (2012). *Fundamentals of information systems security*. Sudbury, MA: Jones and Bartlett Learning.

Kirschbaum, C., & Hellhammer, D. H. (1994). Salivary cortisol in psychoneuroendocrine research: Recent developments and applications. *Psychoneuroendocrinology*, *19*(4), 313–333. doi:10.1016/0306-4530(94)90013-2 PMID:8047637

Kleijnen, M., De Ruyter, K., & Wetzels, M. (2007). An assessment of value creation in mobile service delivery and the moderating role of time consciousness. *Journal of Retailing*, *83*(1), 33–46. doi:10.1016/j.jretai.2006.10.004

Kofod-Petersen, A., & Cassens, J. (2006). Using activity theory to model context awareness. *Lecture Notes in Computer Science*, *3946*, 1–17. doi:10.1007/11740674_1

Kollmann, A., Riedl, M., Kastner, P., Schreier, G., & Ludvik, B. (2007). Feasibility of a mobile phone–based data service for functional insulin treatment of type 1 diabetes mellitus patients. *Journal of Medical Internet Research*, *9*(5), e36. doi:10.2196/jmir.9.5.e36 PMID:18166525

Kolvenbach, S., Grather, W., & Klockner, K. (2004, February). Making community work aware. In *Parallel, Distributed and Network-Based Processing, 2004. Proceedings. 12th Euromicro Conference on* (pp. 358-363). IEEE. 10.1109/EMPDP.2004.1271466

Koster, R. (2013). *Theory of fun for game design.* O'Reilly Media, Inc.

Koutromanos, G., & Avraamidou, L. (2014). The use of mobile games in formal and informal learning environments: A review of the literature. *Educational Media International, 51*(1), 49–65. doi:10.1080/09523987.2014.889409

Kremers, R., & Brassett, J. (2017). Mobile payments, social money: Everyday politics of the consumer subject. *New Political Economy*, 1–16.

Krendl, K. A., & Clark, G. (1994). The impact of computers on learning: Research on in-school and out-of-school settings. *Journal of Computing in Higher Education, 5*(2), 85–112. doi:10.1007/BF02948572

Kuisma, T., Laukkanen, T., & Hiltunen, M. (2007). Mapping the reasons for resistance to Internet banking: A means-end approach. *International Journal of Information Management, 27*(2), 75–85. doi:10.1016/j.ijinfomgt.2006.08.006

Kukulska-Hulme, A. (2010). *Mobile learning for quality education and social inclusion.* Academic Press.

Kukulska-Hulme, A. (2007). Mobile usability in educational contexts: What have we learnt? *The International Review of Research in Open and Distributed Learning, 8*(2). doi:10.19173/irrodl.v8i2.356

Kukulska-Hulme, A., & Shield, L. (2008). An overview of mobile assisted language learning: From content delivery to supported collaboration and interaction. *ReCALL, 20*(3), 271–289. doi:10.1017/S0958344008000335

Kukulska-Hulme, A., & Traxler, J. (2005). *Mobile learning: A handbook for educators and trainers.* Routledge.

Kumar, A., & Gupta, S. (2012). *Health infrastructure in India: Critical analysis of policy gaps in the indian healthcare delivery.* Vivekanand International Foundation.

Kurkovsky, S. (2009). Engaging students through mobile game development. *ACM SIGCSE Bulletin, 41*(1), 44–48. doi:10.1145/1539024.1508881

Kurkovsky, S. (2013). Mobile game development: Improving student engagement and motivation in introductory computing courses. *Computer Science Education, 23*(2), 138–157. doi:10.1080/08993408.2013.777236

Kurtz, R.G.M. (2016). *Resistência à atitude e intenção de adoção do m-learning por professores do ensino superior* (tese de doutorado). Disponível na base de dados da Pontífica Universidade Católica do Rio de Janeiro.

Lane, N. D., Miluzzo, E., Lu, H., Peebles, D., Choudhury, T., & Campbell, A. T. (2010). A survey of mobile phone sensing. *IEEE Communications Magazine*, *48*(9), 140–150. doi:10.1109/MCOM.2010.5560598

Latif, S., Qadir, J., Farooq, S., & Imran, M. A. (2017). How 5g wireless (and concomitant technologies) will revolutionize healthcare? *Future Internet*, *9*(4), 93. doi:10.3390/fi9040093

Latif, S., Rana, R., Qadir, J., Imran, M., & Younis, S. (2017). Mobile health in the developing world: Review of literature and lessons from a case study. *IEEE Access: Practical Innovations, Open Solutions*, *5*, 11540–11556. doi:10.1109/ACCESS.2017.2710800

Laukkanen, T. (2007). Internet vs mobile banking: Comparing customer value perceptions. *Business Process Management Journal*, *13*(6), 788–797. doi:10.1108/14637150710834550

Laukkanen, T., & Cruz, P. (2008, July). Comparing consumer resistance to mobile banking in Finland and Portugal. In *International Conference on E-Business and Telecommunications* (pp. 89-98). Springer.

Laukkanen, T., Sinkkonen, S., Kivijärvi, M., & Laukkanen, P. (2007). Innovation resistance among mature consumers. *Journal of Consumer Marketing*, *24*(7), 419–427. doi:10.1108/07363760710834834

Lazarus, R. S., & Folkman, S. (1984). *Stress*. New York: Appraisal, and Coping.

Lecheta, R. R. (2012). *Google Android para Tablets*. São Paulo: Novatec.

Lecheta, R. R. (2013). *Google Android-3ª Edição: Aprenda a criar aplicações para dispositivos móveis com o Android SDK*. Novatec Editora.

Lee, J., Cho, C. H., & Jun, M. S. (2011). Secure quick response-payment (QR-Pay) system using mobile device. In *The 13th International Conference on Advanced Communication Technology* (pp. 1424–1427). IEEE.

Lee, J. J., & Hammer, J. (2011). Gamification in education: What, how, why bother? *Academic Exchange Quarterly*, *15*(2), 146.

Lei n. 12.249, de 11 de junho de 2010. (n.d.). *Criação do Programa um Computador por Aluno*. Retirado em 28 de setembro de 2017, de http://www.planalto.gov.br/ccivil_03/_ato2007-2010/2010/lei/l12249.htm

Lei n. 14.363 de 25 de janeiro de 2008. (n.d.). *Proibição do uso de telefone celular em escolas públicas do estado de Santa Catarina*. Retirado em 28 de setembro de 2017, de http://www.leisestaduais.com.br/sc/lei-ordinaria-n-14363-2008-santa-catarina-dispoe-sobre-a-proibicao-do-uso-de-telefone-celular-nas-escolas-estaduais-do-estado-de-santa-catarina

Lei n. 7.232, de 29 de outubro de 1984. (n.d.). *Criação do Conselho Nacional de Informática*. Retirado em 28 de setembro de 2017, de http://www.planalto.gov.br/ccivil_03/leis/L7232.htm

Lerner, T. (2013). *Mobile payment*. Springer.

Liébana-Cabanillas, F., Muñoz-Leiva, F., & Sánchez-Fernández, J. (2017). A global approach to the analysis of user behavior in mobile payment systems in the new electronic environment. *Service Business*, 1–40.

Liébana-Cabanillas, F., Sánchez-Fernández, J., & Muñoz-Leiva, F. (2014). The moderating effect of experience in the adoption of mobile payment tools in Virtual Social Networks: The m-Payment Acceptance Model in Virtual Social Networks (MPAM-VSN). *International Journal of Information Management*, *34*(2), 151–166. doi:10.1016/j.ijinfomgt.2013.12.006

LifeMap Solutions. (n.d.). *COPD Navigator - LifeMap Solutions*. Retrieved February 24, 2017, from http://www.lifemap-solutions.com/products/copd-navigator/

Liu, G.-Z., & Hwang, G.-J. (2010). A key step to understanding paradigm shifts in e-learning: Towards context-aware ubiquitous learning. *British Journal of Educational Technology*, *41*(2), E1–E9. doi:10.1111/j.1467-8535.2009.00976.x

Li, Y., & Liu, X. (2017). Integration of IPad-Based M-Learning into a Creative Engineering Module in a Secondary School in England. Tojet. *The Turkish Online Journal of Educational Technology*, *16*, 43–57.

Li, Y., & Wang, L. (n.d.). Using iPad-based mobile learning to teach creative engineering within a problem-based learning pedagogy. *Education and Information Technologies*, *22*, 1–14.

Logan, A. G., McIsaac, W. J., Tisler, A., Irvine, M. J., Saunders, A., Dunai, A., ... Trudel, M. (2007). Mobile phone–based remote patient monitoring system for management of hypertension in diabetic patients. *American Journal of Hypertension*, *20*(9), 942–948. doi:10.1016/j.amjhyper.2007.03.020 PMID:17765133

Lopes, P. F. (2014). Counting App for local observations and space syntax. In A. Moural, P. F. Lopes, & S. Eloy (Eds.), *CLOSE CLOSER: Close do Cities, Closer to People* (p. 24). ISCTE-IUL.

Lóscio, B., Oliveira, H., & Pontes, J. (2011). NoSQL no desenvolvimento de aplicações Web colaborativas. *VIII Simpósio Brasileiro de*. Retrieved from http://www.addlabs.uff.br/sbsc_site/SBSC2011_NoSQL.pdf

Lowe, D. G. (1999). Object Recognition from Local Scale-Invariant Features. *Proceedings of the International Conference on Computer Vision*, 1150–1157. Available at: http://www.cs.ubc.ca/~lowe/papers/iccv99.pdf

Lu, C., Chang, M., Kinshuk, D., Huang, E., & Chen, C.-W. (2011). Usability of context-aware mobile educational game. *Knowledge Management & E-Learning: An International Journal*, *3*(3), 448–477.

Lu, H., Pan, W., Lane, N. D., Choudhury, T., & Campbell, A. T. (2009, June). SoundSense: scalable sound sensing for people-centric applications on mobile phones. In *Proceedings of the 7th international conference on Mobile systems, applications, and services* (pp. 165-178). ACM. 10.1145/1555816.1555834

Lukowicz, P., Kirstein, T., & Tröster, G. (2004). *Wearable systems for health care applications*. Retrieved June 3, 2017, from https://www.researchgate.net/publication/8480044_Wearable_systems_for_health_care_applications

Lu, M. T., Tzeng, G. H., Cheng, H., & Hsu, C. C. (2015). Exploring mobile banking services for user behavior in intention adoption: Using new hybrid MADM model. *Service Business*, *9*(3), 541–565. doi:10.100711628-014-0239-9

Lunardini, F., Basilico, N., Ambrosini, E., Essenziale, J., Mainetti, R., & Pedrocchi, A., … Borghese, N. A. (2017). Exergaming for balance training, transparent monitoring, and social inclusion of community-dwelling elderly. *RTSI 2017 - IEEE 3rd International Forum on Research and Technologies for Society and Industry, Conference Proceedings*. 10.1109/RTSI.2017.8065964

Lyytinen, K., & Yoo, Y. (2002). Ubiquitous computing. *Communications of the ACM*, *45*(12), 63–96.

Mackillop, L. (2014). *Smartphone based management of gestational diabetes*. Retrieved from https://goo.gl/Pq4N9c

Majumdar, A., Kar, S. S., Kumar, G., Palanivel, C., & Misra, P. (2015). mhealth in the prevention and control of non-communicable diseases in India: Current possibilities and the way forward. *Journal of Clinical and Diagnostic Research : JCDR*, *9*(2), LE06.

Males, S., Bate, F., & Macnish, J. (2017). The impact of mobile learning on student performance as gauged by standardised test (NAPLAN) scores. *Issues in Educational Research*, *1*(27), 99–114.

Manikandan, M., & Chandramohan, S. (2016). Mobile wallet: A virtual physical wallet to the customers. *PARIPEX-Indian Journal of Research, 4*(9).

Markovich, S., Markovich, S., Snyder, C., & Snyder, C. (2017). M-Pesa and Mobile Money in Kenya: Pricing for Success. Kellogg School of Management Cases, 1–17.

Martin & Chuck. (2011). *The Third Screen: Marketing to Your Customers in a World Gone Mobile*. London: Nicholas Brealey Publishing.

Martonik, A. (2015). *What's the difference between Android Pay and the new Google Wallet?* Available at http://www.androidcentral.com/whats-difference-between-android-pay-and-new-google-wallet

Martucci, L. A., Andersson, C., Schreurs, W., & Fischer-Hübner, S. (2006). Trusted Server Model for Privacy-Enhanced Location Based Services. In *Proceedings of the 11th Nordic Workshop on Secure IT Systems* (pp. 19–20). Academic Press.

Maslach, C., & Leiter, M. P. (2008). Early predictors of job burnout and engagement. *The Journal of Applied Psychology*, *93*(3), 498–512. doi:10.1037/0021-9010.93.3.498 PMID:18457483

Masulli, F., Rovetta, S., Cabri, A., Traverso, C., Capris, E., & Torretta, S. (2017). An Assistive Mobile System Supporting Blind and Visual Impaired People when Are Outdoor. In *IEEE 3rd International Forum on Research and Technologies for Society and Industry (RTSI)* (pp. 1–6). IEEE.

Matic, A., Osmani, V., Maxhuni, A., & Mayora, O. (2012, May). Multi-modal mobile sensing of social interactions. In *Pervasive computing technologies for healthcare (PervasiveHealth), 2012 6th international conference on* (pp. 105-114). IEEE. 10.4108/icst.pervasivehealth.2012.248689

Melo, T. (2012). *Designer projeta aplicativo como guia de pontos turísticos, em Manaus.* Retrieved June 23, 2017, from http://g1.globo.com/am/amazonas/noticia/2012/12/designer-projeta-aplicativo-como-guia-de-pontos-turisticos-em-manaus.html

Mendes, D., Martins, P., & Paiva, S. (2016). Mobile platform for helping visually impaired citizens using public transportation: A case study in a Portuguese historic center. *International Journal of Emerging Research in Management and Technology*, *5*(6).

Menezes, J., Fernando, J., Carvalho, C., Barbosa, J., & Mansilha, B. (2009). Estudo da Prevalência da Doença Arterial Periférica em Portugal. *Angiologia e Cirurgia Vascular*, *5*(2), 59–68.

Meola, A. (2016). Mobile payments technology and contactless payments explained. *Business Insider.* Available at http://www.businessinsider.com/mobile-payment-technology-contactless-payments-explained-2016-11

mHealth Alliance. (2013). *Mhealth opportunities for non-communicable diseases among the elderly.* Retrieved from https://goo.gl/3Pk1Ks

Miller, S. (2010). *The moral foundations of social institutions: A philosophical study.* Cambridge University Press.

Miluzzo, E., Lane, N. D., Eisenman, S. B., & Campbell, A. T. (2007, October). CenceMe–injecting sensing presence into social networking applications. In *European Conference on Smart Sensing and Context* (pp. 1-28). Springer. 10.1007/978-3-540-75696-5_1

Ministério da Educação. (2006). *Programa Nacional de Informática na educação.* Retirado em 28 de setembro de 2017, https://www.fnde.gov.br/sigetec/relatorios/indicadores_rel.html#Dois

Miniwatts Marketing Group. (2015, Nov. 20). *World internet usage and population statistics.* Author.

Minussi, M. M., & Wyse, A. T. S. (2016). Web-Game educacional para ensino e aprendizagem de Ciências. *Novas Tecnologias na Educação, 14*, 1–10.

Miranda, A. L. (2002). *Da natureza da tecnologia: uma análise filosófica sobre as dimensões ontológica, epistemológica e axiológica da tecnologia moderna.* Curitiba, Brazil: Biblioteca do Centro Federal de Educação Tecnológica do Paraná.

Monjelat, N. (2017). Programming Technologies for Social Inclusion: An experience in professional development with elementary teachers. In *2017 Twelfth Latin American Conference on Learning Technologies (LACLO)* (pp. 1–8). Academic Press. 10.1109/LACLO.2017.8120901

Montrieux, H., Courtois, C., Grove, F., Raes, A., Schellens, T., & Marez, L. (2013). Mobile Learning in Secondary Education: Perceptions and Acceptance of Tablets of Teachers and Pupils. *International Conference Mobile Learning*, 204-208.

Moreira, F., Ferreira, M. J., Santos, C. P., & Durão, N. (2017). Evolution and use of mobile devices in higher education: A case study in Portuguese Higher Education Institutions between 2009/2010 and 2014/2015. *Telematics and Informatics*, *34*(6), 838–852. doi:10.1016/j.tele.2016.08.010

Moreno, J. L. (1953). *Who shall survive?* Academic Press.

Morgan, M. J., Translate, V., & Summers, J. (2008). *Marketing esportivo*. Thomson Learning.

Motorola Mobility, L. L. C. (2017). *Moto 360 Sport - Sports Smartwatch powered by Android Wear - Motorola*. Retrieved March 21, 2017, from https://www.motorola.com/us/products/moto-360-sport

Moural, A., Lopes, P. F., & Eloy, S. (Eds.). (2014). *CLOSE CLOSER: Close do Cities, Closer to People*. Lisboa: ISCTE-IUL.

Murcho, N., Jesus, S., & Pacheco, E. (2009). O mal-estar relacionado com o trabalho em enfermeiros: um estudo empírico. *Actas Do I Congresso Luso-Brasileiro de Psicologia*. Retrieved from https://scholar.google.pt/scholar?q=O+mal-estar+relacionado+com+o+trabalho+em+enfermeiros%3A+um+estudo+empírico&btnG=&hl=pt-PT&as_sdt=0%2C5

Nations, U. (2016). *Identifying social inclusion and exclusion*. Academic Press.

Ndayizigamiye, P., & Hangulu, L. (2017). A Design of a Mobile Health Intervention to Enhance Home- Carers ' Disposal of Medical Waste in South Africa. In *IEEE Global Humanitarian Technology Conference (GHTC)* (pp. 1–6). IEEE. 10.1109/GHTC.2017.8239241

Neri, M. C. (2009). *O Tempo de Permanência na Escola e as Motivações dos Sem-Escola*. Federal University of Rio de Janeiro.

Neuendorf, K. A. (2016). The content analysis guidebook. *Sage (Atlanta, Ga.)*.

Neves, I. A., Semprebom, E., & Lima, A. A. (2011). Copa 2014: Expectativa e Receptividade os Setores Hoteleiro, Gastronômico e Turístico na Cidade de Curitiba. In Simpósio de Administração da Produção, Logística e Operações Internacionais. (SIMPOI). São Paulo, Brazil: Academic Press.

Nielsen, J. (1992). Usability engineering lifecycle. *IEEE Computer*, *25*(3), 12–22.

Nielsen, J. (1995). *10 Usability Heuristics for User Interface Design*. Available at: https://www.nngroup.com/articles/ten-usability-heuristics/

Nielsen, J. (1994). *Usability Engineering*. Cambridge, MA: Morgan Kaufmann Publishers.

Nielsen, J., & Molich, R. (1990). Heuristic evaluation of user interfaces. In *Proceedings of the SIGCHI conference on Human factors in computing systems* (pp. 249–256). Seattle, WA: ACM.

Nike, Inc. (2017). *Nike Running. Nike.com*. Retrieved March 21, 2017, from http://www.nike.com/us/en_us/c/running

Nikolic, I. A., Stanciole, A. E., & Zaydman, M. (2011). *Chronic emergency: Why NCDs matter*. Academic Press.

NIOA. (2007). *National Institute of Aging. Why population Aging Matters – A global Perspective.* Washington, DC: U.S. Department of State.

NIST. (1995). *SP 800-12: An Introduction to Computer Security: The NIST Handbook.* NIST. Retrieved March 3, 2017, from http://csrc.nist.gov/publications/nistpubs/800-12/handbook.pdf

Noronha, N. (2016). *Doentes com insuficiência respiratória acompanhados via smartphone | SAPO Lifestyle.* Retrieved September 24, 2016, from http://lifestyle.sapo.pt/saude/noticias-saude/artigos/doentes-com-insuficiencia-respiratoria-acompanhados-via-smartphone?artigo-completo=sim

Nsamenang, A. B., & Tchombé, T. M. (Eds.). (2012). *Handbook of African educational theories and practices: A generative teacher education curriculum.* HDRC.

Nseir, S., Hirzallah, N., & Aqel, M. (2013). A secure mobile payment system using QR code. In *The 5th International Conference on Computer Science and Information Technology* (pp. 111–114). IEEE. 10.1109/CSIT.2013.6588767

O'Donnell, O. (2007). Access to health care in developing countries: Breaking down demand side barriers. *Cadernos de Saude Publica, 23*(12), 2820–2834. doi:10.1590/S0102-311X2007001200003 PMID:18157324

Oerlemans, W., & Bakker, A. (2014). Burnout and daily recovery: A day reconstruction study. *Journal of Occupational Health.* Retrieved from http://psycnet.apa.org/journals/ocp/19/3/303/

Oinas-Kukkonen, H., & Harjumaa, M. (2009). Persuasive systems design: Key issues, process model, and system features. *Communications of the Association for Information Systems, 24*(1), 28.

Omonedo, P., & Bocij, P. (2017). Potential impact of perceived security, trust, cost and social influence on m-commerce adoption in a developing economy. *WORLD (Oakland, Calif.), 7*(1), 147–160.

Open mHealth. (2015). *Case study: Type 1 diabetes - Open mHealth.* Retrieved May 24, 2017, from http://www.openmhealth.org/features/case-studies/case-study-type-1-diabetes/

Open mHealth. (n.d.). *About Us | Open mHealth.* Retrieved July 24, 2017, from http://www.openmhealth.org/organization/about/

Ourique, L. (n.d.). *Home mobility hazards detected via object recognition in augmented reality.* (forthcoming)

OutSystems. (n.d.). *Low-Code Development Platform for Mobile and Web Apps | OutSystems.* Retrieved January 12, 2017, from https://www.outsystems.com/platform/

Owsley, S. (2011). Soldiers Take First Step in Combating Mozambique's Landmines. *U.S. Africa Command.* Available at: http://www.africom.mil/getArticle.asp?art=6854

Oxford Dictionaries. (n.d.). *Smartphone - definition of smartphone in English.* Retrieved June 29, 2017, from https://en.oxforddictionaries.com/definition/smartphone

Oxford, R. L. (Ed.). (1996). *Language learning motivation: Pathways to the new century* (Vol. 11). Natl Foreign Lg Resource Ctr.

Pachler, N., Bachmair, B., & Cook, J. (2010). Mobile Learning Structures, Agency, Practices (G. Kress, Ed.). Academic Press.

Palaghias, N., Hoseinitabatabaei, S. A., Nati, M., Gluhak, A., & Moessner, K. (2015, June). Accurate detection of real-world social interactions with smartphones. In *Communications (ICC), 2015 IEEE International Conference on* (pp. 579-585). IEEE. 10.1109/ICC.2015.7248384

Palaghias, N., Hoseinitabatabaei, S. A., Nati, M., Gluhak, A., & Moessner, K. (2016). A survey on mobile social signal processing. *ACM Computing Surveys*, *48*(4), 57. doi:10.1145/2893487

Pan, C., Yang, W., Jia, W., Weng, J., & Tian, H. (2009). Management of chinese patients with type 2 diabetes, 1998–2006: The diabcare-china surveys. *Current Medical Research and Opinion*, *25*(1), 39–45. doi:10.1185/03007990802586079 PMID:19210137

Pangaribuan, & Junifer, J. (2014). Diagnosis of diabetes mellitus using extreme learning machine. In *Information technology systems and innovation (ICITSI), 2014 international conference on* (pp. 33–38). Academic Press.

Panzavolta, S., & Lotti, P. (2013). *Serious Games and Inclusion. Special Education Needs Network*. European Commission.

Papastergiou, M. (2009). Digital game-based learning in high school computer science education: Impact on educational effectiveness and student motivation. *Computers & Education*, *52*(1), 1–12. doi:10.1016/j.compedu.2008.06.004

Pareschi, L., Riboni, D., & Bettini, C. (2008). Protecting users' anonymity in pervasive computing environments. In *Pervasive Computing and Communications, 2008. PerCom 2008. Sixth Annual IEEE International Conference on* (pp. 11–19). IEEE.

Park, R. E., & Burgess, E. W. (1921). *Introduction to the Science of Sociology*. Chicago: University of Chicago Press.

Park, Y. (2011). A Pedagogical Framework for Mobile Learning: Categorizing Educational Applications of Mobile Technologies into Four Types. *The International Review of Research in Open and Distributed Learning*, *12*(2), 78–102. doi:10.19173/irrodl.v12i2.791

Park, Y., & Chen, J. V. (2007). Acceptance and adoption of the innovative use of smartphone. *Industrial Management & Data Systems*, *107*(9), 1349–1365. doi:10.1108/02635570710834009

Pascoal, P. (2015). Retrieval of Objects Captured with Kinect One Camera. In *SHREC'15 Track: Eurographics Workshop on 3D Object Retrieval*. Eurographics Workshop on 3D Object Retriev.

Paulino, D., Reis, A., Paredes, H., & Barroso, J. (2016). Usage of mobile devices for monitoring and encouraging active life. *1st International Conference on Technology and Innovation is Sports, Health and Wellbeing*.

Pearce, P. L. (2001). *A relação entre residentes e turistas: literatura sobre pesquisas e diretrizes de gestão. In Turismo Global* (pp. 145–164). São Paulo: Editora SENAC.

Perrotta, C., Featherstone, G., Aston, H., & Houghton, E. (2013). *Game-based learning: Latest evidence and future directions. NFER Research Programme: Innovation in Education*. Slough: NFER.

Pesquisa Nacional por Amostra de Domicílios. (2014). Available at: https://ww2.ibge.gov.br/home/estatistica/pesquisas/pesquisa_resultados.php?id_pesquisa=40

Pfleeger, C., & Shari, L. (2007). *Security in Computing* (4th ed.). Prentice Hall PTR.

Phonthanukitithaworn, C., Sellitto, C., & Fong, M. (2015). User intentions to adopt mobile payment services: A study of early adopters in Thailand. *Journal of Internet Banking and Commerce*.

Picard, R. W., & Healey, J. (1997). Affective wearables. *Personal Technologies*, *1*(4), 231–240. doi:10.1007/BF01682026

Pina, F., Kurtz, R., Ferreira, J., Freitas, A., Silva, J. F., & Giovannini, C. J. (2016). Adoção de m-Learning no Ensino Superior: O Ponto de Vista Dos Professores. *Revista Eletrônica de Administração*, *22*, 279–306.

Pinheiro, N. A. M., Silveira, R. M. C. F., & Bazzo, W. A. (2007). Ciência, tecnologia e sociedade: A relevância do enfoque CTS para o contexto do ensino médio. *Ciência & Educação (Bauru)*, *13*(1), 71–84. doi:10.1590/S1516-73132007000100005

Piot, P., Caldwell, A., Lamptey, P., Nyrirenda, M., Mehra, S., Cahill, K., & Aerts, A. (2016). Addressing the growing burden of non–communicable disease by leveraging lessons from infectious disease management. *Journal of Global Health*, *6*(1), 010304. doi:10.7189/jogh.06.010304 PMID:26955469

Porter, M. E. (2001). *Estratégia competitiva: técnicas para análise de indústrias e da concorrência* (Vol. 2). Rio de Janeiro, Brazil: Campus.

Prasad, L. (2012). *Pervasive Computing goals and its Challenges for Modern Era*. Academic Press.

Preece, J., Rogers, Y., & Sharp, H. (2005). *Design de interação: Além da interação homem-computador*. Porto Alegre, Brazil: Bookman.

Prensky, M. (2001). Digital Natives Digital Immigrants. On the Horizon, 9.

Prensky, M. (2003). Digital game-based learning. *Computers in Entertainment*, *1*(1), 21. doi:10.1145/950566.950596

Presby, B. (2017). *Barriers to reducing the digital-use divide as perceived by middle school principals* (Order No. 10268273). Available from ProQuest Dissertations & Theses Global. (1887122371). Retrieved from https://search.proquest.com/docview/1887122371?accountid=26642

Pressman, S., Fachinger, J., den Exter, M., Grambow, B., Holgerson, S., Landesmann, C., & Titov, M. (2005). *Software engineering: a practitioner's approach*. Palgrave Macmillan.

Primo, A. F. T. (2006). O aspecto relacional das interações na Web 2.0. In Congresso Brasileiro de Ciências da Comunicação (29.: 2006 set.: Brasília). Anais: estado e comunicação [recurso eletrônico]. Brasília, DF: Intercom: Universidade de Brasília.

Proença, P. (2013). *Object Category Recognition through RGB-D Data*. ISCTE-IUL.

Pukkasenung, P., & Chokngamwong, R. (2016). Review and comparison of mobile payment protocol. In Advances in Parallel and Distributed Computing and Ubiquitous Services (pp. 11–20). Springer Singapore. doi:10.1007/978-981-10-0068-3_2

Quinn, C. C., Clough, S. S., Minor, J. M., Lender, D., Okafor, M. C., & Gruber-Baldini, A. (2008). Welldoc™ mobile diabetes management randomized controlled trial: Change in clinical and behavioral outcomes and patient and physician satisfaction. *Diabetes Technology & Therapeutics*, *10*(3), 160–168. doi:10.1089/dia.2008.0283 PMID:18473689

Rabello, E. T., & Passos, J. S. (2010). *Vygotsky e o desenvolvimento humano. Formato do arquivo: Microsoft Powerpoint-Visualização rápida*. Retrieved from www.ceesp.com.br/arquivos/Aula

Rabello, R. R. (2009). Android: um novo paradigma de desenvolvimento móvel. *Revista WebMobile, 18*.

Rahi, S., Ghani, M. A., & Alnaser, F. M. (2017). The Influence of E-Customer Services and Perceived Value on Brand Loyalty of Banks and Internet Banking Adoption: A Structural Equation Model (SEM). *Journal of Internet Banking and Commerce*, *22*(1), 1–18.

Rajalakshmi, R., Arulmalar, S., Usha, M., Prathiba, V., Kareemuddin, K. S., Anjana, R. M., & Mohan, V. (2015). Validation of smartphone based retinal photography for diabetic retinopathy screening. *PLoS One*, *10*(9), e0138285. doi:10.1371/journal.pone.0138285 PMID:26401839

Ranjan, P., & Om, H. (2017). Computational Intelligence Based Security in Wireless Sensor Networks: Technologies and Design Challenges. *Studies in Computational Intelligence Computational Intelligence in Wireless Sensor Networks*, 131-151. doi:10.1007/978-3-319-47715-2_6

Rao, S., & Troshani, I. (2007). A conceptual framework and propositions for the acceptance of mobile services. *Journal of Theoretical and Applied Electronic Commerce Research*, *2*(2).

Raposo, M., Eloy, S., & Dias, M. S. (2017). Revisiting the city, augmented with digital technologies : the SeeARch tool. In *Proceedings of the REHAB 2017 conference*. Available at: http://rehab.greenlines-institute.org/

Raposo, M. (2016). *Ver a Arquitectura através das tecnologias digitais*. Instituto Universitário de Lisboa.

Recife. (2017). *Portal de dados abertos da Prefeitura do Recife*. Retrieved June 23, 2017, from http://dados.recife.pe.gov.br/

Reckziegel, M. (2009). *Entendendo os WebServices*. Academic Press.

Regensteiner, J., Meyer, T., Krupski, W., Cranford, L., Hiatt, W., & Regensteiner, J. G. (1997). Hospital vs. home based exercise rehabilitation for patients with peripheral arterial occlusive disease. *Angiology*, *48*(4), 291–300. doi:10.1177/000331979704800402 PMID:9112877

Reis, A., Barroso, J., & Gonçalves, R. (2013). Supporting Accessibility in Higher Education Information Systems. *Proceedings of the 7th international conference on Universal Access in Human-Computer Interaction: applications and services for quality of life*. 10.1007/978-3-642-39194-1_29

Reis, A., Morgado, L., Tavares, F., Guedes, M., Reis, C., Borges, J., Gonçalves, R., & Cruz, J. (2016b). Gestão de listas de espera para cirurgia na rede hospitalar pública portuguesa - O sistema de informação dos programas de recuperação de listas de espera. *CISTI, 11.ª Conferência Ibérica de Sistemas e Tecnologias de Informação*. DOI:10.1109/CISTI.2016.7521612

Reis, A., Lains, J., Paredes, H., Filipe, V., Abrantes, C., Ferreira, F., ... Barroso, J. (2016a). *Developing a System for Post-Stroke Rehabilitation: An Exergames Approach. In Universal Access in Human-Computer Interaction. Users and Context Diversity* (pp. 403–413). Springer International Publishing. doi:10.1007/978-3-319-40238-3_39

Rema, M., Premkumar, S., Anitha, B., Deepa, R., Pradeepa, R., & Mohan, V. (2005). Prevalence of diabetic retinopathy in urban india: The chennai urban rural epidemiology study (cures) eye study, i. *Investigative Ophthalmology & Visual Science*, *46*(7), 2328–2333. doi:10.1167/iovs.05-0019 PMID:15980218

Research2Guidance. (2013, Mar 4). *Mobile-Health-Market-Report-2013-2017*. Author.

Research, R. P. (2014). *Retinitis Pigmentosa Q&A*. Seoul, South Korea: Elsevier Korea.

Rettig, M. (1994). Prototyping for Tiny Fingers. *Communications of the ACM*, *37*(4), 21–27. doi:10.1145/175276.175288

Ribeiro, F. N., & Zorzo, S. D. (2009). A Quantitative Evaluation of Privacy in Location Based Services. In *Systems, Signals and Image Processing, 2009. IWSSIP 2009. 16th International Conference on* (pp. 1–4). Academic Press. 10.1109/IWSSIP.2009.5367735

Richardson, C., & Rymer, J. R. (2016). The Forrester Wave. *Low-Code Development Platforms*, *Q2*, 18.

Riff, D., Lacy, S., & Fico, F. (2014). *Analyzing media messages: Using quantitative content analysis in research*. Routledge.

Rock, D., & Page, L. J. (2009). *Coaching with the brain in mind: Foundations for practice*. John Wiley & Sons.

Rodrigues, A. L. (2016). A integração pedagógica das tecnologias digitais na formação ativa de professores. *Atas do IV Congresso Internacional das TIC na Educação: Tecnologias digitais e a Escola do Futuro,* 1320-1333.

Rostampoor-Vajari, M. (2012). What Is Sociometry and How We Can Apply It in Our Life. *Advances in Asian Social Science*, *2*(4), 570–573.

Roth, G. A., Johnson, C., Abajobir, A., Abd-Allah, F., Abera, S. F., & Abyu, G. (2017). Global, regional, and national burden of cardiovascular diseases for 10 causes, 1990 to 2015. *Journal of the American College of Cardiology*.

Rouse, M. (2008). *EU Data Protection Directive (Directive 95/46/EC)*. TechTarget. Retrieved June 10, 2017, from: http://whatis.techtarget.com/definition/EU-Data-Protection-Directive-Directive-95-46-EC

Roussos, G. (2006). Ubiquitous computing for electronic business. *Ubiquitous and Pervasive Commerce*, 1-12.

Safeena, R., Date, H., & Kammani, A. (2011). Internet Banking Adoption in an Emerging Economy: Indian Consumer's Perspective. *Int. Arab J. e-Technol.*, *2*(1), 56-64.

Saha, D., & Mukherjee, A. (2003). Pervasive computing: A paradigm for the 21st century. *Computer*, *36*(3), 25–31. doi:10.1109/MC.2003.1185214

Saint-Germain, R. (2005). *Information Security Management Best Practice Based on ISO/IEC 17799*. The Information Management Journal.

Salmony, D. M., & Jin, B. (2016). Pan-European mobile peer-to-peer payment–and beyond. *Journal of Digital Banking*, *1*(1), 85–96.

Samsung, (2017b). *S health I Start a Health Challenge*. Retrieved March 21, 2017, from https://shealth.samsung.com/

Samsung. (2017a). *Samsung Gear S2 - The Official Samsung Galaxy Site*. Retrieved March 21, 2017, from http://www.samsung.com/global/galaxy/gear-s2/

Sandberg, J., Maris, M., & Hoogendoorn, P. (2014). The added value of a gaming context and intelligent adaptation for a mobile learning application for vocabulary learning. *Computers & Education*, *76*, 119–130. doi:10.1016/j.compedu.2014.03.006

Santa Catarina. (2016). *Lei nº 14.363, de 25 de janeiro de 2008. DispÕe Sobre A ProibiÇÃo do Uso de Telefone Celular nas Escolas Estaduais do Estado de Santa Catarina*. Retrieved September 14, 2017, from http://www.leisestaduais.com.br/sc/lei-ordinaria-n-14363-2008-santa-catarina-dispoe-sobre-a-proibicao-do-uso-de-telefone-celular-nas-escolas-estaduais-do-estado-de-santa-catarina

Santos, R. P., & Freitas, S. R. S. (2017). Tecnologias Digitais Na Educação: Experiência do uso de Aplicativos de Celular no Ensino da Biologia. *Cadernos de Educação*, *16*(32), 135–150. doi:10.15603/1679-8104/ce.v16n32p135-150

Schilit, B., Adams, N., & Want, R. (1994, December). Context-aware computing applications. In *Mobile Computing Systems and Applications, 1994. WMCSA 1994. First Workshop on* (pp. 85-90). IEEE. 10.1109/WMCSA.1994.16

Schuster, D., Rosi, A., Mamei, M., Springer, T., Endler, M., & Zambonelli, F. (2013). Pervasive social context: Taxonomy and survey. *ACM Transactions on Intelligent Systems and Technology*, *4*(3), 46. doi:10.1145/2483669.2483679

Schwabe, G., & Göth, C. (2005). Mobile learning with a mobile game: Design and motivational effects. *Journal of Computer Assisted Learning*, *21*(3), 204–216. doi:10.1111/j.1365-2729.2005.00128.x

Sekaran, U., & Bougie, R. (2016). *Research methods for business: A skill building approach*. John Wiley & Sons.

Shaikh, A. A. (2013). Mobile banking adoption issues in Pakistan and challenges ahead. *J. Inst. Bank. Pak*, *80*(3), 12–15.

Shaikh, A. A., Hanafizadeh, P., & Karjaluoto, H. (2017). Mobile banking and payment system: A conceptual standpoint. *International Journal of E-Business Research*, *13*(2), 14–27. doi:10.4018/IJEBR.2017040102

Shaikh, A. A., & Karjaluoto, H. (2015). Mobile banking adoption: A literature review. *Telematics and Informatics*, *32*(1), 129–142. doi:10.1016/j.tele.2014.05.003

Sharples, M., Taylor, J., & Vavoula, G. (2010). A theory of learning for the mobile age. In *Medienbildung in neuen Kulturräumen* (pp. 87–99). VS Verlag für Sozialwissenschaften. doi:10.1007/978-3-531-92133-4_6

Shih, P. C., Han, K., Poole, E. S., Rosson, M. B., & Carroll, J. M. (2015). Use and Adoption Challenges of Wearable Activity Trackers. In *iConference 2015 Proceedings*. iSchools. Retrieved from https://www.ideals.illinois.edu/handle/2142/73649

Shrier, D., Canale, G., & Pentland, A. (2016). *Mobile money & payments: Technology trends*. Academic Press.

Silveira, L. M. D. O. B., & Wagner, A. (2012). A interação família-escola diante dos problemas de comportamento da criança: Estudos de caso. *Psicologia da Educação*, (35): 95–119.

Silver, H. (2015). *The Contexts of Social Inclusion*. Academic Press.

Silverman, D. (Ed.). (2016). *Qualitative research*. Sage.

Six, J., Cornelis, O., & Leman, M. (2014, January). TarsosDSP, a real-time audio processing framework in Java. In *Audio Engineering Society Conference: 53rd International Conference: Semantic Audio*. Audio Engineering Society.

Slama, S., Kim, H.-J., Roglic, G., Boulle, P., Hering, H., Varghese, C., & Tonelli, M. (2016). *Care of non-communicable diseases in emergencies*. London: Lancet.

Sohail, M. S., & Al-Jabri, I. M. (2014). Attitudes towards mobile banking: Are there any differences between users and non-users? *Behaviour & Information Technology*, *33*(4), 335–344. doi:10.1080/0144929X.2013.763861

Son, I., Lee, H., Kim, G., & Kim, J. (2015). The effect of Samsung Pay on Korea equity market: Using the Samsung's domestic supply chain. *Advanced Science and Technology Letters, 114*, 51–55. doi:10.14257/astl.2015.114.10

Spikol, D., & Milrad, M. (2008). Combining physical activities and mobile games to promote novel learning practices. In *Wireless, Mobile, and Ubiquitous Technology in Education, 2008. WMUTE 2008. Fifth IEEE International Conference on* (pp. 31–38). IEEE. 10.1109/WMUTE.2008.37

Spolsky, B. (1986). 11 Overcoming language barriers to education in a multilingual world. *Language and Education in Multilingual Settings, 25*, 182.

Sridar, K., & Shanthi, D. (2017). Design and development of mobile phone based diabetes mellitus diagnosis system by using ann-fp-growth techniques. *Biomedical Research.*

Sripalawat, J., Thongmak, M., & Ngramyarn, A. (2011). M-banking in metropolitan Bangkok and a comparison with other countries. *Journal of Computer Information Systems, 51*(3), 67–76.

Stansfeld, S. A., Fuhrer, R., Shipley, M. J., & Marmot, M. G. (1999). Work characteristics predict psychiatric disorder: Prospective results from the Whitehall II Study. *Occupational and Environmental Medicine, 56*(5), 302–307. doi:10.1136/oem.56.5.302 PMID:10472303

Statista. (2017a). *Statista data.* Available at https://www.statista.com/statistics/277819/paypals-annual-mobile-payment-volume/

Statista. (2017b). *Statista data.* Available at https://www.statista.com/statistics/277841/paypals-total-payment-volume/

Statista. (2017c). *Statista data.* Available at https://www.statista.com/statistics/279957/number-of-mobile-payment-users-by-region/

Steinert, M. E. P., Hardoim, E. L., & Pinto, M. P. P. R. C. (2017). De mãos limpas com as tecnologias. *Revista Sustinere, 4*, 233–252.

Stewart, J., Bleumers, L., Van Looy, J., Mariën, I., All, A., & ... (2013). *The potential of digital games for empowerment and social inclusion of groups at risk of social and economic exclusion: evidence and opportunity for policy.* Joint Research Centre, European Commission.

Stylos, J. (2009). *Making APIs more usable with improved API designs, documentation and tools.* Pittsburgh: Carnegie Mellon University. Retrieved from http://search.proquest.com/openview/5d9b5ea97d780f089afab105565a902c/1?pq-origsite=gscholar&cbl=18750&diss=y

Sung, M., Marci, C., & Pentland, A. (2005). No Title. *Journal of Neuroengineering and Rehabilitation, 2*(1), 17. doi:10.1186/1743-0003-2-17 PMID:15987514

Sung, Y.-T., Chang, K.-E., & Liu, T.-C. (2016). The effects of integrating mobile devices with teaching and learning on students' learning performance: A meta-analysis and research synthesis. *Computers & Education, 94*(Supplement C), 252–275. doi:10.1016/j.compedu.2015.11.008

Swaine, M. D. (2015). Chinese views and commentary on the 'One Belt, One Road' initiative. *China Leadership Monitor, 47*, 1–24.

Taormina-Weiss, W. (2012). *Workplace Stress - Symptoms and Solutions - Disabled World*. Retrieved July 20, 2016, from https://www.disabled-world.com/disability/types/psychological/workplacestress.php

Tapscott, D. (1997). *Economia digital: promessa e perigo na era da intelig{ê}ncia em rede*. São Paulo: Makron Books.

Taylor, S. J., Bogdan, R., & DeVault, M. (2015). *Introduction to qualitative research methods: A guidebook and resource*. John Wiley & Sons.

Team, G. (2016). *GSMA Intelligence. Global Mobile Engagement Index*. Retrieved from https://gsmaintelligence.com/

Teleco. (2016a). *Smartphones no Brasil*. Retrieved from www.teleco.com.br/smartphone.asp

Teleco. (2016b). *Perfil dos Usuários de Internet no Brasil*. Retrieved from www.teleco.com.br/internet_usu.asp

Teleco. (2016c). *Internet no Brasil*. Retrieved from www.teleco.com.br/internet.asp

Temdee, P., & Prasad, R. (2017). *Context-Aware Communication and Computing: Applications for Smart Environment*. Academic Press.

Teng, J., Zhang, B., Li, X., Bai, X., & Xuan, D. (2014). E-shadow: Lubricating social interaction using mobile phones. *IEEE Transactions on Computers, 63*(6), 1422–1433. doi:10.1109/TC.2012.290

Thompson, J. B. (1998). *O advento da interação mediada. A mídia e a modernidade: uma teoria social da mídia*. Petrópolis, RJ: Vozes.

Thrasher, J. (2013). RFID vs. NFC: What's the difference? *RFID Insider*. Available at http://blog.atlasrfidstore.com/rfid-vs-nfc

Tighe, L. (2016). *Teacher perceptions of usefulness of mobile learning devices in rural secondary science classrooms* (Order No. 10239808). Available from ProQuest Dissertations & Theses Global. (1868429722). Retrieved from https://search.proquest.com/docview/1868429722?accountid=26642

To, W. M., & Lai, L. S. (2014). Mobile banking and payment in China. *IT Professional, 16*(3), 22–27. doi:10.1109/MITP.2014.35

Tylor, E. B. (1871). *Primitive culture: researches into the development of mythology, philosophy, religion, art, and custom* (Vol. 2). J. Murray.

Ulrich, R. S., Simons, R. F., Losito, B. D., Fiorito, E., Miles, M. A., & Zelson, M. (1991). Stress recovery during exposure to natural and urban environments. *Journal of Environmental Psychology, 11*(3), 201–230. doi:10.1016/S0272-4944(05)80184-7

UN. (2017). *World Population Prospects: The 2017 Revision*. New York: United Nations.

UNESCO. (2014). *O futuro da aprendizagem móvel*. Paris: Unesco.

UNICEF & WHO. (2015). *Water, sanitation and hygiene in health care facilities: status in low and middle income countries and way forward*. Author.

United Nations Department of Economic and Social Affairs. (2016). New York: UN DESA.

UXPA. (2017). *Usability Body of Knowledge*. UXPA.

Van Eck, R. (2006). Digital game-based learning: It's not just the digital natives who are restless. *EDUCAUSE Review*, *41*(2), 16.

Vassos, T. (1998). *Marketing estratégico na Internet*. Makron Books.

Vicentin, I. C., & Hoppen, N. (2002). Tecnologia da Informação aplicada aos negócios de Turismo no Brasil. *Turismo-Visão E Ação, 4*(11), 61–78.

Vinciarelli, A., Murray-Smith, R., & Bourlard, H. (2010, September). Mobile social signal processing: vision and research issues. In *Proceedings of the 12th international conference on Human computer interaction with mobile devices and services* (pp. 513-516). ACM. 10.1145/1851600.1851731

Volkswagen. (n.d.). *The Fun Theory*. Retrieved December 3, 2017, from http://www.thefuntheory.com/

Vrijkotte, T. G. M., van Doornen, L. J. P., & de Geus, E. J. C. (2000). Effects of Work Stress on Ambulatory Blood Pressure, Heart Rate, and Heart Rate Variability. *Hypertension*, *35*(4), 880–886. doi:10.1161/01.HYP.35.4.880 PMID:10775555

Wang, A. I., Øfsdahl, T., & Mørch-Storstein, O. K. (2008). An evaluation of a mobile game concept for lectures. In *Software Engineering Education and Training, 2008. CSEET'08. IEEE 21st Conference on* (pp. 197–204). IEEE. 10.1109/CSEET.2008.15

Wang, R., & Qian, X. (2010). *OpenSceneGraph 3.0: Beginner's guide*. Packt Publishing Ltd.

Warschauer, M. (2004). *Technology and social inclusion: Rethinking the digital divide*. MIT press.

Weatherford, J. (2009). *The history of money*. Crown Business.

Wee, W. (2012). China's Alipay has 700 million registered accounts, beats PayPal? *Tech in Asia*. Available at https://www.techinasia.com/chinas-alipay-700-million-registered-accounts-beatspaypal

Weiser, M. (1991). The Computer for the 21 st Century. *Scientific American*, *265*(3), 94–105. doi:10.1038cientificamerican0991-94 PMID:1675486

Werbach, K., & Hunter, D. (2012). *For the win: How game thinking can revolutionize your business*. Wharton Digital Press.

Werthner, H., Klein, S., & ... (1999). *Information technology and tourism: a challenging relationship*. Springer-Verlag Wien. doi:10.1007/978-3-7091-6363-4

West, D. M. (2014). *Going mobile: How wireless technology is reshaping our lives*. Brookings Institution Press. Retrieved from http://www.jstor.org/stable/10.7864/j.ctt7zsvqt

WHO. (2009). *Four noncommunicable diseases, four shared risk factors*. Retrieved from https://goo.gl/ixGLRJ

WHO. (2010). *Package of essential noncommunicable (pen) disease interventions for primary health care in low-resource settings*. WHO.

WHO. (2010a). *Back ground paper: Non communicable diseases in low and middle income countries*. Geneva: WHO.

WHO. (2010b). *Global status report on noncommunicable diseases 2010*. Retrieved from https://goo.gl/NOucW

WHO. (2011). *mHealth: New horizons for health through mobile technologies: second global survey on eHealth*. Geneva, Switzerland: WHO Library Cataloguing-in-Publication Data. Retrieved from http://www.who.int/goe/publications/goe_mhealth_web.pdf

WHO. (2013). *Global health workforce shortage to reach 12.9 million in coming decades*. Geneva, Switzerland: World Health Organization.

WHO. (2014a). *Mobile phones help people with diabetes to manage fasting and feasting during Ramadan*. Retrieved from https://goo.gl/spN2E4

WHO. (2014b). *World health statistics*. Retrieved from https://goo.gl/9q2sSA

WHO. (2017a). *Cancer by world health organization*. Retrieved from http://www.who.int/cancer/en/

WHO. (2017b). *Noncommunicable diseases and their risk factors*. Retrieved from http://www.who.int/ncds/introduction/en/

Wild, S. H., Roglic, G., Green, A., Sicree, R., & King, H. (2004). Global prevalence of diabetes: estimates for the year 2000 and projections for 2030: response to Rathman and Giani. *Diabetes Care*, *27*(10), 2569–2569. doi:10.2337/diacare.27.10.2569-a PMID:15451948

Winters, N. (2007). What is mobile learning. *Big Issues in Mobile Learning*, 7–11.

Word Bank. (2017). *Population ages 65 and above (% of total)*. Retrieved July 3, 2017, from http://data.worldbank.org/indicator/SP.POP.65UP.TO.ZS?locations=PT

World Health Organization, . (2002). *Reducing risks, promoting healthy life*. Author.

World, B. (2016). *Population ages 65 and above*. Retrieved from https://goo.gl/wPzRdX

Xavier, O. S. (1990). A sociometria na administração de recursos humanos. *Revista de Administração de Empresas*, *30*(1), 45–54. doi:10.1590/S0034-75901990000100005

Xin, H., Techatassanasoontorn, A. A., & Tan, F. B. (2015). Antecedents of consumer trust in mobile payment adoption. *Journal of Computer Information Systems*, *55*(4), 1–10. doi:10.108 0/08874417.2015.11645781

Xu, Q., Chia, S. C., Mandal, B., Li, L., Lim, J. H., Mukawa, M. A., & Tan, C. (2016). SocioGlass: Social interaction assistance with face recognition on google glass. *Scientific Phone Apps and Mobile Devices*, *2*(1), 1–4. doi:10.118641070-016-0011-8

Yuan, S., Liu, Y., Yao, R., & Liu, J. (2016). An investigation of users' continuance intention towards mobile banking in China. *Information Development*, *32*(1), 20–34. doi:10.1177/0266666914522140

Zhang, R., Zhang, Y., Sun, J., & Yan, G. (2012, March). Fine-grained private matching for proximity-based mobile social networking. In INFOCOM, 2012 Proceedings IEEE (pp. 1969-1977). IEEE. doi:10.1109/INFCOM.2012.6195574

Zhang, X., & Zeng, H. (2017). Mobile Payment Protocol Based on Dynamic Mobile Phone Token. In *2017 Twelfth Latin American Conference on Learning Technologies (LACLO)* (pp. 680–985). Academic Press. 10.1109/ICCSN.2017.8230198

Zhou, T. (2012). Examining mobile banking user adoption from the perspectives of trust and flow experience. *Information Technology Management*, *13*(1), 27–37. doi:10.100710799-011-0111-8

Zhou, T. (2014). An empirical examination of initial trust in mobile payment. *Wireless Personal Communications*, *77*(2), 1519–1531. doi:10.100711277-013-1596-8

About the Contributors

Sara Paiva holds a PhD in Informatics Engineering and has been teaching at the School of Technology and Management for more 10 years. For several years now, she has been dedicated to the area of mobile development (Android and iOS), both at the learner level and at the research and project level. She has more than 20 publications in magazines and conferences and has participated in several research projects. Her main area of work is currently the use of mobile solutions for social inclusion, having worked extensively with the ACAPO association and participated in several seminars on the subject. On the other hand, she has organized special sessions in conferences, special issues in newspapers and books dedicated to the theme of the application of mobile solutions to the area of social inclusion.

* * *

Ahmed Al-Haiqi has finished his PhD in 2015 at the National University of Malaysia (UKM), Malaysia, where he had obtained the Master of Engineering degree in 2010. His current research interests are the smartphone -in particular, Android-security, sensors threats, sensor programming, emerging computer communication technologies and the application of machine learning in several interdisciplinary problems. Previously he had some research in network security, and before many years he used to be a professional database developer. Currently a senior lecturer at Universiti Tenaga Nasional, Malaysia.

Syed Mustafa Ali is healthcare informatician, working towards the use of mHealth technologies to accelerate public health efforts in Pakistan.

António Amaral is an Industrial Engineer (2005) from University of Minho and holds a Ph.D. in Industrial and Systems Engineering also from University of Minho. Worked as an Operations Engineer in Sociedade de Transportes Coletivos do Porto – STCP, SA; Program team member in EISS (Exploration and Information Support System). Member of the technical commission – CT175, ONS APOGEP, for working groups GT4 – Portfolio Management and GT5 – Governance. Worked as an Invited Assistant Professor in the Production and Systems Department at University of Minho and as an Assistant Professor in the Faculty of Business Economics and Social Sciences at Lusófona University of Porto. Presently, works as an Invited Adjunct Professor at the Business School of the Polytechnic Institute of Viana do Castelo, in the disciplinary group of Organisation, Logistics and Marketing. He is also a researcher of the ALGORITMI Research Centre in the Industrial Engineering and Management Line.

Luís Barreto is currently a Professor, Vice-President of the direction board of Business Science Superior School, Polytechnic University of Viana do Castelo (Escola Superior de Ciências Empresariais- Instituto Politécnico de Viana do Castelo), he has been the manager of different R&D projects, namely: NetStart (http://www.netstart.pt), CSI- Cooperation Servicing Innovation and SeeLe (Seeking Learning Evaluation- http://seele.ipvc.pt) and Alto Minho S.mob. He has several scientific papers published and is also a member of the editorial board of different scientific journals. The following subjects are Luís Barreto main interests: Networking Protocols, Wireless Security, Ad-hoc and Wireless Networks, E-learning, Web 2.0, Informal and Formal Learning.

João Barroso is Associate Professor with Habilitation at University of Trás-os-Montes e Alto Douro (UTAD). He earned a doctorate in UTAD in 2002 in Electrical Engineering and held in 2008 the Habilitation Public Examinations in Electrical Engineering. In 2012 he became Associate Professor on Informatics/Accessibility at UTAD. He was Pro-Rector of Innovation and Information Management of UTAD between July 2010 and July 2013. He was part of the creative team from the area of the Department of Computer Engineering, has been Coordinator of the Bachelor of Computer Science, Director of the Master in Computer Science and a member of the Director board of Doctoral Program in Informatics. He is a Senior Researcher at INESC Technology and Science – INESC TEC. His main research interests are in the areas of Digital Image Processing, Human Computer Interaction and Accessibility. He produced more than 170 academic papers, including book chapters, articles and communications in conference proceedings and guided twenty-five graduate work, as well as Master and PhD. He participated in the organization of scientific meetings of various international and national nature, which emphasizes the creation of the conference "DSAI - Software Development for Enhancing Accessibility and Fighting Info- exclusion" in 2006 (http://www.dsai.ws) and the "TISHW - International Conference on Technology and Innovation in Sports, Health and Wellbeing" in 2016 (http://www.tishw.ws).

William Bortoluzzi Pereira received a bachelor's degree in Information Systems from the Centro Universitário Franciscano in 2016. He has experience in the area of Computer Science, with emphasis on Information Technology Management and Context-aware Computing. He is currently a Master's student in Computer Science at the Federal University of Santa Maria.

Filipe Carvalho is a Mathematician with a degree (2002) from University of Minho and holds a PhD in Science with specializations in Mathematics (2013) also from University of Minho. He works as a professor at the Business School of Polytechnic Institute of Viana do Castelo, in the disciplinary group of Mathematics. He is also a researcher of the CMAT Research Centre in the School of Science, Minho University. Apart from other projects he worked in the high school book manuals in Mathematics with ASA editors between 2012 and 2017.

Juliana da Rocha holds a Bachelor of Science degree in Computer Science from Mackenzie Presbyterian University (2016). Is a Support and Test Analyst for international projects at IBM Brazil.

Luís Santos Dias holds a BSc degree in Chemical Engineering from Instituto Superior Técnico. He has past experience in several research and development projects in domains ranging from parallel computing to image compression and remote sensing using super and hyper spectral sensors. He was also a software developer in various companies that operated in the telecommunications and banking sectors. Currently he has a research scholarship at ISTAR-IUL, a research centre of ISCTE-IUL, were he is actively involved in a European R&D project on Ambient Assisted Living.

Sara Eloy is an Assistant Professor at Instituto Universitário de Lisboa (ISCTE-IUL) where she teaches Architectural Computer Aided Design, Computation and Drawing. Her main research interest is shape grammars and the possibilities of using them in real design scenarios. Other areas of research are CAAD, the use of immersive virtual and augmented Reality for the design process, and the analysis of the building space namely considering space perception, and space syntax. She graduated in Architecture (FAUTL 1998) and has a PhD in Architecture (IST UL 2012) where she investigated on a transformation grammar-based methodology for housing rehabilitation. She is the director of the Information Sciences and Technologies and Architecture Research Center (ISTAR-IUL) and was director of the Department of Architecture and Urbanism (DAU) and of the Integrated Master in Architecture at ISCTE-IUL in Lisbon between 2013 and 2016. She has collaborated in several architectural firms in Portugal.

Ivaldir Farias Junior has a D.Sc. (2014) and an M.Sc. (2008) Degree in Computer Science, from the Federal University of Pernambuco (UFPE, Recife, Brazil), in software engineering/project management area. He also has a degree in Data Processing Technology, from the FACIR College. He is a College Instructor at the Estácio Faculty, a researcher at the UFPE University and Consultant in Processes and Software Quality at the Softex Recife. He is a researcher and member of the gp2 research group, from UFPE, focused on project management. He is a member of the Innovation and Research group in Educational Technology and a member of the SEMENTE Nucleus (a system for the elaboration of materials using new technologies), both from UFRPE. He is Reviewer of the IEEE Latin America Magazine and Argentina-Brazil Electronic Magazine of Communication and Information Technologies and editor of the Estácio Recife Electronic Magazine (REER).

Jarbas Junior is a visual/digital designer, master in Information Design. Interested in new media (interface) and innovative and experimental projects. Lecturer at UNICAP (communication degrees - digital games, journalism and advertising) and researcher.

Muhammad Yasir Khan is a research student and currently working on ICT-based solutions for healthcare sector. He received his BS degree in Computer Science from Air University, Multan, Pakistan in 2015. Currently, he is doing M.S. in Computer Science from Information Technology University (ITU) Lahore, Pakistan. His research interest includes machine learning and mobile health.

Siddique Latif received his B.Sc. degree in Electronic Engineering from the International Islamic University, Islamabad, Pakistan in 2014, sponsored by National ICT Scholarship Program. Currently, he is doing M.S. in Electrical Engineering from the National University of Sciences and Technology, Islamabad, Pakistan. His research interests include healthcare, signal processing, and deep machine learning.

Nelson G. de Sá Leitão Júnior is a Doctorate candidate at the Federal University of Pernambuco (UFPE, Recife, Brazil), has an M.Sc. degree in Software Engineering (2016) from the CESAR.EDU Faculty, and a B.S. degree in Computer Science, from the Catholic University of Pernambuco (UNICAP). He is a researcher in Team Communication and Distributed Software Development (DSD) areas and a Software Engineer at Recife Center for Advanced Studies of Recife (CESAR). He has experience with Team Leadership, Requirements Elicitation, Architecture Conception, and Programming, under Software Development Projects.

João Carlos Lima graduated in Electrical Engineering from the Federal University of Santa Maria (UFSM), Master in Computer Science from the Federal University of Rio Grande do Sul (UFRGS) and Doctor of Knowledge Engineering from the Federal University of Santa Catarina (UFSC). Currently Associate Professor at UFSM. He works in the areas of Context-awareness, Database and Mobile and Pervasive Computing.

Pedro Faria Lopes has a degree in electronics engineering and information technology. MSc and PhD were done using electronics, digital imaging, computer animation, multimedia, programming, writing. Multimedia made me study literature, theater, improvisation and music. Pedro created and directed a Multimedia Center. Directed digital videos for e-/b-learning for university students. Since 1999 Pedro supervise's pedagogical computer games in Mathematics, Environment preservation, Biology, English, Gestural Sign Language for 5 years old to young adults. Pedro is involved in technology creation (including Apps) to support Architects in customizable low cost house creation, space syntax, simulations and house rehabilitation. Present/past activities: Associate Professor at ISCTE-IUL, invited professor at Instituto Superior Técnico, Université de Genève, Universidade de Coimbra, Universidade Atlântica, Theater and Cinema University; evaluator to the European Commission, consultant/worked with Biblioteca Nacional de Portugal, Ministério da Educação, Câmara Municipal de Lisboa, INDEG, Atitude Virtual, Pim Pam Pum, Sonofilme, Thomson Digital Image, Amélia Muge, Raul de Orofino.

Karen Lotthammer is an undergraduate in Information and Communication Technologies Federal University of Santa Catarina.

Zeni Marcelino is a teacher at Otávio Manoel Anastácio School, Araranguá/SC - Brazil. BsC in Pedagogy.

Valéria Martins holds a Bachelor's Degree in Computer Science from the Paulista Júlio de Mesquita Filho State University (1995), Brazil, a master's degree in Computer Science from the Federal University of São Carlos (2000), Brazil and a PhD in Electrical Engineering from the Polytechnic School of the University of São Paulo (2011), Brazil. She has a postdoctoral degree from the Federal University of Itajubá (2014), Brazil. She is an assistant professor in courses of Computer Science area of Mackenzie Presbyterian University and is an associate professor in the Post-Graduate Program in Developmental Disorders. She has experience in Computer Science, with emphasis on Human-Computer Interaction, working mainly in the following topics: Augmented Reality, Virtual Reality, Natural Interfaces (voice interfaces, eye tracker, gesture interfaces, brain-computer interface), acting mainly in Informatics in Education and Health Informatics.

Lázaro Ourique is currently a PhD Student at ISTAR-IUL. Lázaro previously graduated with a MSc. in Architecture and Urban Design (FA-UL 2014) and with a European MSc. in Planning and policies for the City, Environment and Landscape (FA-UL & IUAV 2014) under the theme "For a New Urbanity – Vendas Novas, Regeneration and Reintegration of the Industrial Hub". Lázaro is currently a Research Intern at the Information Sciences, Technologies and Architecture Research Center (ISTAR-IUL) at ISCTE-IUL in Lisbon, Portugal and has previously worked at ADETTI-IUL (2014-2015), at Elemento Arquitectura (2014) and collaborated with Assistant Professor Francisco Serdoura (FA-UL 2012-2013) and Assistant Professor Carlos Ferreira (FA-UL 2010-2011). Lázaro has acquired extensive experience in the areas of Space Syntax, Space Use and Space Perception and became well versed in areas like Virtual Reality, Augmented Reality, Eye-tracking, Physiological Data Collection and the Monitoring of Human Subject Experiments.

Hugo Paredes received B.Eng. and Ph.D. degrees in Computer Science from the University of Minho, Braga, Portugal, in 2000 and 2008, and the Habilitation title from the University of Trás-os-Montes e Alto Douro (UTAD), Vila Real, Portugal in 2016. He was software engineer at SiBS, S.A. and software consultant at Novabase Outsoursing, S.A. Since 2003, he has been at UTAD, where he is currently Assistant Professor with Habilitation, lecturing on systems integration and distributed systems. Currently he is the Director of the Masters in Informatics Engineering at UTAD. He is a Senior Researcher at Institute for Systems and Computer Engineering, Technology and Science – INESC TEC. His main research interests are in the domain of Human Computer Interaction, including Collaboration and Accessibility topics. He is a member of the J.UCS board of editors, was guest editor of three Special Issues in journals indexed by the Journal Citation Reports and collaborates with the steering committee of the DSAI International Conference. He has authored or co-authored more than 100 refereed journal, book chapters and conference papers. He is one of the inventors of a granted patent and a patent pending request. He participated in thirteen national projects and three international projects, eight of them with public funding and six with private funding.

Dennis Paulino concluded Bachelor degree in Informatic Engineer at UTAD in June 2016. Since September 2016, he is currently taking the Master´s Degree in Informatic Engineer at UTAD and collaborating as a research fellow at INESC TEC, working in the project NanoSTIMA (Macro-to-Nano Human Sensing: Towards Integrated Multimodal Health Monitoring and Analytics).

Tiago Pedro is a young passionate student and researcher in the fields of Architecture and Computer Science. He started his career as an Architecture student, earned his Bachelor on 2014 and is now working to finish his thesis on MSc Computer Science Engineering in the fields of Computer Graphics, Virtual Reality and Human-Computer Interaction. In late 2013 he started collaborating on ISTAR-IUL and there he developed his skills related to Computer Science. Through this experience he was able to contribute to several projects that resulted in 8 papers published as a co-author.

Teresa Pereira is currently an assistant lecturer at the Superior School of Business Studies of Polytechnic Institute of Viana do Castelo. She accomplished the graduation in Mathematics and Computer Science (5 years program) at University of Minho in 2002 and obtained the MSc degree in Information Technologies (pre-Bologna) in 2007. In 26.July of 2012 she got a Ph.D degree in Technologies and Information Systems at University of Minho, Portugal. From 2002 to 2004 she worked as a researcher in the OmniPaper project (IST-2001-32174) funded under 5th FWP (Fifth Framework Program). Teresa research interests include Semantic Web, Information Management, Ontologies, Security Management Information Systems; Security Risk Management, Cybersecurity.

Adnan Qayyum did his B.S. in Electrical (Computer) Engineering form COMSATS Institute of Information Technology, Wah in 2014 and M.S. in Computer Engineering (Signal and Image Processing) from University of Engineering and Technology (UET), Taxila in 2016. Currently he is working as research associate in the Department of Electrical Engineering, Information Technology University (ITU), Lahore. His research interests include mobile health, machine learning, deep learning and medical imaging.

Isabela Silva is a researcher at the Remote Experimentation Laboratory - RExLab, at the Federal University of Santa Catarina. Master's student at the Information and Communication Technologies Graduate Program at Federal University of Santa Catarina. BsC in Information and Communication Technologies at Federal University of Santa Catarina.

Joaquim G. Pereira Silva is currently an Adjunct Professor in the School of Technology at the Polytechnic Institute of Cávado and Ave, Portugal. He holds a Dipl. Eng. in the field of Systems and Informatics Engineering and an M.Sci. in the field of Industrial Engineering. Currently, he is developing research studies for completing a PhD degree in the field of Computer Science. He teaches subjects related with Information Systems, Software Engineering and Decision Support Systems to undergraduate and post-graduate studies. He worked for about 10 years as a information systems manager in industry before initiating the academic career. His research work has covered the areas of information systems, moving object databases and data mining.

Juarez Bento da Silva has a degree in Business Administration from the Catholic University of Rio Grande do Sul (1991), specialization in Networking and Telecommunications at the UFSC(1999), Masters in Computer Science from UFSC (2002), PhD in Engineering and Knowledge Management, UFSC(2007). He is currently Associate Professor at UFSC and coordinator of the Remote Experimentation Lab (RExLab). With 36 years of professional experience and of these 12 years were working in industries in areas electrical and electronic / computer on maintenance activities and development and 26 years working in Higher Education Institutions (HEI) of which 14 years as a teacher. He is currently Associate Professor at the Federal University of Santa Catarina (UFSC) and coordinator of the Remote Experimentation Lab (RExLab). He has experience in Computer Science with emphasis in Computer Systems and acting on the following topics: remote experimentation, 3D virtual worlds, embedded systems, computers in education and digital inclusion. The main interest in research and current focus is the development and use of new information and communication technologies in teaching and learning. Currently we seek to develop and implement new devices for remote testing and work on the integration of hardware (remote experiments) in 3D virtual worlds, using open-source project.

Karmel Silva has a Master's degree in Information and Communication Technology at the Federal University of Santa Catarina. He is currently a Fellow of the Coordination of Improvement of Higher Level Personnel, working in the Remote Experimentation Laboratory with team management, research development and participation in projects. Has a degree in Psychology and specialization in Health Management, with emphasis in Organizational Psychology. It mainly works in the following subjects: remote experimentation, m-learning and digital contents.

Liping Sun is a professor at School of Science, Harbin University of Science and Technology, China. She obtained her Ph.D. degree from Harbin Institute of Technology in 2014. Her research interests include Lie (super) algebra and education in mathematics.

Marcelo Mendonça Teixeira concluded a Postdoctoral at the Facultat of Ciències de la Comunicació of Universitat Autònoma of Barcelona (Barcelona, Spain) and a Postdoctoral in the Department of Statistics and Informatics of the Federal Rural University of Pernambuco (UFRPE, Recife, Brazil) from 2013 to 2014. He is a Researcher at the Innovation and Research group in Educational Technology, and the SEMENTE Nucleus (a system for the elaboration of materials using new technologies), both from UFRPE University. He has a Scholarship by the Foundation for the Support of Science and Technology of Pernambuco (FACEPE) and the CAPES, from 2014 to 2016. He is a Specialist, Master, and Doctor of Educational Technology, at the University of Minho (Braga, Portugal), with the completed degrees from 2007 to 2013. He is also a Professor and IT Consultant.

Muhammad Usman received BS degree in Electronics Engineering from International Islamic University, Islamabad in 2014. He has been working at Nayatel Pvt. Ltd. Islamabad as Operations Engineer since 2014 and doing his research work at Medical Image Processing Research Group (MIPRG) COMSATS Institute of Information, Islamabad. His research interests include computer vision, medical Image processing, machines learning and robotics.

Isadora Vasconcellos e Souza received a bachelor's degree in Computer Science from the Federal University of Santa Maria in 2015. She has experience in the area of Computer Science with emphasis on Mobile Computing and Context-aware Computing. She is currently a Master's student in Computer Science at the Federal University of Santa Maria.

Loren Viegas is an undergraduate student at Information and Communication Technologies course at Federal University of Santa Catarina – Brazil.

Liguo Yu received his PhD degree in Computer Science from Vanderbilt University in 2004. He is an associate professor at Indiana University South Bend. He received his MS degree from Institute of Metal Research, Chinese Academy of Science in 1995. He received his BS degree in Physics from Jilin University in 1992. Before joining IUSB, he was a visiting assistant professor at Tennessee Tech University. During his sabbatical of Spring 2014, he was a visiting faculty at New York Institute of Technology-Nanjing. His research interests include software coupling, software maintenance and software evolution, empirical software engineering, and open-source development.

Index

T

U

W

Printed in the United States
By Bookmasters